Systems and Control: Foundations & Applications

Marian V. Iordache
Panos J. Antsaklis

Supervisory Control of Concurrent Systems

A Petri Net Structural Approach

Birkhäuser
Boston • Basel • Berlin

Marian V. Iordache
School of Engineering
  and Engineering Technology
LeTourneau University
Longview, TX 75607-7001
USA

Panos J. Antsaklis
Department of Electrical Engineering
University of Notre Dame
Notre Dame, IN 46556
USA

Mathematics Subject Classification: 68Q60, 68Q85, 93C65

Library of Congress Control Number: 2006922103

ISBN-10 0-8176-4357-5   e-ISBN 0-8176-4488-1
ISBN-13 978-0-8176-4357-7

Printed on acid-free paper.

Printed in the United States of America.   (EB)

9 8 7 6 5 4 3 2 1

www.birkhauser.com

Preface

Increasing complexity in engineering projects together with requirements on reduced design time and low costs raise difficult challenges in industry. Thus, effective tools for correct by construction design or for design verification are very much needed. In practice, design and verification are done hierarchically, at several levels of abstraction. This book addresses the design of tools for correct by construction synthesis at the discrete-event level of abstraction. The approach of this book is to use Petri nets as discrete-event models and structural methods for the synthesis of supervisors enforcing the design specifications. This approach promises significant computational benefits.

Discrete-event systems are a high-level representation of dynamical systems that exclude continuous dynamics details and timing information. Discrete-event modeling has been used to represent a variety of systems, from computer programs and logic circuits to manufacturing systems and traffic control systems. Notably, purely continuous dynamic systems can also be abstracted as discrete-event systems.

In the supervision of discrete-event systems, the goal is to design supervisors that ensure that only the behaviors consistent with the specification may occur in the system. The supervisor can be seen as an enhancement of the design, and is implemented by additional programming code or hardware. This book addresses the formal development of tools that design supervisors automatically, given a Petri net model and a specification. Apart from supervision design tools, a designer must do by hand the mistake-prone part of coordinating the discrete-event modules such that specification constraints are satisfied. On the other hand, when supervision design tools are used, the focus of the designer is shifted towards developing a formal specification of the desired properties of the system. For instance, a software developer could use such tools to generate the code that ensures that parallel threads of a program do not deadlock each other.

Petri net models arise naturally in the context of concurrency and discrete-event systems. Petri nets have been used in applications from various fields, such as flexible manufacturing, process control in the chemical industry, software specification, communication protocols, asynchronous digital circuits, and robotics. Compared to automata, Petri nets have been appreciated for offering a compact, higher level representation of concurrent systems. Note that reachability analysis can be used in order to obtain an automaton equivalent to a Petri net. However, since automata represent explicitly the reachable states of a system, an automaton representation of a Petri net model may not be finite.

This book focuses on methods that exploit the structure of Petri nets for supervisor design. This structural approach avoids reachability analysis and promises to alleviate or, for certain problems, altogether remove the state-space explosion problem. This benefit is often achieved at the cost of the suboptimality of the design. Indeed, structural analysis cannot provide all details available in a reachability analysis. Compared to the supervision of discrete-event systems based on automaton models, the supervisor design for Petri net models is equivalent only if reachability analysis is used. Thus, structural methods provide an interesting alternative to the traditional supervision of discrete-event systems, which is based on automaton models.

In this book we present structural methods for the design of Petri net supervisors. Following much of the literature, we focus on specifications represented by conjunctions of linear constraints, specifically by systems of linear inequalities in terms of the Petri net marking. This class of specifications has grown out of the need to express mutual exclusion constraints and has been proven to be useful in many applications. Further, as we show in this book, many problems involving more general specifications, such as languages or disjunctions of linear constraints, can be written in terms of the linear inequality specifications on transformed Petri nets.

Much of the material in this book deals with methods for the design of supervisors in both centralized and decentralized settings. The important and difficult problem of liveness enforcement is also addressed: a structural method for liveness enforcement is presented. We conclude with additional results that explore the application of Petri net methods to the supervision of concurrent hybrid systems. Note that hybrid systems are systems having both discrete-event dynamics and continuous dynamics.

After the introduction in Chapter 1, a concise introduction to Petri nets is presented in Chapter 2. More Petri net concepts are defined in the following chapters, as needed. Chapter 3 introduces the supervision problem for Petri nets and contains an overview of the methods available in the literature. Specifically, the chapter focuses on structural methods dealing with a particular type of specifications, which are described by systems of linear inequalities in the Petri net marking. In Chapter 4 it is shown how some important classes of specifications can be reduced to the marking inequality specifications on transformed Petri nets. Notably, language specifications and constraints described by disjunctions of inequalities are treated here, under appropriate assumptions.

Chapter 5 considers the problem of decentralized or distributed supervision. Here, the problem is to design decentralized/distributed supervisors that ensure that a global specification on a system is satisfied. The design problem is approached in several settings: when the supervisors cannot communicate; when the communication between supervisors is unrestricted; and when the communication is constrained. The goal here is also to minimize a communication cost. Chapters 6 and 7 deal with liveness specifications. Note that a system is live when no deadlocks are possible. Important concepts and results relating the structure of Petri nets to deadlocks are presented in Chapter 6. These are used in Chapter 7 in a structural method for the design of liveness enforcing supervisors.

Chapters 8 and 9 consider the supervision of concurrent hybrid systems. The approach here is to reduce the problem to the supervision of Petri nets. The approach is hierarchical, involving low-level controllers of hybrid systems, and a higher level discrete-event supervisor, which issues commands to the low-level controllers. The discrete-event supervisor is designed based on the Petri net abstraction of the closed loop of controllers and hybrid systems. The control and abstraction of hybrid systems is described in Chapter 9, while supervisor design for abstractions is discussed in Chapter 8.

The authors have made a significant effort to ensure that the book is readable. All relevant background is explained in the book, and numerous examples are used to illustrate the material. The index at the end contains references to all concepts defined in the book. This book is rigorous, with results being formally stated and proven. The book is also practical, as it includes fully developed methods, ready to be implemented in software. A MATLAB® toolbox has already been developed containing many of the methods presented in this work. The toolbox is available on the web, at http://www.letu.edu/people/marianiordache. Many of the examples given in this book have been solved using this software toolbox.

This book represents a novel contribution to the field. The literature already contains some books on this topic that provide valuable material, which is complemented and expanded in the present book. The authors believe the ideas presented here may help researchers and developers from various engineering fields reduce the computational burden of verification or supervision problems for discrete-event models. This book can also be used by graduate students and for advanced courses in discrete-event systems.

LeTourneau University
University of Notre Dame
April 2005

Marian V. Iordache
Panos J. Antsaklis

Contents

Symbols

$\mathbb{N}$	the set of nonnegative integers
$\mathbb{N}^*$	the set of positive integers
$\mathbb{Z}$	the set of integers
$\mathbb{R}$	the set of real numbers
$\geq$	$x \geq y$, where $x, y \in \mathbb{R}^n$, if $x(i) \geq y(i)$ for $i = 1 \ldots n$
	$A \geq B$, where $A, B \in \mathbb{R}^{m \times n}$, if $A(i,j) \geq B(i,j)$ for $i = 1 \ldots m$, $j = 1 \ldots n$
$\setminus$	set minus: $A \setminus B = \{x : x \in A, x \notin B\}$
$\subset$	$A \subset B$ if A is a proper subset of the set B
$\subseteq$	$A \subseteq B$ if A is a proper subset of the set B or if $A = B$
A^T	the transpose of the matrix A
$\lfloor x \rfloor$	the largest integer less than or equal to the real number x
$\lceil x \rceil$	the smallest integer greater than or equal to the real number x
$\|x\|$	the support of a vector x
$|x|$	the absolute value of the real number x
$|A|$	$[|A(i,j)|]_{i,j}$ for a matrix A
$|X|$	the number of elements of the set X
2^X	the set of all subsets of the set X
$\bullet p$	the set of transitions entering the place p
$\bullet t$	the set of places that are input to the transition t
$p \bullet$	the set of transitions departing from the place p
$t \bullet$	the set of places that are output to the transition t
D	usually the incidence matrix of a Petri net: $D = D^+ - D^-$
D^+	the input matrix (taken with respect to incoming transition arcs)
D^-	the output matrix (taken with respect to outgoing transition arcs)
F	usually the set of transition arcs of a Petri net
μ	a marking

μ_i $\qquad$ $\mu(p_i)$

$\mu(A)$ $\qquad$ $\sum_{p \in A} \mu(p)$, where A is a set of places

$\mu[t$ $\qquad$ the marking μ enables the transition t

$\mu[t > \mu'$ $\quad$ μ enables the transition t and μ' is reached by firing t

$\mu \xrightarrow{t} \mu'$ $\quad$ μ enables the transition t and μ' is reached by firing t

$\mu[\sigma > \mu'$ $\quad$ μ enables the firing sequence σ and μ' is reached by firing σ

$\mu \xrightarrow{\sigma} \mu'$ $\quad$ μ enables the firing sequence σ and μ' is reached by firing σ

$\mu|_{\mathcal{N}}$ $\qquad$ the marking μ restricted to the places of the net $\mathcal{N}$

$\mu|_S$ $\qquad$ the marking μ restricted to the places in the set S

$\mathcal{N}$ $\qquad$ a Petri net structure

$(\mathcal{N}, \mu_0)$ $\quad$ a Petri net with initial marking μ_0

p_i $\qquad$ the place whose marking is on the row i of the marking vector

P $\qquad$ usually the set of places of a Petri net

q $\qquad$ usually a firing vector

$\mathcal{R}(\mathcal{N}, \mu_0)$ the set of reachable markings in the Petri net $(\mathcal{N}, \mu_0)$

σ $\qquad$ usually a firing sequence

Σ^* $\qquad$ the set of all finite strings of elements $\alpha \in \Sigma$, including the empty string λ

t_i $\qquad$ the transition that corresponds to column i of the incidence matrix

T $\qquad$ usually the set of transitions of a Petri net

W $\qquad$ usually the weight function

Ξ $\qquad$ a supervisor

*Supervisory Control
of Concurrent Systems*

1

Introduction

1.1 Contribution and Background

Technological advances in our world have increasingly required new techniques for the synthesis and verification of complex systems. Typically, real-world systems involve both continuous and discrete/logical signals. Further, it is often the case that in a system there are several activities carried out at the same time. This work addresses the supervisory control of concurrent systems, that is, systems that include subsystems operating in parallel (i.e., at the same time). The supervisory control problem consists of designing a supervisor that coordinates the activities of the subsystems such that the overall system satisfies given specifications. This work could be relevant for applications within fields such as automated manufacturing, robotics, communication networks, traffic control systems, and others.

The goal of this work is to provide systematic ways to generate supervisors that coordinate the operation of the subsystems of concurrent systems, so that given specifications are satisfied. For systems that can be readily represented in a logical form as *discrete event systems* (DESs), new methodologies are proposed for the synthesis of supervisors. For more general systems involving both continuous and discrete/logical dynamics, a two-level design approach is proposed. Design at the higher level involves logical specifications and is carried out using DES methodologies. Design at the lower level involves *hybrid systems* methodologies, as this level considers both the continuous and the discrete dynamics of each subsystem. The goal of the lower level design is to obtain controllers for each of the subsystems and to abstract the controlled subsystem to a logical model to be used at the higher level design. The goal of the higher level design is to generate a supervisor that, by selecting appropriate inputs for the subsystem controllers, ensures that the requirements on the (global) system are satisfied.

In this work, Petri nets are used to represent DESs. Compared to finite automata, which are extensively used in the DES framework, Petri nets offer a compact representation of DES, as they do not represent explicitly the state space of the system. Further, Petri nets model concurrency, as they allow more than one event to occur at the same time. This work focuses on the structural approach to the supervision of Petri nets, presenting several contributions of the authors to this theory, including:

- Results relating the structure of Petri nets to liveness. Liveness is the quality of a system that from any system state any system action (event) can eventually occur again.
- Structural methods for the design of liveness enforcing supervisors. A liveness enforcing supervisor restricts the operation of a system to ensure liveness.
- Structural methods for the design of supervisors for a wide class of specifications and supervision settings.
- A structural approach to the decentralized supervision of Petri nets.

This work approaches the hybrid systems as well, and in particular the concurrent hybrid systems. Hybrid systems are systems that involve both continuous and DES dynamics. There are also several contributions of the authors to the theory of hybrid systems that are presented in this book, including:

- A design framework in which the supervisor design problem for concurrent hybrid systems is decomposed into a DES (Petri net) supervisor synthesis problem, and a hybrid system controller synthesis problem.
- An abstraction-based approach intended for the extraction of DES models, and related computational methods applying for linear discrete-time mode dynamics.

DESs are systems with dynamics driven by event occurrences. In the context of control systems, the control of DESs has been studied first by Ramadge and Wonham [159]. In this framework, supervisors are designed to ensure that a desired specification is satisfied. The specification is given as a language that the controlled DES should achieve. Then, based on the observation of the occurrence of observable events, the supervisor is to disable the controllable events that would eventually lead to strings not in the specification language. The DESs in [159] as well as in much of the literature are represented as automata. In this book, the DESs are represented as Petri nets. Petri nets originate from the work of C.A. Petri [150, 151, 152], who formulated a general theory for discrete concurrent systems. Among the many varieties of Petri nets in the literature, this book considers the Petri nets defined in the survey paper of Murata [145]. This type of Petri net is also known as the place/transition (P/T) net, as in the textbook by Reisig [160]. Unlike automata, Petri nets can represent events occurring at the same time. The supervisory control of Petri nets has also been studied, and surveys can be found in [70, 88]. In the context of the supervision of Petri nets, two classes of specifications have been considered: forbidden state specifications [57, 72, 124, 125, 141, 198, 214] and language specifications [33, 124]. Among the supervision techniques for the enforcement of forbidden state specifications, the supervision based on place invariants (SBPI) is a very efficient technique [57, 141, 142, 214]. This technique can enforce specifications consisting of linear constraints on the state (marking) of the Petri net. The SBPI has a central role in this work, as many of the DES methods that are proposed here either use or generalize the SBPI. Thus, on one hand, the SBPI is applied for the liveness-enforcing methods presented in this book. On the other hand, it is shown that the SBPI approaches can be extended to deal with more general specifications, specifically disjunctions of linear constraints and language specifications, and also extended to more general settings of concurrency and partial controllability/observability. The extension of the SBPI to decentralized supervision is also considered in this book.

In decentralized supervision, instead of a centralized supervisor coordinating the operations of the components of a system, several supervisors are used, each in charge of a part of the system. Decentralization can reduce the cost and increase the speed of control. For instance, a central supervisor coordinating the operation of several components at different physical locations would need communication channels allowing it to receive information from local sensors and send commands to local actuators. However, by means of local supervisors associated with each component, the communication requirements could be eliminated or reduced, and in this way both the cost of control and the delays due to communication constraints would be reduced. The main challenge in decentralized supervision is to decompose a global specification into subspecifications to be enforced by each supervisor. Sometimes, a solution may not even be possible without allowing either some communication between supervisors or a central coordinator. The design of decentralized supervisors in the DES context has been focused on automata models and has been considered in numerous papers, such as [167, 166, 200, 216]. A survey can also be found in [164]. This book presents methods for the design of decentralized supervisors for Petri net models. The methods address both decentralized supervision with communication and decentralized supervision without communication. In the case of the supervision with communication, the design can be carried out by minimizing a communication cost function.

A concurrent system may have deadlock states. Deadlock may arise when the subsystems of the concurrent system are interdependent. A deadlock is a state in which the system (or a part of the system) has no feasible sequence of operations to continue its execution. Therefore, the system (or the part of the system that is involved) halts when it reaches a deadlock state. Liveness is the quality of a system that has no deadlocks. A survey of deadlock prevention results on Petri net models can be found in [70]. Deadlock prevention has been studied first in the context of computer operating systems. A survey of the early results can be found in [36]. Deadlock prevention has also been studied in the context of resource allocation in flexible manufacturing systems. In this context, numerous papers have used Petri net modeling, such as [13, 14, 48, 74, 147]. A general feature of these papers is that they use restricted classes of Petri net models. Further, they typically assume that the system is fully controllable and fully observable. This book considers liveness enforcement without any assumptions on the Petri net structure. Partial controllability and observability is also allowed. Unlike most previous results, the supervisor design approach presented here is independent of the initial state. These benefits come at the cost of a design procedure that does not have guaranteed termination.

Hybrid systems are systems with both continuous and discrete-event dynamics. In the emerging area of hybrid systems, significant research effort has been carried out, beginning in the 1990s [62, 7, 3, 8, 6]. Hybrid systems surveys include [10, 11], and tutorials include [9, 168, 122]. The major hybrid systems approaches can be found in [5]. The hybrid automata [2] represent the most popular class of hybrid system models. They extend the DES automata by adding continuous dynamics. Petri nets have also been extended to the hybrid systems framework, such as in [65, 109]. Other references on hybrid Petri nets can be found in the survey paper [11] and the special issue [69] with the references therein. Note that hybrid Petri nets are not used in this book. Rather, the supervisor design problem for concurrent hybrid systems is approached by decomposing it into a problem of hybrid system controller design and a problem of Petri net supervisor design.

Many of the methods presented in this book have been implemented in software as functions of a MATLAB toolbox [82]. Thus, many of the examples of the book have been solved using these functions. The toolbox can be downloaded from the webpages of the authors.

1.2 Outline of the Book

Chapter 2: This chapter introduces the Petri nets. The most important concepts are defined in this chapter, while other specific concepts are introduced in subsequent chapters, as necessary. A comparison of Petri nets to automata is also included in this chapter.

Chapter 3: This chapter introduces the supervision of Petri nets. The presentation includes an overview of the literature methods. The chapter focuses on the supervision based on place invariants (SBPI) [141, 142, 214], a structural method of supervision that is referred to throughout the book.

Chapter 4: This chapter extends the SBPI to more general specifications and supervision settings. First, we consider an extended class of linear constraints, involving the marking of the Petri net, the firing vector, and the Parikh vector. It is shown that this extended class of constraints can describe any control place connected to a Petri net. This class of constraints describes the P-type languages of free-labeled Petri nets. Design methods for disjunctions of constraints of this type are developed in a general supervision setting, allowing for concurrency and for a form of partial controllability and observability that is weaker than that of labeled Petri nets. Next, the chapter presents an approach for the enforcement of specifications expressed by general P-type languages (not just the P-type languages of free-labeled Petri nets). Finally, the chapter presents a method to obtain Petri net supervisors from specifications expressed by disjunctions of linear marking constraints. Some of the material of this chapter corresponds to results communicated by the authors in [83, 97, 89] and the technical reports [84, 88].

Chapter 5: The SBPI is extended here to a decentralized setting, in which decentralized supervisors are to achieve a global specification on the state of the system. A decentralized admissibility concept is introduced. It is shown that decentralized admissible specifications can be enforced as easily as the centralized admissible specifications of the SBPI. Specifications that are not decentralized admissible can become admissible by enabling communication between supervisors. An algorithm is provided for the design of decentralized supervisors with communication. Finally, an integer linear programming approach is proposed in order to deal with cases in which the communication is restricted or unavailable, and the specification is decentralized *inadmissible*. The authors have published earlier versions of the material of this chapter in [85, 86, 77] and the technical report [81]. An extensive review of the work on the decentralized control of DES is also included in this chapter.

Chapter 6: This chapter contributes to the theory of Petri nets with new results on liveness, liveness of a subset of transitions, and deadlock in Petri nets. Liveness is seen as a particular case of T-liveness, which means that all transitions in a set T are live. The first results of this chapter characterize the relation between supervisors enforcing liveness or T-liveness with supervisors preventing deadlock. Then a class of Petri net subnets is introduced. This class of subnets is used to extend two well-known results in the Petri net literature. Specifically, the result relating deadlock to siphons is generalized to a powerful necessary condition for deadlock and to a sufficient condition for deadlock. Further, Commoner's theorem is also extended. The final part of the chapter shows how these new results can be used for deadlock prevention, least restrictive total-deadlock prevention, and least restrictive T-liveness enforcement. The authors have published much of this material in [80]. This material is also covered in part by the technical reports [91, 92].

Chapter 7: This chapter presents a procedure that can be used for deadlock prevention or T-liveness enforcement. T-liveness means that the transitions in the given set T are live. T-liveness enforcement corresponds to full liveness enforcement when T equals the total set of transitions. Rather than assuming a given initial marking, this procedure generates at every iteration a convex set of admissible initial markings. In the case of full liveness enforcement and under certain conditions also in the case of T-liveness enforcement, the convex set of each iteration includes the set of markings for which liveness/T-liveness can be enforced. The T-liveness enforcing version of the procedure has the following property. If the procedure terminates, the final convex set contains only markings for which T-liveness can be enforced. Then, the supervisor keeping the Petri net marking in this convex set can be easily designed using the SBPI. The deadlock prevention version of the procedure is typically not guaranteed to produce a $\mathcal{T}$-liveness enforcing supervisor. However, it has the benefit of faster convergence. This chapter focuses on the fully controllable and observable Petri nets. The next chapter considers the partially controllable and observable case. The authors have included many of the results of this chapter in [94, 79, 95, 96] and the technical reports [91, 92]. Earlier versions of this work have appeared in [93], the thesis [78], and the technical report [90].

Chapter 8: The procedure introduced in the previous chapter is extended here in several directions. First, the procedure is extended to deal with partial controllability and observability. Further, the procedure accepts additional constraints on the Petri net state. Such constraints can be used to help the procedure converge. Finally, the procedure is also extended to handle the case when the set of transitions $\mathcal{T}$ cannot be made live. In this case, the $\mathcal{T}$-liveness version of the procedure guarantees liveness only for a subset $\mathcal{T}'$ of $\mathcal{T}$. The material of this chapter corresponds to results published in [94, 95, 96] and the technical reports [91, 92].

Chapter 9: This chapter extends the DES framework toward the description of concurrency in hybrid systems. Specifically, several implicit assumptions of the DES modeling are removed. Thus, it is no longer assumed that a supervisor is allowed to keep at will, for arbitrarily long times, the system in a certain state. Rather, a supervisor should be designed such that it enables at least one transition moving the system from such an "unstable" state when the state is reached. Transition uncontrollability is also further refined, to distinguish between inability to force a firing and inability to disable a firing. Then, several supervisory control problems are formulated in this framework. These problems can be solved using methods of the traditional DES framework.

Chapter 10: This chapter presents hybrid system methods that can be used in the process of abstracting concurrent hybrid systems to the DES setting of Chapter 9. The focus is on hybrid systems with discrete-time dynamics. In particular, an efficient method for the computation of a class of controlled invariant sets is proposed. The authors have presented this method also in [87].

2

An Introduction to Petri Nets

Discrete-event systems (DESs) are systems with dynamics driven by the occurrence of events. Petri nets and automata denote two major classes of DES models. This chapter begins with an introduction to automata. Then, Petri nets are presented. We conclude with a comparison of Petri nets and automata.

2.1 Automata

We begin with an example. Consider a vending machine from which three types of products can be purchased: iced tea (t), water (w), and lemonade (l). The operation of the machine can be graphically represented as shown on the left side of Fig. 2.1. When the machine is ready, it accepts coins. Once the required amount is entered, the machine waits for the customer to select one of the three drinks and then delivers it. This behavior can be modeled by the automaton shown on the right of Fig. 2.1 as follows. The automaton has a set of events $\Sigma = \{i, t, w, l, d, r\}$, where i occurs when the coins are inserted, $t/w/l$ when the customer selects tea/water/lemonade, d when the machine dispenses the product, and r when the machine is ready for a new order. Further, the automaton has a set of six states: $Q = \{q_0, q_1, \ldots, q_5\}$, corresponding to the machine waiting for coins, waiting for a selection, dispensing the selected product, and reinitializing.

Formally, a **deterministic automaton** (which we will also call a **state machine**) is a tuple $G = (Q, \Sigma, \delta, s, F)$, where Q is the set of states, Σ the set of events, $\delta : Q \times \Sigma \to Q$ the (partial) transition function, s the initial state, and F the set of final states. Note that δ is a partial function when $\delta(q, \alpha)$ is not defined for all $(q, \alpha) \in Q \times \Sigma$. When $\delta(q, \alpha)$ is defined, the equality $q' = \delta(q, \alpha)$ indicates there is a transition from q to q' which is labeled by α. For instance, in Fig. 2.1, $\delta(q_1, w) = q_3$ but $\delta(q_5, w)$ is undefined, since the event w cannot be accepted by the machine at the state q_5. Further, assuming the machine starts in the state q_5, the initial state of the automaton is $s = q_5$.

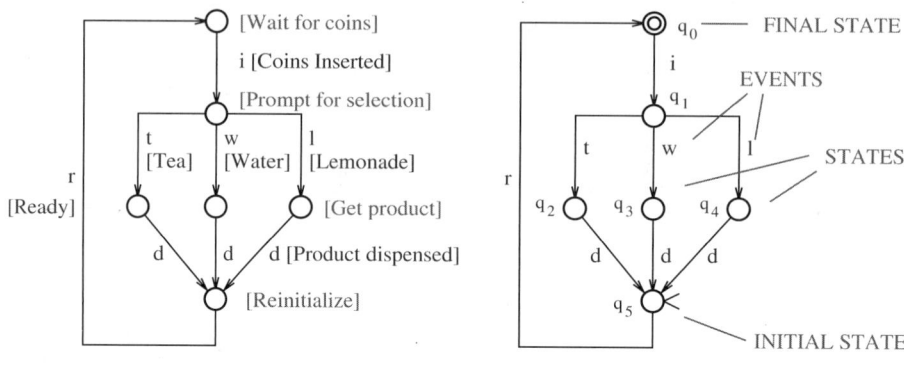

VENDING MACHINE   **AUTOMATON MODEL**

Fig. 2.1. DES example.

F denotes a set of states that should be reached after a complete execution; in our example, we can choose $F = \{q_0\}$. Note that G is said to be **finite** if Q is finite.

The behavior of a system modeled by an automaton can be characterized by the sequences of events that are possible in the automaton, starting from the initial state. For instance, the sequence r, i, t is possible in our example, but r, t, w is not possible, since the automaton cannot execute the event t at state q_0. Formally, a sequence of events σ is a string $\alpha_1\alpha_2\alpha_3 \ldots$, where $\alpha_i \in \Sigma$ for all i. Let Σ^* denote the set of all finite sequences of events, including λ, the empty string. To simplify notation, δ is extended to $\delta : Q \times \Sigma^* \to Q$, where $\delta(q, \lambda) = q$ and $\delta(q, \sigma\alpha) = \delta(\delta(q, \sigma), \alpha)$, for $\sigma \in \Sigma^*$ and $\alpha \in \Sigma$. Note that $\delta(q, \sigma)$ is defined only if all $\delta(q, \sigma_i)$ are defined, where σ_i denotes the prefix of length i of σ (e.g., $\sigma_2 = \alpha_1\alpha_2$ when $\sigma = \alpha_1\alpha_2\alpha_3 \ldots$). The **language** of a deterministic automaton G is defined as $\mathcal{L}(G) = \{\sigma \in \Sigma^* : \delta(s, \sigma) \text{ is defined}\}$. Moreover, the **marked language** of G is defined as $\mathcal{L}_m(G) = \{\sigma \in \Sigma^* : \delta(s, \sigma) \in F\}$.

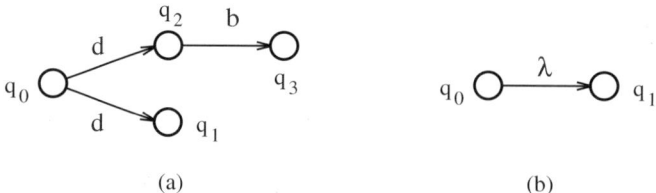

(a)   (b)

Fig. 2.2. Extensions for nondeterminism.

Note that in the context of the supervisory control, the results in the literature refer usually to deterministic automata. Nondeterministic automata extend the deterministic automata models by allowing different transitions with the same label to start from the same state (Fig. 2.2(a)). Some authors allow nondeterministic automata to also contain λ (empty string) labels (Fig. 2.2(b)). The situation of Fig. 2.2(a) is interpreted as follows: when the automaton is in state q_0 and d occurs, the state switches nondeterministically (i.e., randomly) to one of q_1 or q_2. Further, the situation of Fig. 2.2(b) corresponds to a transition that can occur at any time, apart from the occurrence of any event.

The **nondeterministic automaton** is defined the same way as the deterministic one, except that now $\delta : Q \times (\Sigma \cup \{\lambda\}) \to 2^Q$, where 2^Q denotes the set of subsets of Q. As before, δ is extended to sequences of events, and we have $\mathcal{L}(G) = \{\sigma \in \Sigma^* : \delta(s, \sigma) \neq \emptyset\}$ and $\mathcal{L}_m(G) = \{\sigma \in \Sigma^* : \delta(s, \sigma) \cap F \neq \emptyset\}$. Notably, $\sigma \in \mathcal{L}_m(G)$ if σ leads to a final state in one of the nondeterministic executions of G. For instance, in Fig. 2.2(a), if $s = q_0$ and $F = \{q_1\}$, then $\mathcal{L}(G) = \{d, db\}$ and $\mathcal{L}_m(G) = \{d\}$. Thus, $d \in \mathcal{L}_m(G)$ even though no final state is reached when q_0 switches to q_2 instead of q_1.

2.2 Petri Nets

2.2.1 Ordinary Petri Nets

Petri net models arise naturally in the context of systems in which there are several activities carried out at the same time. For instance, in a factory a robot can unload a part at the same time another part is drilled by a drilling machine. Petri nets consist of a structure and a state, where the state is called marking. The marking changes when transitions occur. Formally, an **ordinary Petri net structure** is the tuple $\mathcal{N} = (P, T, F)$, where P is the **set of places**, T the **set of transitions**, and $F \subseteq (P \times T) \cup (T \times P)$ is the set of **transition arcs**. A **marking** μ of the structure $\mathcal{N}$ is a map $\mu : P \to \mathbb{N}$. An ordinary Petri net structure $\mathcal{N}$ with **initial marking** μ_0 is called ordinary Petri net and will be denoted by $(\mathcal{N}, \mu_0)$. Note that in this book *Petri nets* will denote an extension of the ordinary Petri nets, as shown in section 2.2.2. However, for convenience, we will call ordinary Petri nets simply Petri nets, unless the context requires the distinction to be made. Further, sometimes we will call Petri nets the Petri net structures.

Petri nets have a graphical representation that is very useful for tutorial purposes. Fig. 2.3 shows an example. Note that places are represented by circles and transitions by short line segments. Further, transition arcs are represented as arrows between places and transitions. Thus, Fig. 2.3 represents the Petri net with $P = \{p_1, p_2, p_3, p_4, p_5\}$, $T = \{t_1, t_2, t_3, t_4\}$, and $F = \{(t_4, p_1), (p_1, t_1), (p_2, t_1), (p_3, t_1), (t_1, p_4), (p_4, t_3), (p_4, t_2), (t_3, p_5), (t_2, p_5), (t_2, p_2)\}$.

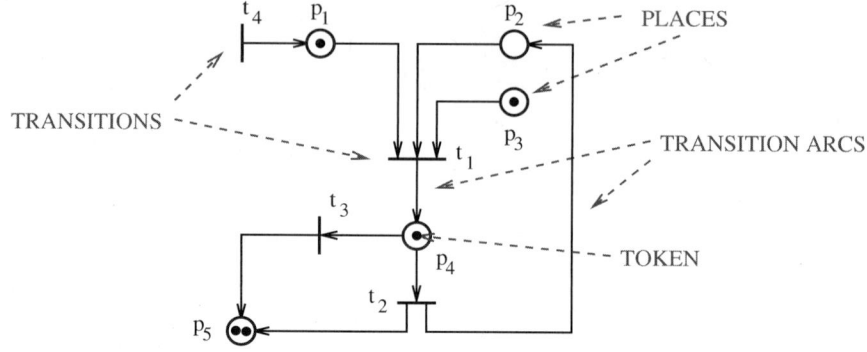

Fig. 2.3. The graphical representation of Petri nets.

It is useful to consider a marking μ both as a map and as a vector. The **marking vector** is defined as $[\mu(p_1), \mu(p_2), \ldots, \mu(p_n)]^T$, where $p_1, p_2, \ldots, p_n$ are the places of the net. The same symbol μ will be used to denote a marking vector. In the literature, an equivalent way of saying that place p has the marking $\mu(p)$ is that p has $\mu(p)$ **tokens**. The marking can also be represented graphically. As an example, consider again Fig. 2.3. The tokens are represented by small dark circles. Let's denote $\mu(p_i)$ by μ_i. Then, $\mu_1 = 1$, $\mu_2 = 0$, $\mu_3 = 1$, $\mu_4 = 1$, and $\mu_5 = 2$. Thus, the marking vector is $\mu = [1, 0, 1, 1, 2]^T$.

The structure of a Petri net can be described by preset/postset operations. The **preset** of a place p is the set of input transitions of p: $\bullet p = \{t \in T : (t, p) \in F\}$. The **postset** of a place p is the set of output transitions of p: $p\bullet = \{t \in T : (p, t) \in F\}$. Similar definitions apply for transitions. They are also extended for sets of places or transitions; for instance, if $A \subseteq P$, then $\bullet A = \bigcup_{p \in A} \bullet p$ and $A\bullet = \bigcup_{p \in A} p\bullet$. As an example, in Fig. 2.3 we have $\bullet p_1 = \{t_4\}$, $p_4\bullet = \{t_2, t_3\}$, $t_2\bullet = \{p_2, p_3\}$, and $\bullet\{p_2, p_3\} = \{t_2\}$. Note that $\bullet p_3 = \emptyset$, $\bullet t_4 = \emptyset$, and $p_5\bullet = \emptyset$. We say that p is a **source place** if $\bullet p = \emptyset$ and a **sink place** if $p\bullet = \emptyset$. Source and sink transitions are similarly defined.

The operation of a Petri net is as follows. The marking μ **enables** the transition t if $\forall p \in \bullet t$: $\mu(p) \geq 1$. When μ enables t and t **fires**, the marking is changed. Let μ' be the new marking obtained by firing t. The marking μ' satisfies:

$$\mu'(p) = \begin{cases} \mu(p) + 1 & \text{if } p \in t\bullet \setminus \bullet t, \\ \mu(p) - 1 & \text{if } p \in \bullet t \setminus t\bullet, \\ \mu(p) & \text{otherwise.} \end{cases}$$

The notation $\mu \xrightarrow{t} \mu'$ is used to express that firing t at μ results in the new marking μ'. As an example, the transitions enabled in Fig. 2.3 are t_2, t_3, and t_4. The transition t_1 is not enabled because $\mu_2 = 0$ and $p_2 \in \bullet t_1$. Examples of transition firings can be seen in Fig. 2.4.

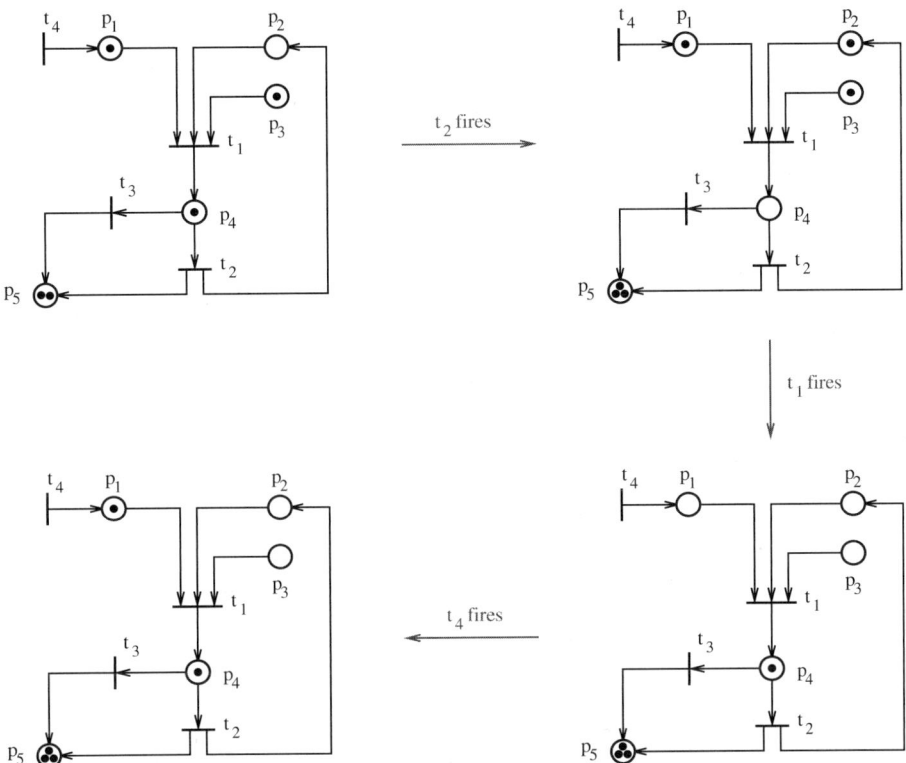

Fig. 2.4. Transition firing examples.

A Petri net modeling illustration is given next. Consider a computer network in which the users can send files to print to a central print server. The print requests from the network are placed in a queue, which is read by the print server. The operations performed by the print server are shown in Fig. 2.5. The server can use two printers: $LP1$ and $LP2$. Each of the two printers is to process only one job at a time. When the server begins processing a request, it waits for one of the printers to become available or selects one of the two printers and waits for it to become available. When the selected printer is available, the request is sent to the printer. Finally, the server notifies the user when the printing job is completed. Assume that the print server processes at most two requests at a time. The Petri net model is shown in Fig. 2.6. The Petri net is shown at the initial marking, for which the server and the printers are idle, waiting for requests. Note that $\mu_1 = 2$ indicates the server can only process two requests at a time.

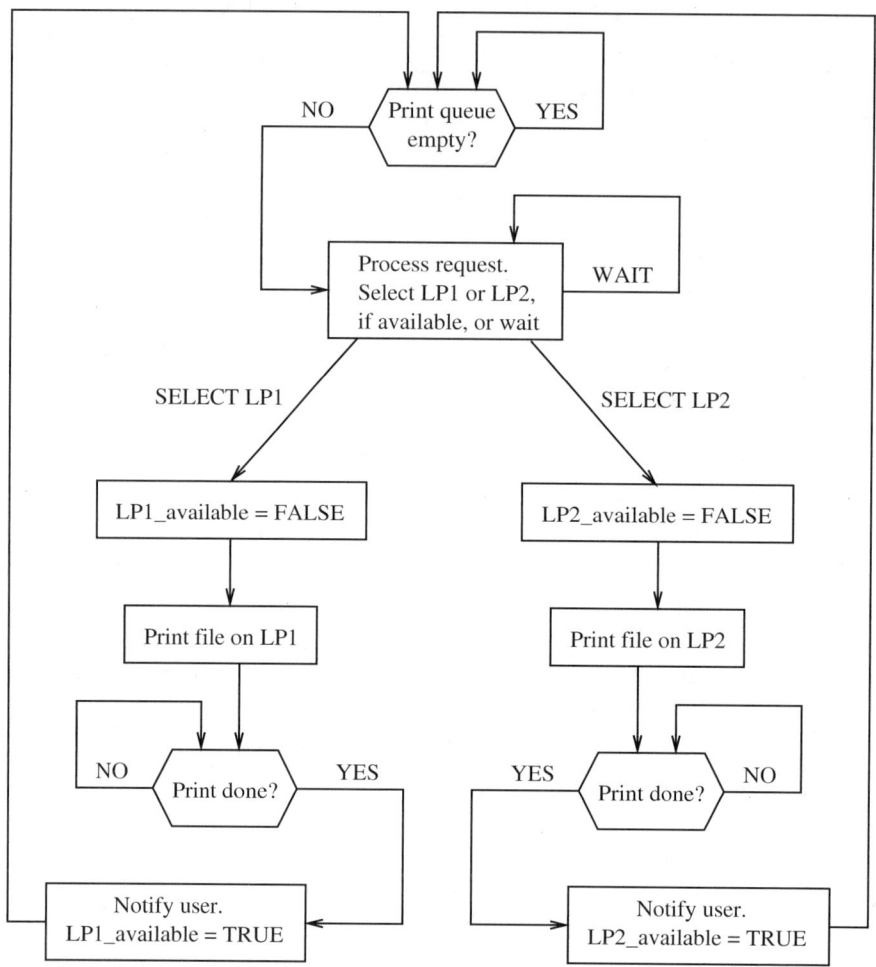

Fig. 2.5. A print server illustration.

2.2.2 Generalized Petri Nets

In some problems it is useful to have multiple arcs between a place and a transition (Fig. 2.7). This need is modeled in the generalized Petri nets, by introducing a weight function. Thus, the Petri net structure definition changes as follows. A **Petri net structure** is a quadruple $\mathcal{N} = (P, T, F, W)$ where P is the **set of places**, T is the **set of transitions**, $F \subseteq (P \times T) \cup (T \times P)$ is the set of **transition arcs**, and $W : F \to \mathbb{N} \setminus \{0\}$ is the **weight function**. Note also that sometimes it is convenient to represent Petri nets in terms of input and output matrices D^+ and D^-, as $\mathcal{N} = (P, T, D^-, D^+)$. This representation is equivalent to $\mathcal{N} = (P, T, F, W)$ and will be discussed later in this section.

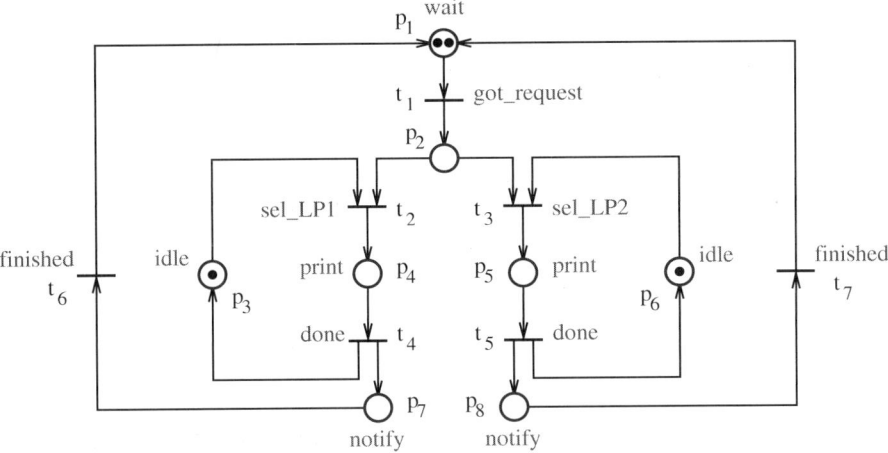

Fig. 2.6. Petri net model of the print server example.

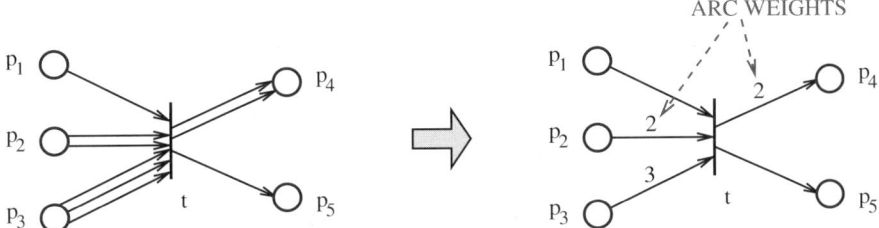

Fig. 2.7. Arc weights.

Note that the graphical representation of weights different than 1 is illustrated in Fig. 2.7. For instance, in Fig. 2.7 we have $W(p_1, t) = 1$, $W(p_3, t) = 3$, and $W(t, p_4) = 2$.

The firing rules in generalized Petri nets are as follows. The marking μ **enables** the transition t if $\forall p \in \bullet t$, $\mu(p) \geq W(p, t)$. When μ enables t and t **fires**, the marking is changed as follows. Let μ' be the next reached marking. The marking μ' satisfies

$$\mu'(p) = \begin{cases} \mu(p) & \text{if } p \notin \bullet t \cup t\bullet, \\ \mu(p) + W(t, p) & \text{if } p \in t\bullet \setminus \bullet t, \\ \mu(p) - W(p, t) & \text{if } p \in \bullet t \setminus t\bullet, \\ \mu(p) - W(p, t) + W(t, p) & \text{if } p \in \bullet t \cap t\bullet. \end{cases}$$

An example is shown in Fig. 2.8.

Given a Petri net $\mathcal{N} = (P, T, F, W)$ with m places and n transitions, let $P = \{p_1, p_2, \ldots, p_m\}$ and $T = \{t_1, t_2, \ldots, t_n\}$. The **incidence matrix** is an

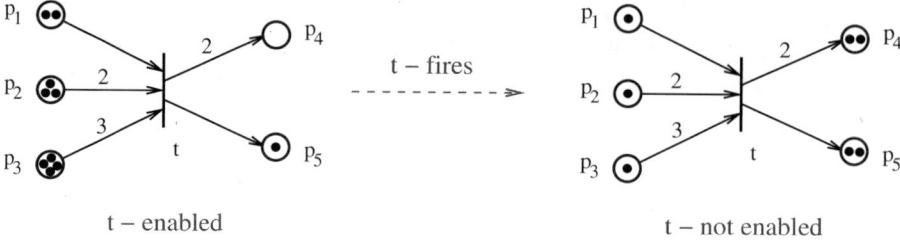

t – enabled   t – not enabled

Fig. 2.8. Firing transitions in generalized Petri nets.

$m \times n$ matrix defined by $D = D^+ - D^-$, where the elements d_{ij}^+ and d_{ij}^- of D^+ and D^- are

$d_{ij}^+ = W(t_j, p_i)$ if $(t_j, p_i) \in F$ and $d_{ij}^+ = 0$ otherwise;
$d_{ij}^- = W(p_i, t_j)$ if $(p_i, t_j) \in F$ and $d_{ij}^- = 0$ otherwise.

The matrices D^+ and D^- are called the **input matrix** and the **output matrix**, respectively. They provide an alternative representation of a Petri net structure as $\mathcal{N} = (P, T, D^-, D^+)$. It is convenient to view D^+ and D^- also as functions: $D^- : P \times T \to \mathbb{N}^{|P| \times |T|}$ and $D^+ : P \times T \to \mathbb{N}^{|P| \times |T|}$. The relation to the $\mathcal{N} = (P, T, F, W)$ representation is as follows: $(p, t) \in F \Leftrightarrow D^-(p, t) \neq 0$ and $(t, p) \in F \Leftrightarrow D^+(p, t) \neq 0$; also, $(p, t) \in F \Rightarrow D^-(p, t) = W(p, t)$ and $(t, p) \in F \Rightarrow D^+(p, t) = W(t, p)$.

A **self-loop** is a pair consisting of a place p and transition t such that $p \in \bullet t \cap t \bullet$. In other words, p is both an "input" and an "output" to t. For instance, the pair of p_1 and t_1 in Fig. 2.9 is a self-loop. When a place p or a transition t participates in a self-loop, we will also say that p *(or t) has a self-loop*.

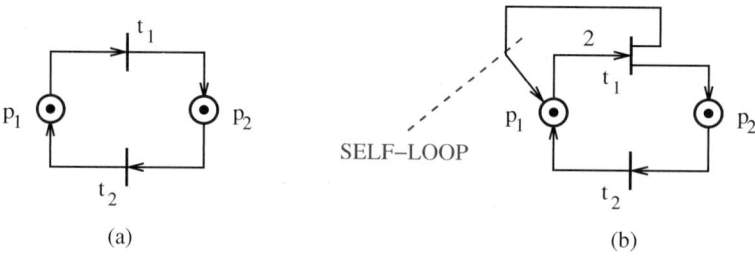

(a)   (b)

Fig. 2.9. Self-loops.

While the input and output matrices can describe completely a Petri net structure, note that the incidence matrix is in general not sufficient. For instance, the two Petri nets of Fig. 2.9 have the same incidence matrix:

$$D = \begin{bmatrix} -1 & 1 \\ 1 & -1 \end{bmatrix} \tag{2.1}$$

though they have different input and output matrices:

$$D_a^+ = \begin{bmatrix} 0 & 1 \\ 1 & 0 \end{bmatrix}, \ D_b^+ = \begin{bmatrix} 1 & 1 \\ 1 & 0 \end{bmatrix}, \ D_a^- = \begin{bmatrix} 1 & 0 \\ 0 & 1 \end{bmatrix}, \ D_b^- = \begin{bmatrix} 2 & 0 \\ 0 & 1 \end{bmatrix}, \tag{2.2}$$

where the subscripts a and b distinguish the matrices of the Petri net (a) and (b). However, the incidence matrix D determines uniquely a Petri net structure without self-loops: we have $D^- = \max(-D, 0)$ and $D^+ = \max(D, 0)$, where max is taken element by element. Thus, D^+ is given by the positive elements of D and D^- by the negative elements of D.

The incidence matrix allows an algebraic description of the marking change of a Petri net:

$$\mu_k = \mu_{k-1} + D q_k, \tag{2.3}$$

where q_k is called the **firing vector** and its elements are all zero excepting $q_{k,i} = 1$, where i corresponds to the transition t_i that fired. Note that t_i is enabled if and only if

$$\mu_{k-1} \geq D^- q_k. \tag{2.4}$$

Equation (2.3) is also known as the *state equation* of Petri nets, as it relates the marking (the state variable) to the firing vector (the control input). As an example, in Fig. 2.9(a), the firing vector associated with the firing of t_2 is $q = [0, 1]^T$. Note that (2.4) is satisfied for $\mu_{k-1} = [1, 1]^T$ and $q_k = q$, as t_2 is enabled.

A **firing sequence** σ is a sequence of transitions $t_{i_1} t_{i_2} t_{i_3} \ldots t_{i_k} \in T$, $k = 1, 2, 3 \ldots$. A firing sequence may be finite or infinite. Given a marking μ, we say that σ is **enabled** if we can fire each of $t_{i_1}, t_{i_2}, t_{i_3}, \ldots$, in this order. For instance, $\sigma = t_1 t_2 t_2$ is enabled in Fig. 2.9(a), but not in Fig. 2.9(b).

Denoting by $t_1, t_2, \ldots, t_n$ the elements of T, the **firing count vector** of a firing sequence σ is an $n \times 1$ vector x such that each element x_i of x indicates how many times the transition t_i appears in σ. For instance, $\sigma = t_1 t_2 t_2$ in Fig. 2.9(a) corresponds to $x = [1, 2]^T$. If x is the firing count vector of a transition sequence leading the Petri net from the marking μ_i to μ_k, then

$$\mu_k = \mu_i + D x. \tag{2.5}$$

A closely related concept is what we call the **Parikh vector**. The Parikh vector, denoted by v, is a state vector defined by $v = v_0 + x$, where v_0 is the initial value of v and x is the firing count vector corresponding to the sequence of transitions fired since the initialization of the system. Thus, if $v_0 = 0$, the marking μ and the Parikh vector v are related by

$$\mu = \mu_0 + D v. \tag{2.6}$$

2.2.3 Additional Concepts

The marking μ' is **reachable** from μ if there is a sequence of markings $\mu_1, \ldots, \mu_k$, $\mu_k = \mu'$, and a sequence of transitions $\sigma = t_{i_1}, \ldots, t_{i_k}$ such that $\mu \xrightarrow{t_{i_1}} \mu_1 \ldots \xrightarrow{t_{i_k}} \mu'$. This is also written as $\mu \xrightarrow{\sigma} \mu'$. The **set of reachable markings** of a Petri net $(\mathcal{N}, \mu)$ (i.e., the set of markings reachable from the initial marking μ) will be denoted by $\mathcal{R}(\mathcal{N}, \mu)$.

A vector x is called a **place invariant** if $x^T D = 0$. A vector x is called a **transition invariant** if $Dx = 0$. In view of (2.6), if x is a place invariant, then $x^T \mu = x^T \mu_0$ for all reachable markings. Note also that in (2.6), if $v = x$, x has nonnegative elements, and x is a transition invariant, then $\mu_k = \mu_0$. For instance, as seen from (2.1), $x = [1, 1]^T$ is both a place and a transition invariant. Thus, $\mu_1 + \mu_2 = 2$ for all reachable markings, and firing the sequence $t_1 t_2$ leads to a marking identical to the initial marking.

A Petri net $(\mathcal{N}, \mu_0)$ is **safe** if $\forall \mu \in \mathcal{R}(\mathcal{N}, \mu_0)$, $\forall p \in P$, $\mu(p) \leq 1$. A Petri net $(\mathcal{N}, \mu_0)$ is **bounded** if there is an integer k such that $\forall \mu \in \mathcal{R}(\mathcal{N}, \mu_0)$, $\forall p \in P$, $\mu(p) \leq k$. Further, $(\mathcal{N}, \mu_0)$ is **unbounded** if not bounded. Note that $(\mathcal{N}, \mu_0)$ is bounded if and only if $\mathcal{R}(\mathcal{N}, \mu_0)$ has a finite number of elements. A Petri net $\mathcal{N}$ is **structurally bounded** if $(\mathcal{N}, \mu_0)$ is bounded for all initial markings μ_0. As examples, the Petri net of Fig. 2.10 is bounded and structurally bounded, the Petri net of Fig. 2.3 is unbounded and not structurally bounded, and the Petri net of Fig. 2.12(c) is bounded but not structurally bounded.

The reachable space $\mathcal{R}(\mathcal{N}, \mu_0)$ of a Petri net $(\mathcal{N}, \mu)$ can be expressed by the **reachability graph**. The reachability graph is a directed graph whose nodes are the markings of $\mathcal{R}(\mathcal{N}, \mu_0)$ and whose arcs are labeled by the transitions of $\mathcal{N}$. The graph is such that for all $\mu, \mu' \in \mathcal{R}(\mathcal{N}, \mu_0)$, there is an arc from μ to μ' that is labeled by t if and only if $\mu \xrightarrow{t} \mu'$. An example of a reachability graph is shown in Fig. 2.10.

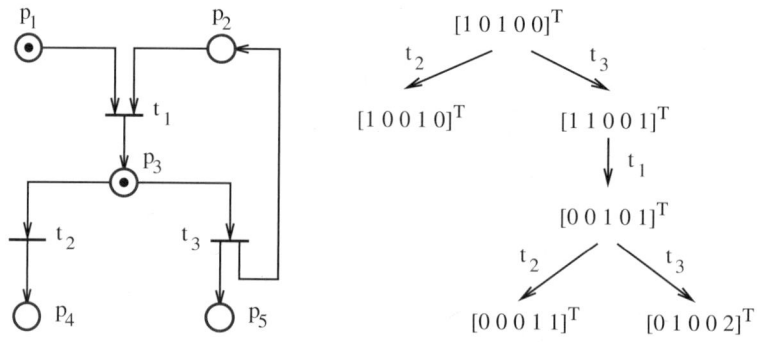

Fig. 2.10. Reachability graph example.

Note that the reachability graph is finite when $(\mathcal{N}, \mu_0)$ is bounded. When $(\mathcal{N}, \mu_0)$ is not bounded, we can still represent the infinite reachable space by a finite **coverability graph**. A coverability graph has a set of nodes $M \subset \{\mathbb{N} \cup \omega\}^m$, where ω denotes infinity, with the property that $\forall \mu \in \mathcal{R}(\mathcal{N}, \mu_0)$ $\exists \mu' \in M \colon \mu' \geq \mu$. The following algorithm, adapted from [99], can be used for the construction of coverability graphs.

ALGORITHM 2.1   Constructing Coverability Graphs

1. *Let $M = \{\mu_0\}$.*

2. *Select a pair (μ, t), $\mu \in M$ and $t \in T$, that has not been considered before, such that μ enables t. If no such pair exists, stop.*

3. *Let μ' be the marking reached by firing t. For all $\mu'' \in M$ on the path from μ_0 to μ', if $\mu' \geq \mu''$, then set $\mu'(p) = \omega$ for all places p such that $\mu'(p) > \mu''(p)$.*

4. *Let $M = M \cup \{\mu'\}$, add the arc (μ, t, μ') to the graph, and go to step 2.*

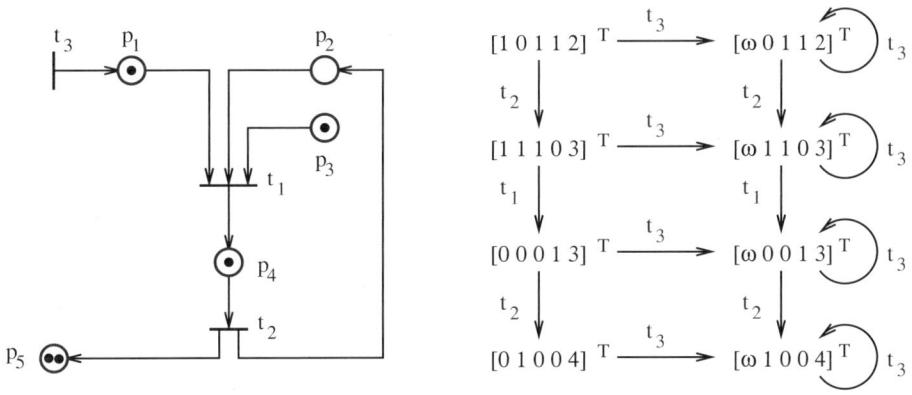

Fig. 2.11. Coverability graph example.

Note that if a Petri net is bounded, the algorithm above produces the reachability graph. An example of a coverability graph is shown in Fig. 2.11.

A Petri net $(\mathcal{N}, \mu_0)$ is said to be **deadlock-free** if for any reachable marking μ there is an enabled transition. $(\mathcal{N}, \mu)$ is in **deadlock** if no transition is enabled at marking μ. For instance the Petri net of Fig. 2.12(b) is in deadlock, but the Petri net of Fig. 2.12(c) is deadlock-free.

Let $(\mathcal{N}, \mu_0)$ be a Petri net. A transition t is said to be **live** if $\forall \mu \in \mathcal{R}(\mathcal{N}, \mu_0)$ $\exists \mu' \in \mathcal{R}(\mathcal{N}, \mu)$ such that t is enabled by μ'. A transition t is **dead** at marking μ if no marking $\mu' \in \mathcal{R}(\mathcal{N}, \mu)$ enables t. $(\mathcal{N}, \mu_0)$ is said to be **live** if every transition is live. For instance, in Fig. 2.12(c) t_2 and t_4 are live, while t_1 and t_3 are dead.

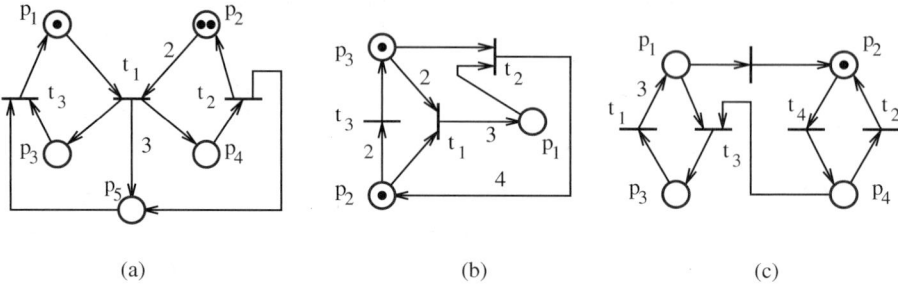

(a)   (b)   (c)

Fig. 2.12. Petri nets illustrating deadlock-related concepts.

Traps and *siphons* are two concepts that are very often used in the dead-lock/liveness analysis of Petri nets. A nonempty set of places $S \subseteq P$ is called a **siphon** if $\bullet S \subseteq S\bullet$ and **trap** if $S\bullet \subseteq \bullet S$. In particular, $S = P$ may be a siphon. An **empty siphon** with respect to a Petri net marking μ is a siphon S such that $\sum_{p \in S} \mu(p) = 0$. The attribute "empty" refers to the fact that S has no tokens. A siphon has the property that if for some marking it is empty, it will stay so for all subsequent reachable markings. A trap has the property that if at some marking it has one token, then for all subsequent reachable markings it will have at least one token. In Fig. 2.12(a), $\{p_1, p_3\}$ and $\{p_2, p_4\}$ are traps. S is a **minimal siphon** if there is no other siphon S' (by defini-tion, $S' \neq \emptyset$) such that $S' \subset S$. $\{p_1, p_3\}$ in Fig. 2.12(c) is a minimal siphon. Given a Petri net with an initial marking, we say that a siphon is **controlled** if for all reachable markings it contains at least one token; given a Petri net structure and a set of initial markings we also say that a siphon is *controlled* if the siphon is controlled for each of the initial markings. A siphon that is not controlled is said to be **uncontrolled**.

2.2.4 Special Structures

Several special Petri net structures are defined below. Examples can be found in Fig. 2.13.

A **state machine** (SM) is an ordinary Petri net $\mathcal{N}$ such that each transi-tion has one input place and one output place: $|\bullet t| = |t \bullet| = 1 \; \forall t \in T$.

A **marked graph** (MG) is an ordinary Petri net $\mathcal{N}$ such that each place has one input transition and one output transition: $|\bullet p| = |p \bullet| = 1 \; \forall p \in P$.

A **free-choice** (FC) net is a Petri net $\mathcal{N}$ with the property that $\forall p_1, p_2 \in P$ $(p_1 \neq p_2)$, $p_1 \bullet \cap p_2 \bullet \neq \emptyset \Rightarrow |p_1 \bullet| = |p_2 \bullet| = 1$.

An **extended free-choice** (EFC) net is a Petri net $\mathcal{N}$ with the property that $\forall p_1, p_2 \in P,\ p_1 \bullet \cap p_2 \bullet \neq \emptyset \Rightarrow p_1 \bullet = p_2 \bullet$.

An **asymmetric choice** (AC) net is a Petri net $\mathcal{N}$ with the property that $\forall p_1, p_2 \in P,\ p_1 \bullet \cap p_2 \bullet \neq \emptyset \Rightarrow p_1 \bullet \subseteq p_2 \bullet$ or $p_2 \bullet \subseteq p_1 \bullet$.

2.2.5 Concurrency

The Petri nets discussed so far include a form of concurrency, as they can model subsystems that operate at the same time. For instance, referring back to the print server example of section 2.2.1, the Petri net of Fig. 2.6 models the situation in which two print requests are processed in parallel. However, note that in the Petri net literature, the notion of concurrency refers to the possibility of transitions firing *at the same time*. This possibility has not been considered in the previous sections, and so is presented here. We detail the various concurrency settings and the extension of the firing rules for concurrent firings of transitions.

The **no concurrency assumption** describes the situation in which only one transition may fire at a time. This is the assumption we have had so far in our introduction to Petri nets. Under the no concurrency assumption, the firing vector q satisfies $q \in \{0,1\}^{|T|}$ and $\sum_{t \in T} q(t) = 1$, where the entry with $q(t) = 1$ indicates the transition that is to fire. Further, recall that when $q(t) = 1$ for some t, t is enabled if and only if the marking μ satisfies

$$\mu \geq D^- q. \tag{2.7}$$

In the **concurrency assumption**, groups of transitions may fire at the same time. In this case, $q \in \{0,1\}^{|T|}$ and $\{t : q(t) = 1\}$ identifies the transitions t that are to be fired at the same time. We will say that q fires when the transitions with $\{t : q(t) = 1\}$ fire at the same time. By definition q is **enabled** (i.e., q may be fired) when (2.7) is satisfied. Note that the marking μ' that is reached by firing q is the marking reached by firing each $t \in \{t : q(t) = 1\}$, one at a time. Thus, μ' is also described by the state equation

$$\mu' = \mu + Dq. \tag{2.8}$$

We express by $\mu \xrightarrow{q} \mu'$ that μ' is obtained by firing q at μ. As an example, consider the Petri net of Fig. 2.9(a), where the incidence matrix D is given in (2.1) and the output matrix D^- is D_a^- in (2.2). Note that $q = [1,1]^T$ is enabled at $\mu = [1,1]^T$, and firing it results in $\mu' = \mu$. However, $q = [1,1]^T$ is not enabled at $\mu = [2,0]^T$.

In the **transition-bag assumption**, groups of transitions may fire at the same time and the transitions in a group may be fired several times each, at the same firing instance. Thus, $q \in \mathbb{N}^{|T|}$ and for each t, $q(t)$ indicates how many times t is fired. Again, by definition, q is enabled when (2.7) is satisfied. Further, firing q results in the same marking as that obtained after firing $q(t)$ times each transition $t \in \{t : q(t) \neq 0\}$. The marking μ' reached by firing q is

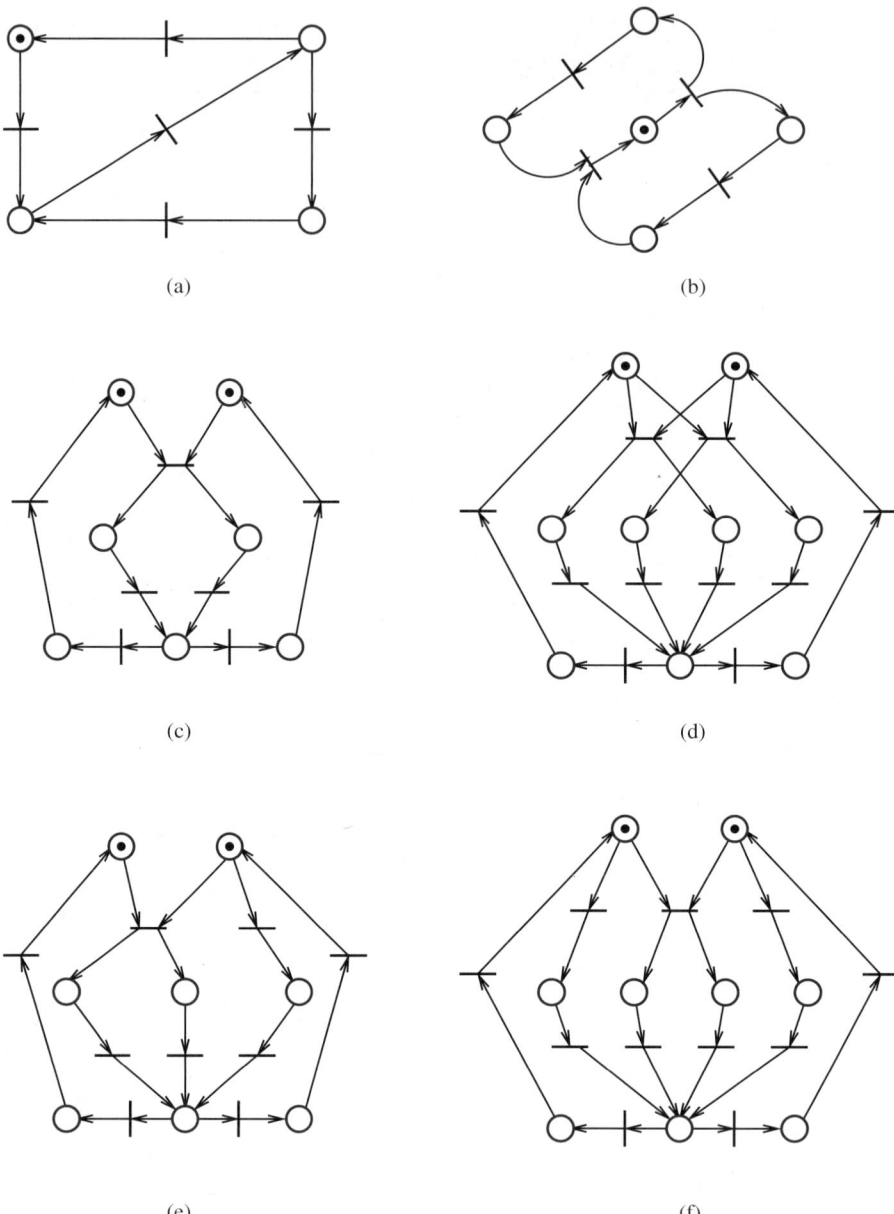

Fig. 2.13. Examples of: (a) state machine, (b) marked graph, (c) free-choice net, (d) extended free-choice net that is not free-choice, (e) asymmetric choice net that is not extended free-choice, (f) Petri net that is not an asymmetric choice net.

expressed by (2.8). As an example, consider again the Petri net of Fig. 2.9(a). Note that at the marking $\mu = [1,1]^T$ the firing vector $q = [1,1]^T$ is enabled, but $q' = [2,2]^T$ is not. Further, $q = [2,0]^T$ is enabled at $\mu = [2,0]^T$.

Following [181], we can incorporate the no concurrency, concurrency, and transition-bag assumptions into a more general setting in which we require $q \in \Delta$, for a given $\Delta \subseteq \mathbb{N}^{|T|}$. The enabling and firing rules are the same as before.

In this book, unless otherwise implied by the context, we assume the no concurrency assumption.

2.3 Petri Nets versus Automata

We have seen that the transitions of automata are labeled by events in a set Σ, where the same event may label two or more different transitions. Petri nets in which the transitions are also labeled by events are called *labeled Petri nets*. A **labeled Petri net** is a Petri net enhanced with a labeling function $\rho : T \to 2^{\Sigma} \cup \{\lambda\}$, where Σ is the set of events, ρ the labeling function, and λ the null event. When each transition $t \in T$ is labeled by a single event $\alpha \in \Sigma$ and there are no two transitions with the same label, the Petri net is said to be **free-labeled**. Unless otherwise implied by the context, Petri nets will be assumed to be free-labeled, where ρ is defined by $\rho(t) = \{t\}$ $\forall t \in T$ for $\Sigma = T$. With regard to languages, we have distinguished between the language and the marked language of an automaton. There are several ways in which Petri net languages can be defined [149]. This book will refer the P-type languages of labeled Petri nets in the context of the no concurrency setting. The **P-type language** of a Petri net $(\mathcal{N}, \rho, \mu_0)$ consists of all sequences of events $w = \rho(\sigma)$ generated by the firing sequences σ enabled at μ_0. Note that for $\sigma = t_1 t_2 t_3 \ldots$, $\rho(\sigma)$ denotes the sequence $\rho(t_1)\rho(t_2)\rho(t_3)\ldots$.

2.3.1 Model Conversions

Petri nets can be transformed into automata by means of the reachability graph. The reachability graph, introduced for free-labeled Petri nets, can easily be extended for labeled Petri nets. Extensions dealing with Petri nets under concurrency assumptions are also possible. For instance, the reachability graph of Fig. 2.10 can be seen as the automaton of Fig. 2.14, which has the set of states $Q = \{[1,0,1,0,0]^T, [1,0,0,1,0]^T, [0,0,0,1,1]^T, [1,1,0,0,1]^T, [0,0,1,0,1]^T, [0,1,0,0,2]^T\}$, the set of events $\Sigma = \{t_1, t_2, t_3\}$, and the initial state $s = [1,0,1,0,0]^T$.

Formally, the equivalent automaton of a Petri net $(\mathcal{N}, \mu_0)$ can be expressed as $G = (Q, \Sigma, \delta, s, F)$ with $Q = \mathcal{R}(\mathcal{N}, \mu_0)$, $\Sigma = \bigcup_{t \in T} \rho(t) \setminus \{\lambda\}$, $s = \mu_0$, and δ such that $\mu' \in \delta(\mu, \alpha)$ if and only if $\mu \xrightarrow{t} \mu'$ and $\alpha \in \rho(t)$. Thus, the equivalent automaton is the reachability graph expressed in the automata terminology.

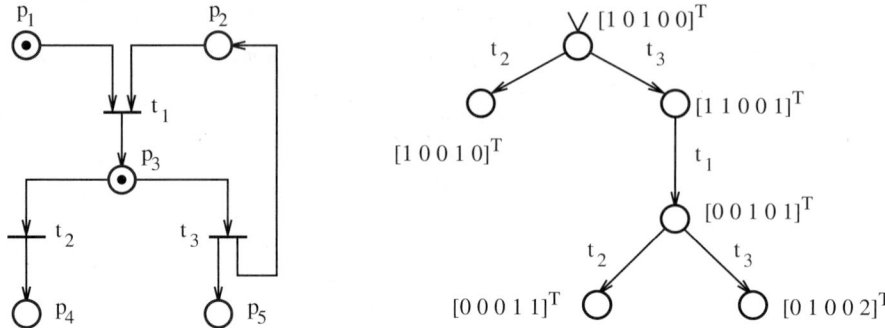

Fig. 2.14. Petri net and equivalent automaton.

Note that the reachability graph corresponds in general to a nondeterministic automaton, unless certain assumptions are made on the labeling of the Petri net. In particular, for free-labeled Petri nets, the reachability graph corresponds to a deterministic automaton.

Recall that the reachability graph of a Petri net is finite if and only if the Petri net is bounded. Thus, an unbounded Petri net (while having a finite Petri net representation) cannot be represented with a finite automaton. Further, even when the Petri net is bounded, the reachability graph may have a very large number of nodes, when compared to the number of places and transitions of the Petri net. For instance, in Fig. 2.9(a), the number of nodes is 2^n for initial markings μ with $\mu_1 + \mu_2 = n$. Situations in which the reachability graph has a small number of nodes or when the number of nodes is comparable to the number of transitions and places of a Petri net are rather unusual. For instance, the reachability graph has only one node when the Petri net is in deadlock at the initial marking. Examples of live Petri nets with small reachability graphs are also possible (Fig. 2.15).

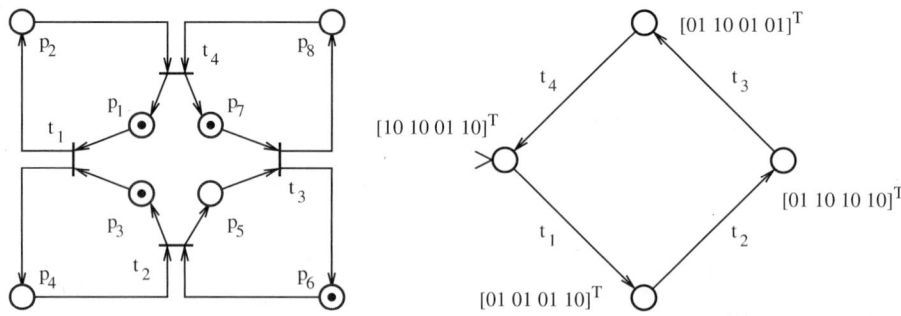

Fig. 2.15. Live Petri net with "small" reachability graph.

Finite automata can easily be converted to labeled Petri nets, as illustrated in Fig. 2.16. The resulting Petri net is a state machine. However, converting finite automata to free-labeled Petri nets is a more difficult problem, which may not always have a solution. For instance, Fig. 2.17 shows an automaton that does not have an equivalent free-labeled Petri net. Indeed an equivalent free-labeled Petri net would enable both c and d after the firing of ab, since the same marking is reached by firing ab and ba.

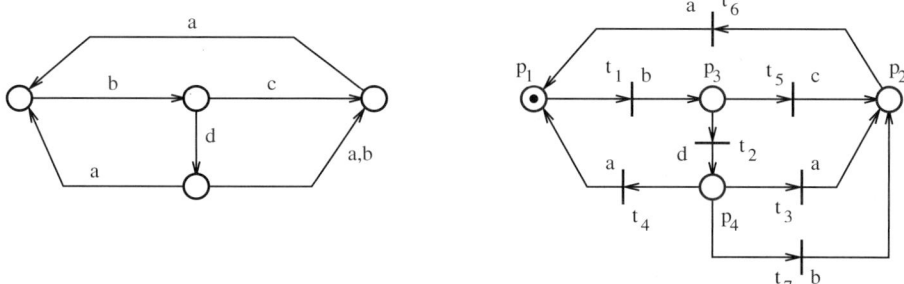

Fig. 2.16. Automaton and equivalent labeled Petri net.

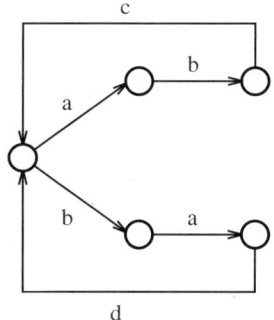

Fig. 2.17. An automaton that does not have an equivalent free-labeled Petri net.

2.3.2 Comparison

Compared to automata, Petri nets offer compact representations of concurrent systems. Further, the modeling of concurrent systems is usually more natural using Petri nets. For instance, modeling the print server problem of section 2.2.1 with automata may appear more involved, due to concurrency: two print jobs may be processed at the same time. We will see that a possible systematic way to model concurrent systems with automata is to first find the Petri net model, and then build the reachability graph.

To compare Petri nets and automata, we consider systems that consist of several subsystems and compare the sizes of the Petri net and automaton models of the overall system. We begin by defining the composition operations, for automata first and then for Petri nets.

For the sake of generality we will refer to the notation of nondeterministic automata, as deterministic automata can be seen as a special class of the nondeterministic automata. Given an automaton $G = (Q, \Sigma, \delta, s, F)$, let $Ac(G)$ denote the automaton obtained by removing from G the states $q \in Q$ that are unreachable, that is, the states $q \in Q$ such that $\not\exists \sigma \in \Sigma^* : q \in \delta(s, \sigma)$. $Ac(G)$ is known as the *accessible part* of G.

The **parallel composition** of automata, also known as the **synchronous composition**, is defined as follows. Let $G_1 = (Q_1, \Sigma_1, \delta_1, s_1, F_1)$ and $G_2 = (Q_2, \Sigma_2, \delta_2, s_2, F_2)$ be two automata. For the sake of generality, assume G_1 and G_2 are nondeterministic. The parallel composition is an automaton G denoted as $G = G_1 \| G_2$, where

$$G = Ac(Q_1 \times Q_2, \Sigma_1 \cup \Sigma_2, \delta, (s_1, s_2), F_1 \times F_2) \tag{2.9}$$

and the transition function δ is defined as follows. For $i = 1, 2$, let $\overline{\delta}_i(q, \alpha) = \delta_i(q, \alpha) \; \forall \alpha \in \Sigma_i$ and $\overline{\delta}_i(q, \alpha) = \{q\} \; \forall \alpha \in (\Sigma_1 \cup \Sigma_2 \cup \{\lambda\}) \setminus \Sigma_i$. Then

$$\delta((q_1, q_2), \alpha) = \overline{\delta}_1(q_1, \alpha) \times \overline{\delta}_2(q_2, \alpha). \tag{2.10}$$

In (2.10), note that for any set A we have $A \times \emptyset = \emptyset$ and $\emptyset \times A = \emptyset$. Note also that $\lambda \notin \Sigma_1 \cup \Sigma_2$. A composition example is shown in Fig. 2.18, in which we compose $G_1 = (\{A, B\}, \{a, b, d\}, \delta_1, A, B)$ with $G_2 = (\{1, 2\}, \{a, b\}, \delta_2, 1, 2)$. The composition follows from the definition. The initial state is $(A, 1)$, as both A and 1 are initial states. The final state is $(B, 2)$, as both B and 2 are final states. There is a transition labeled by the event a between $(A, 1)$ and $(B, 2)$, as there is a transition labeled by a from A to B, and from 1 to 2. The transition between $(B, 2)$ and $(B, 1)$ is obtained the same way. The transition from $(B, 2)$ to $(A, 2)$ appears as the event d leads from B to A and $d \notin \Sigma_1$.

The relation between the language of the parallel composition and the language of the composed automata is as follows. Given two sets of events Σ and Σ_1, a **projection** is a map $P : \Sigma^* \to \Sigma_1^*$ such that $P(\sigma)$ removes from a sequence $\sigma \in \Sigma^*$ all events that are not in Σ_1. Formally, P satisfies that $P(\lambda) = \lambda$, $\forall \alpha \in \Sigma \cap \Sigma_1 : P(\alpha) = \alpha$, $\forall \alpha \in \Sigma \setminus \Sigma_1 : P(\alpha) = \lambda$, and $\forall \sigma \in \Sigma^* \forall \alpha \in \Sigma : P(\alpha\sigma) = P(\alpha)P(\sigma)$. (Since λ is the null symbol, $\lambda\lambda = \lambda$, $\alpha\lambda = \alpha$, $\lambda\alpha = \alpha$.) The inverse of P is $P^{-1} : \Sigma_1^* \to \Sigma^*$, satisfying $P^{-1}(\sigma_1) = \{\sigma \in \Sigma^* : P(\sigma) = \sigma_1\}$. Let P_1 and P_2 be the projections from $(\Sigma_1 \cup \Sigma_2)^*$ to Σ_1^* and Σ_2^*, respectively. Then

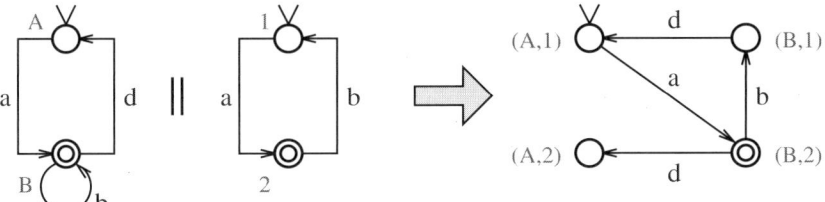

Fig. 2.18. Parallel composition of automata.

$$\mathcal{L}(G_1 \| G_2) = P_1^{-1}(\mathcal{L}(G_1)) \cap P_2^{-1}(\mathcal{L}(G_2)), \tag{2.11}$$
$$\mathcal{L}_m(G_1 \| G_2) = P_1^{-1}(\mathcal{L}_m(G_1)) \cap P_2^{-1}(\mathcal{L}_m(G_2)). \tag{2.12}$$

The size of the composed model is as follows. Assume all states of $Q_1 \times Q_2$ are reachable from the initial state. If G_1 has m_1 states and G_2 has m_2 states, then $G_1 \| G_2$ has $m_1 m_2$ states. This shows why the state space "explodes" when we compose several automata: the number of states of the parallel composition is the product of the number of states of the automata that are composed. Next, for an automaton of state transition δ, let's count each (q, α, q') with $q' \in \delta(q, \alpha)$ as an event-transition.[1] Let's calculate the number of event-transitions of $G_1 \| G_2$. Let $\nu_1 : \Sigma_1 \cup \{\lambda\} \to \mathbb{N}$, such that for all $\alpha \in \Sigma_1 \cup \{\lambda\}$, $\nu_1(\alpha)$ equals the number of event-transitions of G_1 labeled by α. Let ν_2 be similarly defined for $\mathcal{N}_2$. Then, the number of event-transitions of $G_1 \| G_2$ is

$$NET \leq m_1 \left(\nu_2(\lambda) + \sum_{\alpha \in \Sigma_2 \setminus \Sigma_1} \nu_2(\alpha) \right) + m_2 \left(\nu_1(\lambda) + \sum_{\alpha \in \Sigma_1 \setminus \Sigma_2} \nu_1(\alpha) \right)$$
$$+ \sum_{\alpha \in \Sigma_1 \cap \Sigma_2} \nu_1(\alpha) \nu_2(\alpha), \tag{2.13}$$

where we have equality when all states of $Q_1 \times Q_2$ are reachable.

For Petri nets, the parallel composition can be defined in terms of structures. Let $\mathcal{N}_1 = (P_1, T_1, D_1^-, D_1^+, \rho_1)$ and $\mathcal{N}_2 = (P_2, T_2, D_2^-, D_2^+, \rho_2)$ be two labeled structures, where $\rho_1 : T_1 \to \Sigma_1 \cup \{\lambda\}$ and $\rho_2 : T_2 \to \Sigma_2 \cup \{\lambda\}$. The composition results in a structure $\mathcal{N} = (P, T, F, W, \rho)$ such that $\mathcal{N}$ contains the transitions of T_1 that have a λ label or labels not found on the transitions of T_2, the transitions of T_2 that have a λ label or labels not found on the transitions of T_1, and transitions t that model synchronizations of pairs $(t^1, t^2) \in T_1 \times T_2$ that share common labels different from λ. The formal description is introduced by means of the following algorithm, illustrated also on the example of Fig. 2.19.

---

[1] In the graphical representation of an automaton, an arc between states may correspond to several event-transitions, if the arc is labeled by multiple events.

ALGORITHM 2.2   Parallel Composition of Petri Nets

1. Let $P = P_1 \cup P_2$ and $T = \emptyset$.

2. Let $L_1 = \bigcup_{t \in T_2} \rho_2(t) \setminus \{\lambda\}$ and $L_2 = \bigcup_{t \in T_1} \rho_1(t) \setminus \{\lambda\}$.

3. For $i = 1, 2$, let $T_{i,-} = \{t \in T_i : \rho_i(t) \setminus L_i \neq \emptyset\}$.

4. For $i = 1, 2$, $\forall t \in T_{i,-}$, let $T = \{t\} \cup T$, $\rho(t) = \rho_i(t) \setminus L_i$, set $D^-(p, t) = D_i^-(p, t)$ and $D^+(p, t) = D_i^+(p, t)$ $\forall p \in P_i$, and set $D^-(p, t) = 0$ and $D^+(p, t) = 0$ $\forall p \in P \setminus P_i$.

5. For each pair of transitions $(t^1, t^2) \in T_1 \times T_2$ such that $(\rho_1(t^1) \cap \rho_2(t^2)) \setminus \{\lambda\} \neq \emptyset$, create a new transition t: $T = \{t\} \cup T$, $\rho(t) = (\rho_1(t^1) \cap \rho_2(t^2)) \setminus \{\lambda\}$, and set $D^-(p, t) = D_i^-(p, t)$ and $D^+(p, t) = D_i^+(p, t)$ $\forall p \in P_i$ and $i = 1, 2$.

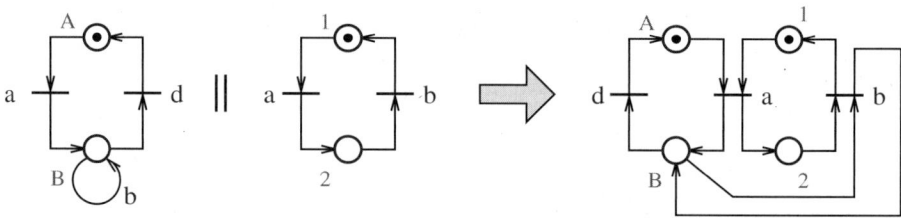

Fig. 2.19. Parallel composition of Petri nets.

Note that if $\mathcal{N}_1$ has m_1 places and $\mathcal{N}_2$ has m_2 places, then $\mathcal{N}$ has $m_1 + m_2$ places. Thus, unlike the automata case, the composition results in a linear increase of the number of places. Let's denote each pair (t, α) of a Petri net, where $\alpha \in \rho(t)$ is a label of t, as an event-transition. Further, let $\nu_1 : \Sigma_1 \cup \{\lambda\} \to \mathbb{N}$, such that for all $\alpha \in \Sigma_1 \cup \{\lambda\}$, $\nu_1(\alpha)$ equals the number of event-transitions of $\mathcal{N}_1$ labeled by α. Let ν_2 be similarly defined for $\mathcal{N}_2$. Then, the number of event-transitions of $\mathcal{N}$ is

$$NET = \nu_2(\lambda) + \sum_{\alpha \in \Sigma_2 \setminus \Sigma_1} \nu_2(\alpha) + \nu_1(\lambda) + \sum_{\alpha \in \Sigma_1 \setminus \Sigma_2} \nu_1(\alpha)$$
$$+ \sum_{\alpha \in \Sigma_1 \cap \Sigma_2} \nu_1(\alpha)\nu_2(\alpha). \tag{2.14}$$

By comparing (2.14) and (2.13), we see that the number of transitions of the composed Petri net does not grow as fast as that of the composed automata. In fact, for Petri nets only synchronizations may cause an increase in the number of transitions that is not linear in terms of the size of the composed models. Further, as seen in the reachability graph construction, a Petri net model of a system may be significantly more compact than its equivalent automaton model of the system. This makes the difference between (2.14) and (2.13) even more dramatic.

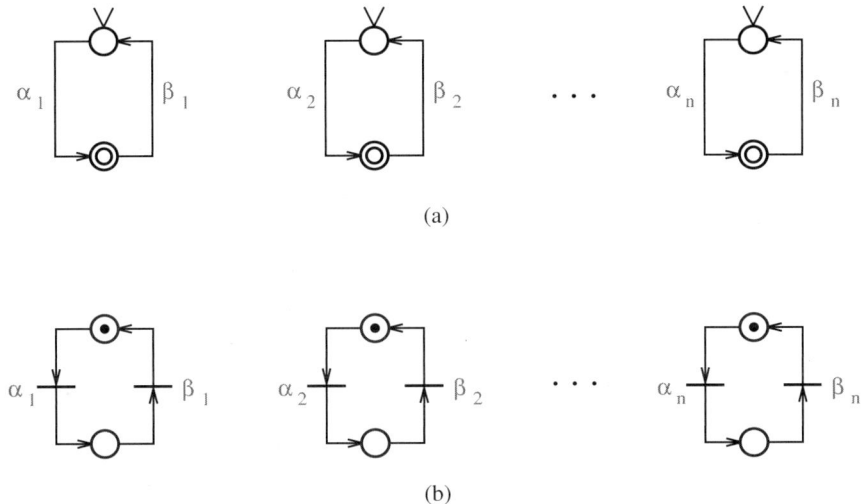

(a)

(b)

Fig. 2.20. Example for parallel composition of automata versus Petri nets.

As an example, consider a system consisting of n independent subsystems, each with two states, shown in Fig. 2.20(a). Thus, all events α_i and β_i are assumed to be distinct. Each subsystem can be modeled by a Petri net of two places and two transitions. The composition of the n subsystems results in a Petri net with $2n$ places and $2n$ transitions (Fig. 2.20(b)). However, the global automaton model has 2^n states and $n2^n$ transitions!

Petri nets and automata can be further compared by the modeling effort they require. Let's consider again a system consisting of several subsystems, each modeled as a finite automaton. Each finite automaton can be seen as a Petri net state machine, and by composing the subsystems using Petri net composition, we obtain a global Petri net model $(\mathcal{N}, \mu_0)$. Note that the reachability graph of $(\mathcal{N}, \mu_0)$ is the same as the parallel composition of the automata representing the subsystems. We can conclude that the Petri net composition is an intermediary result of the automata composition, and hence less computational effort is required for Petri net modeling. Recall also that the conversion of Petri nets to automata tends to be computationally complex, as a "small" Petri net model may correspond to an automaton with a "huge" number of states.

In this book, we deal with the supervision of discrete-event systems. Thus, based on a discrete-event model, a supervisor is designed. The approaches based on Petri net models and automata models can be compared as follows. A "fast" design method for automata having, for instance, polynomial complexity, may or may not be faster than a "slow" design method for Petri nets having, for instance, exponential complexity. This conclusion is not surprising, once we realize that the complexity of an automata-based approach depends on the size of the automata, which may be exponential in terms of the size of the Petri net models. Thus, computationally efficient methods for Petri nets promise computational savings, when compared to automata approaches. Further, automata approaches can be seen as a special case of Petri net approaches, since a Petri net can be converted to an automaton.

Two types of Petri net problems that may not be approachable in the automata setting are as follows. First, problems involving unbounded Petri nets correspond to problems involving automata with an infinite number of states. Second, problems in which the initial marking of the Petri net is a parameter instead of a given variable, may not be approachable in the automata setting. The reason is that the automata approach may need to deal with automata with infinite number of states, since the potential number of initial markings of the Petri net may not be bounded. In this book we will consider arbitrary Petri nets, which are not necessarily bounded, and structural approaches that do not depend on the initial marking.

2.4 Bibliographical Notes

Petri nets have been extensively studied in the literature. Survey papers and books include [41, 145] and [53, 149, 160]. Petri nets and their extensions have been used in various contexts, including distributed algorithms [161, 162], program specification [135, 29], communication protocols [37, 187, 23], manufacturing [46, 45, 155, 218], digital circuits [212, 211], and robotics [136].

In this book we have followed the Petri net definition from the survey by Murata [145]. Minor variations of this definition are used in some of the literature. The Petri nets of this book are also known as P/T nets [160].

3

The Supervision of Petri Nets

3.1 Introduction

The goal of this chapter is to introduce the supervision of Petri nets. A *supervisor* is a DES that is used to monitor and control another DES, which is called *the plant*, such that a given specification is satisfied. The supervisor is different from a controller in the sense that a controller dictates the input applied to the system, while the supervisor only restricts the set of inputs that can be applied to the system. The set of inputs is restricted dynamically, based on the observation of the plant. Ideally, a supervisor should forbid only the inputs that may unavoidably lead to the violation of the specification. The composition of a supervisor with a plant is a DES called *the closed loop*.

When the plant is a Petri net, a supervisor restricts the operation of the net by restricting the set of enabled transitions that may fire at a given state of the Petri net. The specification may consist of requirements concerning the reachable markings or permissible firing sequences.

A possible way to apply the supervision of Petri nets in the real world is as follows. Given a plant with discrete-event-driven dynamics, a Petri net model is extracted. We call the extracted model the *plant Petri net*. In the plant Petri net, the occurrence of an event in the physical plant is modeled by the firing of a transition. The (observable) transition firings of the plant are the information available to the supervisor. A transition firing may trigger a change in the state of the supervisor, while a change in the supervisor state can change the set of plant events disabled by the supervisor. Physically, the event disablement can be done by restricting the range of the inputs of the plant.

The design of a supervisor typically involves the following requirements:

1. If possible, the closed loop should be representable as a Petri net.
2. If the exact implementation of the specification is impossible, transform the specification to a more restrictive form that is implementable.
3. If possible, find a least restrictive supervisor.

When the first requirement is satisfied, Petri net methods can be used for the analysis of the closed loop. The second requirement matters in the case of systems with uncontrollable and/or unobservable transitions. With regard to the third requirement, note that a least restrictive supervisor may not always exist.

Several supervision settings are possible. They are presented in section 3.3. Throughout the book, we will often refer to a supervision setting in which transitions can be individually controlled if controllable, and individually observed if observable. Thus, in this setting, the supervisor observes the firings of the observable transitions and, depending on its state, inhibits firings of controllable transitions. Note that no state parameters of the plant Petri net (such as the marking) are directly available to the supervisor, with the exception of the initial marking. The initial marking is assumed to be known, as it represents the initial state of the plant.

The main supervision approach referred to in this book is *the supervision based on place invariants*. It is described in section 3.2. There are various supervision settings that can be distinguished mainly based on the type of partial controllability and observability. They are defined in section 3.3. The concepts of feasible and admissible specifications are introduced in section 3.4. Methods that extend the supervision based on place invariants to partial controllability and observability are discussed in section 3.5. An overview of the methods for the supervision of Petri nets is presented in section 3.6.

3.2 Supervision Based on Place Invariants

This section briefly reviews an effective approach used for the enforcement of linear marking constraints of the form $L\mu \leq b$. The approach is called supervision based on place invariants (SBPI). Literature references describing this approach include [57, 141, 142, 214]. The case when all transitions of the plant Petri net are controllable and observable is presented first. Then, an extension to Petri nets with uncontrollable and unobservable transitions is also presented.

3.2.1 Fully Controllable and Observable Petri Nets

The control problem considered here is to enforce a set of n_c linear constraints

$$L\mu \leq b. \tag{3.1}$$

In other words, we desire the supervisor to prevent the system from reaching markings which do not satisfy (3.1). The notation is as follows: L is an integer $n_c \times m$ matrix (n_c: the number of constraints, m: the number of places of the given Petri net), b is an integer column vector, and μ is the Petri net marking. Let μ_c be a vector of n_c nonnegative slack variables, defined as

$$\mu_c = b - L\mu. \tag{3.2}$$

Let D be the incidence matrix and $\mu_{c0} = b - L\mu_0$, where μ_0 is the initial marking of the Petri net. Then, it can be verified that the least restrictive supervisor can be implemented by a Petri net with the same set of transitions, having the marking μ_c, the initial marking μ_{c0}, and the incidence matrix

$$D_c = -LD. \tag{3.3}$$

That is, the closed-loop Petri net (the supervisor plus the plant Petri net) has the incidence matrix $D_{cl} = [D^T, D_c^T]^T$ and the marking $\mu_{cl} = [\mu^T, \mu_c^T]^T$. This result is summarized in the following theorem [141, 214].

Theorem 3.1. *Let a Petri net with incidence matrix D and initial marking μ_0 be given. A set of n_c linear constraints $L\mu \leq b$ are to be imposed. If $L\mu_0 - b \leq 0$ then a Petri net supervisor with incidence matrix $D_c = -LD$ and initial marking $\mu_{c0} = b - L\mu_0$ enforces the constraint $L\mu \leq b$ when included in the closed-loop system $D_{cl} = [D^T, D_c^T]^T$. Furthermore, the supervision is least restrictive.*

The places of the supervisor Petri net are called **control places**. Note that each control place is in one of the place invariants described by (3.2). As an example, in Fig. 3.2, the places C_1 and C_2 are the control places of a supervisor enforcing $2\mu_1 + \mu_3 \geq 1$ and $2\mu_2 + \mu_3 \geq 1$.

3.2.2 Partially Controllable and Observable Petri Nets

This section describes the extension of the approach of section 3.2.1 to Petri nets with uncontrollable and unobservable transitions. In a Petri net, a transition is **uncontrollable** if the supervisors are not given the ability to directly inhibit it. Otherwise, the transition is **controllable**. A transition is **unobservable** if the supervisors are not given the ability to directly detect its firing. Otherwise, the transition is **observable**. The firing of uncontrollable (unobservable) transitions corresponds to the occurrence of uncontrollable (unobservable) events. In our paradigm the supervisors observe transition firings, not markings. For instance, consider the Petri net of Fig. 3.1. First assume that t_1 is controllable and t_2 is uncontrollable. This means that t_2 cannot be directly controlled. So, in case (a) t_2 cannot be directly inhibited; it will eventually fire. However in case (b) t_2 can be indirectly prevented from firing by inhibiting t_1. Now assume that t_2 is unobservable and t_3 is observable. This means that we cannot detect when t_2 fires. In other words, the state of a supervisor is not changed by firing t_2. However we can indirectly detect that t_2 has fired, by detecting the firing of t_3.

In section 3.2.1, the Petri net supervisor is implemented in the form of control places connected to the plant Petri net. To deal with uncontrollable and unobservable transitions, it is necessary to ensure that no control place

(a)   (b)

Fig. 3.1. Defining the controllability and observability of transitions.

ever attempts to inhibit an uncontrollable transition enabled in the plant Petri net, and no control place marking is varied by firing unobservable transitions enabled in the closed-loop Petri net. To this end, it is sufficient to have the set of constraints satisfy the following relations of [141, 142]:

$$LD_{uc} \leq 0, \tag{3.4}$$

$$LD_{uo} = 0, \tag{3.5}$$

where $D_{uc} = D(\cdot, T_{uc})$ and $D_{uo} = D(\cdot, T_{uo})$ are the restrictions of the incidence matrix D to the columns corresponding to the set of uncontrollable transitions T_{uc} and to the columns corresponding to the set of unobservable transitions T_{uo}. All sets of constraints $L\mu \leq b$ satisfying (3.4)–(3.5) may be enforced as in section 3.2.1. The relation (3.4) ensures that there is no arc from a control place of the supervisor to an uncontrollable transition. Relation (3.5) ensures that there is no arc between a control place of the supervisor and an unobservable transition.

As will be seen in section 3.6, there are various methods that can be used for constraints $L\mu \leq b$ which do not satisfy (3.4)–(3.5). In this book, of special interest are the methods that replace $L\mu \leq b$ by another set of constraints $L_a\mu \leq b_a$ such that

$$L_a\mu \leq b_a \Rightarrow L\mu \leq b \tag{3.6}$$

and L_a satisfies (3.4)–(3.5). The relation (3.6) should be understood this way: for all markings μ of interest (such as the markings reachable from a set of initial markings), (3.6) is satisfied. The methods that obtain an admissible form $L_a\mu \leq b_a$ based on the specification $L\mu \leq b$ will be called methods for constraint transformation. Constraint transformation methods are discussed later, in section 3.5. It should be noticed that the least restrictive solution to the problem of enforcing a set of constraints $L\mu \leq b$ may not correspond to a set of constraints of the form $L_a\mu \leq b_a$ [57]. Thus, constraint transformation methods are typically suboptimal.

As an example, consider the Petri net of Fig. 3.2(a), where t_1 is unobservable. We have

$$D = \begin{bmatrix} -1 & 1 & 0 \\ -1 & 0 & 1 \\ 2 & -1 & -1 \end{bmatrix}, \quad D_{uo} = \begin{bmatrix} -1 \\ -1 \\ 2 \end{bmatrix}.$$

The matrix D_{uc} is empty. Assume that we want to enforce (3.1) for

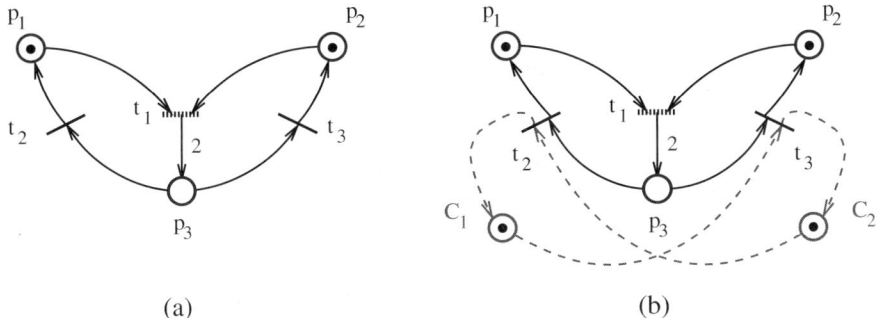

Fig. 3.2. (a) Petri net; (b) closed loop.

$$L = - \begin{bmatrix} 1 & 0 & 1 \\ 0 & 1 & 1 \end{bmatrix}, \quad b = - \begin{bmatrix} 1 \\ 1 \end{bmatrix}.$$

Then we can transform (3.1) to the constraints $L_a \mu \leq b_a$, for

$$L_a = - \begin{bmatrix} 2 & 0 & 1 \\ 0 & 2 & 1 \end{bmatrix}, \quad b_a = - \begin{bmatrix} 1 \\ 1 \end{bmatrix}.$$

By enforcing $L_a \mu \leq b_a$ with the approach of section 3.2.1 we obtain the closed-loop Petri net of Fig. 3.2(b). The transition arcs of the supervisor are represented with dashed lines. The control places are C_1 and C_2; they correspond to the first and second rows of $L_a D$, respectively.

3.3 Supervision Settings

There are various supervision settings depending on the type of concurrency setting and the way uncontrollability and unobservability are modeled. Further, supervision can be *deterministic* or *nondeterministic*. The concurrency settings were introduced in section 2.2.5 on page 19. So far, we have presented the SBPI in the no concurrency setting. The properties of the method do not change in the more general concurrency settings [181]. Three settings of controllability and observability are especially important in this book:

- individually controllable and observable transitions,
- labeled Petri nets,
- double-labeled Petri nets.

In addition, we will also describe the *controlled Petri net* (CtlPN) setting, which has been used much in the literature.

Individually controllable and observable transitions: This is the setting we have used so far. In this setting, the set of transitions T is partitioned in

$T = T_c \cup T_{uc}$ and $T = T_o \cup T_{uo}$, where T_c (T_o) is the set of controllable (observable) transitions and T_{uc} (T_{uo}) is the set of observable (unobservable) transitions. Thus, a supervisor has the ability to control only the transitions $t \in T_c$ and to observe only the firings of $t \in T_o$.

Labeled Petri nets: In this setting the plant is modeled by labeled Petri nets, instead of free-labeled Petri nets. Thus, the set of events Σ is partitioned in $\Sigma = \Sigma_c \cup \Sigma_{uc}$ and $\Sigma = \Sigma_o \cup \Sigma_{uo}$, where Σ_c (Σ_o) is the set of controllable (observable) events and Σ_{uc} (Σ_{uo}) is the set of observable (unobservable) events. Recall, a **labeled Petri net** is a Petri net enhanced with a labeling function $\rho : T \rightarrow 2^{\Sigma} \cup \{\lambda\}$, where Σ is the set of events, ρ the labeling function, and λ the null event. Note that a transition labeled by multiple events can be modeled by several identical transitions, each labeled by a single event. Thus, without loss of generality, we will assume $\rho : T \rightarrow \Sigma \cup \{\lambda\}$.

In this setting, when a transition t fires, an event $e \in \rho(t)$ is generated. If $e \in \Sigma_c$ ($e \in \Sigma_o$), the supervisor is able to disable (observe) this event. Note that t is disabled by the supervisor only when all events $e \in \rho(t)$ are disabled. Compared to the setting of individually controllable and observable transitions, one difference is that two transitions t_1 and t_2 may produce the same event when fired. Here, a supervisor controls/observes transitions indirectly, by disabling/observing events.

This setting corresponds to the Ramadge–Wonham setting used for automata [159]. However, while the latter is usually associated with the no concurrency assumption, we do not assume any particular concurrency setting for labeled Petri nets.

Double-labeled Petri nets: A **double-labeled Petri net** is a labeled Petri net having separate labels for observation and control. Thus, the Petri net is enhanced with two labeling functions: $\rho : T \rightarrow 2^{\mathcal{K}} \cup \{\lambda\}$ and $o : T \rightarrow \mathcal{O} \cup \{\lambda\}$, where ρ labels transitions with subsets of control events $e \in \mathcal{K}$, and o labels transitions with observation events $\alpha \in \mathcal{O}$. Thus, $\mathcal{K}$ ($\mathcal{O}$) is the set of control (observation) events. The meaning of the two labellings is as follows: a transition $t \in T$ is control enabled when there is an event $e \in \rho(t)$ that is enabled. Further, when t fires, the event $o(t)$ is generated. Note that as in the case of labeled Petri nets, we will assume without loss of generality that $\rho : T \rightarrow \mathcal{K} \cup \{\lambda\}$ (each transition is labeled by a unique control event).

The sets $\mathcal{K}$ and $\mathcal{O}$ can be partitioned in $\mathcal{K} = \mathcal{K}_c \cup \mathcal{K}_{uc}$ and $\mathcal{O} = \mathcal{O}_o \cup \mathcal{O}_{uo}$, where $\mathcal{K}_c$ ($\mathcal{O}_o$) is the set of controllable (observable) events and $\mathcal{K}_{uc}$ ($\mathcal{O}_{uo}$) is the set of observable (unobservable) events. Thus, a supervisor is given the ability to disable control events $e \in \mathcal{K}_c$ and observe events $\alpha \in \mathcal{O}_o$.

Controlled Petri nets (CtlPNs): Following [70], a **controlled Petri net (CtlPN)** is a triple $\mathcal{N}^c = (\mathcal{N}, \mathcal{C}, \mathcal{B})$, where $\mathcal{N} = (P, T, F)$ is an ordinary Petri net, $\mathcal{C}$ is a finite set of **control places**, $\mathcal{C} \cap P = \emptyset$, and $\mathcal{B} \subseteq \mathcal{C} \times T$ is a set of directed arcs. As expected, given a marking μ of $\mathcal{N}$, a transition t is enabled by the plant, or *state enabled*, when for all places $p \in P$, $(p, t) \in F \Rightarrow \mu(p) \geq 1$. A **control** for a CtlPN is a function $u : \mathcal{C} \rightarrow \{0, 1\}$. Given a control

u, a transition t is *control enabled* when for all control places c, $(c, t) \in \mathcal{B} \Rightarrow u(c) = 1$. Of course, a transition can be fired only when it is both control and state enabled. In the case of concurrency, the control allows all control-enabled transitions to be fired simultaneously.

Note that firing a transition has no effect on the control (there are no tokens flowing out of the control places). This feature emphasized the difference between the control places of the CtlPNs and the control places of the supervision based on invariants. The latter are completely identical with the places of Petri net places.

In this setting, a transition is controllable if connected to a control place. Compared to the setting of individually controllable transitions, note that here it may not be possible to selectively control-enable arbitrary groups of controllable transitions. For instance, if $(c, t_1), (c, t_2) \in \mathcal{B}$ and there are no other control places connected to c, either both t_1 and t_2 are control enabled or neither of them are. That is, it is not possible to control-enable only t_1 or only t_2.

Defining supervisors: Based on the controllability and observability settings that we have considered, a general and formal definition of supervisors can be presented. Note that this book does not use CtlPNs. The results of this book could be applied to problems involving CtlPN models by converting the CtlPN models to the double-labeled setting. The other two settings can be seen as a particular case of the double-labeled setting. Thus, the individually controllable and observable transition setting corresponds to $\mathcal{K} = \mathcal{O} = T$ and $\rho(t) = o(t) = \{t\} \; \forall t \in T$. Further, the labeled Petri net setting corresponds to $\mathcal{K} = \mathcal{O} = \Sigma$ and $\rho(t) = o(t) \; \forall t \in T$. Therefore, in what follows we will refer to the double-labeled setting.

In the no concurrency case, a firing sequence can be represented by a sequence of transitions $\sigma = t_1 t_2 \ldots$. To account for concurrency, a firing sequence is represented by a sequence of firing vectors $\sigma = q_1 q_2 \ldots$. As we have seen in the section on the SBPI, a supervisor may be defined for several initial states of a system, in which case its behavior depends also on the initial state of the system. Typically, the initial marking is the initial state of a system. In some cases, the initial state may also include the initial value of some other parameters, such as the Parikh vector in section 4.2. Denoting by $\mathcal{M}$ the set of initial states of the plant, by Q the set of firing vectors, and by Q^* the set of sequences σ, a **deterministic supervisor** is a map $\Xi : \mathcal{M} \times Q^* \to Q$. For all $x \in \mathcal{M} \times Q^*$, $\Xi(x)$ represents the set of supervisor-enabled firing vectors: a firing vector q is enabled if and only if $q \leq \Xi(x)$. A **nondeterministic supervisor** is a map $\Xi : \mathcal{M} \times Q^* \to 2^Q$. For all $x \in \mathcal{M} \times Q^*$, an element $A \in \Xi(x)$ is randomly selected to represent the set of supervisor-enabled firing vectors: q is enabled if and only if $q \leq A$.

The formal definitions of supervisors presented so far do not take into account the feasibility issues arising from the limited controllability and observability of the system. Therefore, we define next feasible supervisors. In the no concurrency case, firing $\sigma = t_1 t_2 \ldots$ results in a sequence of events $x = o(t_1) o(t_2) \ldots$. The sequence x is observed as a sequence $x_o = \alpha_1 \alpha_2 \ldots$, $\alpha_i \in \mathcal{O}_o$ for $i = 1, 2, \ldots$, where x_o is obtained from x by deleting all unobservable events. In the case of concurrency, when $\sigma = q_1 q_2 \ldots$ is fired, the observation of σ consists of a sequence of observation vectors $x_o = \omega_1 \omega_2 \ldots$, defined as follows. Given a firing vector q, the observation vector of q is a vector $\omega \in \mathbb{N}^{|\mathcal{O}_o|}$ indexed by the events of $\mathcal{O}_o$, such that $\forall \alpha \in \mathcal{O}_o$, $\omega(\alpha) = \sum_{t \in o^{-1}(\alpha)} q(t)$. In the no concurrency case, based on the observation of the system, subsets of controllable events can be selectively enabled. In the case of concurrency, it is also useful to control the number of times enabled events can be concurrently fired. Thus, the controllable events that are enabled are described by a control vector $\gamma \in \mathbb{N}^{|\mathcal{K}_c|}$ indexed by the events of $\mathcal{K}_c$. By applying the control vector γ to the system, only the firing vectors q that satisfy $\forall e \in \mathcal{K}_c$, $\gamma(e) \geq \sum_{t \in \rho^{-1}(e)} q(t)$ are allowed to fire. Let Γ denote the set of control vectors γ, Ω the set of observation vectors ω, and Ω^* the set of sequences of observation vectors $\omega_1 \omega_2 \ldots$. Let $\pi_o : Q^* \to \Omega^*$ denote the function that maps a firing sequence $\sigma = q_1 q_2 \ldots$ into the sequence of observation vectors. A *feasible* supervisor should be implementable by observing only observation vectors and controlling only controllable events. A formal definition follows.

Definition 3.2. *Given a Petri net $\mathcal{N}$, let $\mathcal{M}$ be the set of all initial states. A deterministic supervisor Ξ is* **feasible** *if there is a map $\overline{\Xi} : \mathcal{M} \times \Omega^* \to \Gamma$ such that $\forall s \in \mathcal{M}$, $\forall \sigma \in Q^*$, $\gamma = \overline{\Xi}(s, \pi_o(\sigma)) \Rightarrow \Xi(s, \sigma) = \{q \in Q : \forall e \in \mathcal{K}_c, \gamma(e) \geq \sum_{t \in \rho^{-1}(e)} q(t)\}$. Further, a nondeterministic supervisor Ξ is* **feasible** *if there is a map $\overline{\Xi} : \mathcal{M} \times \Omega^* \to 2^\Gamma$ such that $\forall s \in \mathcal{M}$, $\forall \sigma \in Q^*$, $\Xi(s, \sigma) = \{S_\gamma : \gamma \in \overline{\Xi}(s, \pi_o(\sigma))\}$, where $S_\gamma = \{q \in Q : \forall e \in \mathcal{K}_c, \gamma(e) \geq \sum_{t \in \rho^{-1}(e)} q(t)\}$.*

The significance of nondeterministic supervisors can be illustrated as follows. Referring to the notation of Definition 3.2, a supervisor Ξ_1 is at least as restrictive as Ξ_2 if for all $x \in \mathcal{M} \times \Omega^*$, $\overline{\Xi}_1(x) \leq \overline{\Xi}_2(x)$. For some specifications, it may be that there is a reachable state s at which two control vectors γ_1 and γ_2 could be applied, $\gamma_1 \not\leq \gamma_2$, $\gamma_1 \not\geq \gamma_2$, and applying any control satisfying $\gamma \geq \gamma_1$ and $\gamma \geq \gamma_2$ may break the specification. Thus, a least restrictive deterministic supervisor enforcing the specification cannot be defined, as a supervisor applying γ_1 at s cannot be compared with one applying γ_2 at s. This obstacle can be removed by allowing a supervisor to make a nondeterministic choice from a set of possible control vectors. Thus, a nondeterministic supervisor could be defined to choose randomly one of γ_1 or γ_2 at the state s. Note that there is no need to define nondeterministic supervisors in the no concurrency setting.

3.4 Admissible and Feasible Sets of Constraints

This section introduces two concepts: admissible sets of constraints and feasible sets of constraints. The presentation is done in the context of the SBPI, in the setting of individually controllable and observable transitions with no concurrency. These definitions are extended to concurrency, double-labeled Petri nets, and more general specifications in section 4.2. Note also that the concept of admissibility used in this work is different from that of [141, 142] (while still being related).

We define as *feasible* constraints those constraints that can be exactly implemented by a feasible supervisor. In other words, the feasible constraints have the property that there is a supervisor that inhibits only enabled transitions that lead to states violating the constraints, never inhibits uncontrollable transitions, and does not detect the firings of unobservable transitions.

Formally, let $\mathcal{N}$ be a Petri net of set of transitions T, set of uncontrollable transitions $T_{uc} \subseteq T$, and set of unobservable transitions $T_{uo} \subseteq T$. Furthermore, let $T_o = T \setminus T_{uo}$ and $T_c = T \setminus T_{uc}$. Referring to the definition of a supervisor from section 3.3, in this context, $\mathcal{M}$ is the set of initial markings. The feasibility of the constraints $L\mu \leq b$ is defined first with respect to a single initial marking μ_0. (Thus, $\mathcal{M} = \{\mu_0\}$.) Let's write $(\mu, \sigma) \in \mathcal{R}_s(\mathcal{N}, \mu_0, \Xi)$ if $\mu_0 \xrightarrow{\sigma} \mu$ and firing σ from μ_0 is allowed in the closed loop of $(\mathcal{N}, \mu_0)$ and Ξ.

A set of constraints $L\mu \leq b$ is **feasible** with respect to $(\mathcal{N}, \mu_0)$ if $L\mu_0 \leq b$ and there is a feasible supervisor Ξ such that

1. $\forall(\mu, \sigma) \in \mathcal{R}_s(\mathcal{N}, \mu_0, \Xi)$: $L\mu \leq b$.
2. $\forall(\mu, \sigma) \in \mathcal{R}_s(\mathcal{N}, \mu_0, \Xi)$: $t \notin \Xi(\mu_0, \sigma) \Rightarrow (\mu \xrightarrow{t} \mu' \land L\mu' \not\leq b)$.

A set of constraints $L\mu \leq b$ is **feasible** with respect to $\mathcal{N}$ if feasible with respect to $(\mathcal{N}, \mu_0)$ for all initial markings μ_0 such that $L\mu_0 \leq b$.

We are also interested in the sets of constraints $L\mu \leq b$ for which the method of section 3.2.1 results in a supervisor that does attempt to disable uncontrollable transitions or observe unobservable transitions. The constraints $L\mu \leq b$ with this property are said to be *admissible*. Note that admissibility implies feasibility. While feasibility is a property of a set of constraints with respect to a given Petri net, admissibility is a property of a set of constraints with respect to a given Petri net and a given supervision method. Formally, a set of constraints $L\mu \leq b$ is **admissible** with respect to $(\mathcal{N}, \mu_0)$ if $L\mu_0 \leq b$ and for all markings μ_k, $k \geq 0$, reachable from μ_0 through any path of consecutively reached markings $\mu_0 \to \mu_1 \to \ldots \mu_k$ such that $L\mu_i \leq b \; \forall i = 0 \ldots k$, μ_k satisfies the following:

1. $\forall t \in T_{uc}$: $\mu_k \xrightarrow{t} \mu \Longrightarrow L\mu \leq b$;
2. $\forall t \in T_{uo}$: $(\mu_k \xrightarrow{t} \mu \land L\mu \leq b) \Longrightarrow L\mu = L\mu_k$.

A set of constraints $L\mu \leq b$ is **admissible** with respect to $\mathcal{N}$ if admissible with respect to $(\mathcal{N}, \mu_0)$ for all initial markings μ_0 such that $L\mu_0 \leq b$.

The two requirements in the definition of admissible constraints have the following role: Consider the control places that result from applying the method of section 3.2.1. The first requirement ensures that these control places never disable plant-enabled uncontrollable transitions. The second requirement ensures that the marking of these control places is not affected by the firing of closed-loop-enabled unobservable transitions.

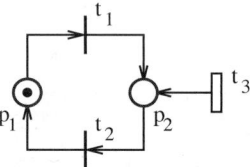

Fig. 3.3. Illustration for inadmissible but feasible constraints.

While all admissible constraints are feasible, the converse is not true. To see this, let's consider again the Petri net of Fig. 3.2(a), with t_1 uncontrollable and unobservable, and t_2 and t_3 controllable and observable. The constraint $\mu_1 + \mu_3 \geq 1$ is not admissible. However, note that $2\mu_1 + \mu_3 \geq 1$ is both admissible and feasible. As $\mu_1 + \mu_3 \geq 1$ and $2\mu_1 + \mu_3 \geq 1$ are equivalent, that is, the set of legal markings specified by $\mu_1 + \mu_3 \geq 1$ equals that of $2\mu_1 + \mu_3 \geq 1$, we can conclude that $\mu_1 + \mu_3 \geq 1$ is feasible, even though not admissible. Another example is the Petri net $(\mathcal{N}, \mu_0)$ of Fig. 3.3, in which t_1 and t_2 are controllable and observable, while t_3 is controllable and unobservable. The constraint $\mu_1 + \mu_2 \leq 1$ is feasible, as a supervisor will only have to disable t_3. However, it is not admissible.

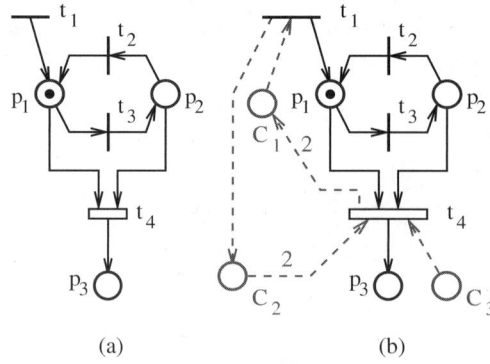

(a)   (b)

Fig. 3.4. Implementation of an admissible set of constraints.

An admissible set of constraints can be enforced by performing just a few matrix multiplications (section 3.2.1). However, checking whether a set of constraints is admissible or not may involve reachability analysis, which is computationally expensive. For this reason the sufficient condition of relations (3.4) and (3.5) is valuable, as it provides a computationally simple test to verify that a set of constraints is admissible. This test insures that the supervisor based on place invariants does not have arcs to the uncontrollable transitions and has no arcs to or from unobservable transitions. Note that the constraints (3.4) and (3.5) are not necessary for $L\mu \leq b$ to be admissible with respect to a Petri net $(\mathcal{N}, \mu_0)$ or a Petri net structure $\mathcal{N}$. For instance, consider the Petri net structure of Fig. 3.4(a), where all transitions are controllable and observable, except for t_4, which is uncontrollable and unobservable. Note that the set of constraints

$$\mu_1 + \mu_2 \leq 1, \tag{3.7}$$

$$-\mu_1 - \mu_2 \leq -1, \tag{3.8}$$

$$\mu_3 \leq 0 \tag{3.9}$$

is admissible with respect to the Petri net structure, in spite of the fact that (3.4) and (3.5) are not satisfied. The set of constraints is admissible because there is no marking satisfying the constraints such that t_4 is plant enabled. Indeed, this is enough to guarantee that there is no reachable state of the closed loop (Fig. 3.4(b)) at which the supervisor should disable or observe the transition t_4.

3.5 Transformations to Admissible Constraints

This section overviews several approaches that can transform inadmissible constraints to an admissible form. In a transformation to admissible constraints, we desire to replace a set of inadmissible constraints $L\mu \leq b$ by a set of admissible constraints $L_a\mu \leq b_a$ such that the new set of constraints is at least as restrictive as the original set of constraints (see relation (3.6)). As mentioned in section 3.2, the conditions (3.4)–(3.5) are sufficient for admissibility in free-labeled Petri nets in the setting of individually controllable and observable transitions. We present here methods that find $L_a\mu \leq b_a$ subject to (3.6) and (3.4)–(3.5) and discuss their extension to more general controllability and observability settings.

The design of admissible constraints has been approached in [141, 142] using the following parameterization:

$$L_a = R_1 + R_2 L, \tag{3.10}$$

$$b_a = R_2(b + \mathbf{1}) - \mathbf{1}, \tag{3.11}$$

where R_1 is an integer matrix with nonnegative elements and R_2 is a diagonal matrix with positive integers on the diagonal. This parameterization is used as a sufficient condition for (3.6). Thus, the approach is to find L_a and b_a subject to (3.10)–(3.11), (3.4), and (3.5). This corresponds to a linear integer programming problem for which, sometimes, solutions may be found using an efficient matrix row operation algorithm [141, 142, 18]. Note that this integer programming formulation of the problem allows introducing additional requirements of interest. For instance, communication constraints and a minimum-communication objective were used in a distributed version of this problem [86]. While the approach of [141, 142] is computationally efficient, it is also suboptimal. That is, a solution may not be found even when solutions exist, and if one is found, it may not be the least restrictive solution. A source of suboptimality is that the computation is not constrained to ensure that if L'_a and b'_a are another solution to (3.10)–(3.11), (3.4), and (3.5), then $L_a\mu \le b_a \not\Rightarrow L'_a\mu \le b'_a$.

The approach of [141, 142] can been improved in several ways. First, it should be noticed that it is difficult to express by linear inequalities the requirement that $L_a\mu \le b_a$ should be as permissive as possible. However, it is easy to constrain the computation of L_a and b_a to guarantee some weaker properties: (a) that a set of markings of interest is included in $\{\mu : L_a\mu \le b_a\}$ and (b) that a set of firing count vectors x is included in $\{x : D_c x \ge 0\}$, where D_c is the incidence matrix of the closed loop. These simple extensions can be found in [86]. As noticed in [19], the admissible constraints $L_a\mu \le b_a$ satisfying (3.6) may not have a unique supremal element. Thus, further work has been done by the authors of [18] towards finding the supremal constraints $L_a\mu \le b_a$ subject to (3.10)–(3.11), (3.4), and (3.5) by means of a parameterization.

Another way to control the selection of L_a and b_a is by means of observation and control costs. Thus, in [20], the optimal design of supervisors is considered, where optimality here is with respect to control and observation costs. Here, instead of having sets of uncontrollable and unobservable transitions T_{uc} and T_{uo}, we have maps $z_c : T \to \mathbb{R}^+$ and $z_o : T \to \mathbb{R}^+$, associating control and observation costs to each transition. The setting of [20] is general, as we can still consider some transitions as uncontrollable/unobservable by associating with them very large control or observation costs. The design problem of [20] is solved by an integer programming approach, using (3.10)–(3.11) and admissibility conditions equivalent to (3.4) and (3.5).

The optimal design of supervisors with respect to the admissibility constraints (3.4) and (3.5) is approached also in chapter 8 of [181]. The proposed method applies to specifications $L\mu \le b$ in which for all rows of L, all elements on a row have the same sign. Note that the solution is given in the form of a disjunction of constraints.

Still another approach appears in [31]. The setting of [31] assumes full observability. Essentially, given the constraint $l\mu \le c$ with $l \in \mathbb{N}^m$ and $c \in \mathbb{N}$, $l\mu \le c$ is replaced with the disjunction

$$\bigvee_{l_i \in SD_{min}(l)} [l_i\mu \le c], \qquad (3.12)$$

where $SD_{min}(l)$ is the set of minimal integer vectors x satisfying $x \geq l$ and $xD(\cdot, T_{uc}) \leq 0$. In particular, $l\mu \leq c$ is replaced with the single admissible constraint $l_1\mu \leq c$ when $SD_{min}(l)$ is the singleton $\{l_1\}$. Under conditions given in [34, 30], the resulting supervisor is least restrictive. It is interesting to notice that some of the assumptions of [31] can be dropped. Indeed, (3.12) is still a valid supervisor even if $l \in \mathbb{Z}^m$ and $c \in \mathbb{Z}$ (as opposed to $l \in \mathbb{N}^m$ and $c \in \mathbb{N}$). Further, partial observability can be incorporated by defining $SD_{min}(l)$ as the set of minimal integer vectors x satisfying $x \geq l$, $xD(\cdot, T_{uc}) \leq 0$, and $xD(\cdot, T_{uo}) = 0$.

Since (3.4)–(3.5) are only sufficient conditions for admissibility, it is clear that the solutions found based on them may not be optimal, in the sense that they may be more restrictive than necessary. An interesting result is that if both the initial marking μ_0 and the free term b of the constraints $L\mu \leq b$ are parameters, then (3.4)–(3.5) are also necessary if we are to ensure that $L\mu \leq b$ is feasible at all μ_0 and b such that $L\mu_0 \leq b$. This result is not very surprising if we take into account that admissibility was defined with respect to the method of section 3.2.1 and that (3.4)–(3.5) are structural conditions, as they do not take into account the initial marking μ_0 of the plant and the initial marking $b - L\mu_0$ of the supervisor of section 3.2.1.

Theorem 3.3. *The set of constraints $L\mu \leq b$ is feasible for all μ_0 and $b \geq L\mu_0$ if and only if L satisfies (3.4)–(3.5).*

Proof. This is a consequence of Theorem 5.7. $\qquad\square$

The methods presented in this section apply to the setting of individually controllable and observable transitions. However, they can be extended to the other controllability and observability setting, based on the observation that the form of the structural conditions (3.4)–(3.5) changes little. In the setting of labeled Petri nets, the structural conditions for admissibility are:

$$\forall t \in T, \ \rho(t) \in \Sigma_{uc} \cup \{\lambda\} \Rightarrow LD(\cdot, t) \leq 0, \tag{3.13}$$

$$\forall t \in T, \ \rho(t) \in \Sigma_{uo} \cup \{\lambda\} \Rightarrow LD(\cdot, t) = 0, \tag{3.14}$$

$$\forall t_1, t_2 \in T, \ \rho(t_1) = \rho(t_2) \Rightarrow LD(\cdot, t_1) = LD(\cdot, t_2). \tag{3.15}$$

Compared to (3.4)–(3.5), the constraint (3.15) has been added. Given any two transitions t_1 and t_2 with the same label, (3.15) ensures that firing t_1 changes the marking of the supervisor (3.2)–(3.3) the same way as the firing of t_2. Thus, the supervisor cannot distinguish between the firing of t_1 and t_2. In the case of double-labeled Petri nets, the conditions for admissibility are that each constraint $l\mu \leq c$ of the set of constraints $L\mu \leq b$ satisfies

$$\forall t \in T, \ \rho(t) \in \mathcal{K}_{uc} \cup \{\lambda\} \Rightarrow lD(\cdot, t) \leq 0, \tag{3.16}$$

$$\forall t \in T, \ o(t) \in \mathcal{O}_{uo} \cup \{\lambda\} \Rightarrow lD(\cdot, t) = 0, \tag{3.17}$$

$$\forall t_1, t_2 \in T, \ o(t_1) = o(t_2) \Rightarrow lD(\cdot, t_1) = lD(\cdot, t_2), \tag{3.18}$$

$$\forall t_1, t_2 \in T, \forall \alpha \in \mathcal{K}, \rho(t_1) = \rho(t_2) = \alpha \Rightarrow$$
$$lD(\cdot, t_1) = lD(\cdot, t_2) \vee [lD(\cdot, t_1) \leq 0 \wedge lD(\cdot, t_2) \leq 0]. \qquad (3.19)$$

Note that the constraint (3.19) is sufficient to guarantee that any two transitions with the same label are either both disabled or both enabled in the closed loop. Let's note that the admissibility constraints (3.4)–(3.5) and (3.13)–(3.15) both have the same form $LA \leq 0$, where A is an integer matrix. Thus, the methods discussed in this section for constraint transformation can be easily extended to the setting of labeled Petri nets.

Due to the constraint (3.19), the conditions for admissibility are no longer linear in the setting of double-labeled Petri nets. The conditions require each constraint $l_i \mu \leq c_i$ of $L\mu \leq b$ to satisfy a disjunction $\bigvee_{j=1}^{q}[l_i B_j \leq 0]$, where $B_1 \ldots B_q$ are integer matrices. By bringing $\bigwedge_i \bigvee_{j=1}^{q}[l_i B_j \leq 0]$ to the disjunctive normal form, the admissibility conditions for double-labeled Petri nets have the form $\bigvee_{i=1}^{r}[LA_i \leq 0]$, where $A_1 \ldots A_r$ are integer matrices.

It is possible to use the methods that rely on admissibility conditions of the form $LA \leq 0$ even in the setting of double-labeled Petri nets. This is how. First, for all constraints $l\mu \leq c$ of $L\mu \leq b$, find for every $j = 1, 2, \ldots, q$ a solution $l_a \mu \leq c_a$ satisfying $l_a B_j \leq 0$ and (3.6). Then, select the "best" solution $l_a \mu \leq c_a$ out of the q cases $i = 1, 2, \ldots, q$. Finally, take $L_a \mu \leq b_a$ as the conjunction of the constraints $l_a \mu \leq c_a$ that were selected for each constraint $l\mu \leq c$ of $L\mu \leq b$.

3.6 Overview of Supervisory Methods

The supervisory control of DESs has been introduced by Ramadge and Wonham [159]. In the work of Ramadge and Wonham, automata are used to represent DESs. In the context of DESs represented as Petri nets, two types of specifications have been considered: language specifications and state specifications. The language specifications describe a language that the closed loop should achieve, as in the Ramadge and Wonham work. On the other hand, state specifications are most often used to indicate a set of *forbidden states (markings)* of the plant that a supervisor should avoid. This chapter has considered the specifications $L\mu \leq b$, which are a particular type of the state specifications. Note that language specifications are more general than state specifications. However, note also that language specifications are most often considered in a no concurrency setting, in which only one event (transition firing) can occur at a time. In [124], in a no concurrency framework, it is shown that any language specification can be realized by a state specification when the plant is enhanced with a (possibly infinite) memory.

Typically, there are two versions of partial controllability and two versions of partial observability in the Petri net literature. The common views of partial controllability correspond to the settings of the individually controllable transitions [124, 125, 141, 142] and of the controlled Petri nets [114, 72, 115].

As mentioned in section 3.3, the difference between these two settings is that in the second one the plant allows the supervisor to disable certain given sets of controllable transitions, as opposed to the ability to disable individually each controllable transition. The first view is the special case of the second when the control sets are singletons. About partial observability, it should be noticed first that this issue is often ignored in the Petri net literature. Based on the papers that do consider partial observability, we can identify two main views of partial observability, arising from two distinct views on the operation of the supervisors. The first view corresponds to the setting of individually observable transitions, in which a supervisor observes the occurrence of the transitions of the plant that are observable. This view follows [159] and appears, for instance, in [141, 142]. The second view is that a supervisor observes the state (marking) of the plant through an observation map. This view appears, for instance, in [59, 32, 124]. Note that it is possible to construct a state observer based on the observation of transitions [59]. Among papers dealing with other settings of partial controllability and observability, we mention [55, 56, 116] for labeled Petri nets.

Three main types of concurrency are considered in the literature on the supervision of Petri nets. The *no concurrency* assumption means that only one event (transition firing) may occur at a time. No concurrency is assumed, for instance, in [124, 125]. The *concurrency* assumption means that only distinct events (firings of distinct transitions) may occur at the same time. This concurrency assumption appears typically in the context of controlled Petri nets, for instance, in [114, 72, 126]. The *transition-bag* assumption is the most permissive concurrency setting, placing no restriction on the events that may occur at the same time and their multiplicity. In [181], dealing with specifications $L\mu \leq b$, arbitrary concurrency settings are described by a set Δ of firing vectors, where Δ describes the firing vectors permitted to fire in a given Petri net structure. Results valid under the various concurrency settings are obtained in [181].

Various approaches have been proposed for the supervision of Petri nets. A survey can be found in [70]. It appears that the work in this field began with the results of Ichikawa et al. [76, 75]. In [76], the specification is to reach a target state (target marking) from the initial state (initial marking), and then to stay there. The paper solves this problem for a class of ordinary Petri nets called structurally conflict-free Petri nets. The solution finds a set of control places that ensure this specification is achieved. The article [75] considers an extended class of Petri nets and an additional specification, requiring a certain firing sequence to be achieved. The solution is to find an appropriate initial marking or to control the firings of certain transitions. Partial controllability is not considered, in the sense that the design of the supervisor is not constrained by inability to control certain transitions. With regard to partial observability, note that for certain approaches or under certain firing assumptions (e.g., assuming that the firing time of an enabled transition is bounded) no observation is necessary. The presentation of the results has been done under the concurrency assumption. However, the results appear to be valid in other concurrency settings as well.

An approach to the enforcement of forbidden state specifications in Petri nets appears in [114, 72, 115]. This work assumes the setting of controlled Petri nets, the marking fully observable, and the concurrency assumption. The Petri net models are assumed to be bounded. Further, in [72, 115] the authors restrict the models to safe Petri nets in the class of cyclic marked graphs. The specification is given in the form of a set of forbidden states. In [72, 115], the specification corresponds to a particular type of constraints $L\mu \leq b$ in which the elements of L are binary. Note that in the context of safe cyclic marked graphs, this specification can describe arbitrary sets of states.

In [114], the maximal control-invariant set is computed using the algorithm of [158]. Then, the paper shows that the optimal deterministic supervisor may not be unique, due to concurrency. Finally, a method for the computation of all deterministic optimal supervisors is presented. Note that a maximally permissive supervisor can be defined in the context of nondeterministic supervisors [70]. The results of [114] are further developed in [72] for marked graphs and nondeterministic supervision. In [72], the construction of the reachability graph is avoided. Instead, predicates based on the structure of the marked graph are used to derive the maximally permissive control. This approach is simplified in [115]. In [73], conditions on the forbidden state set are derived to ensure that liveness can be enforced. Under the conditions given in [73], liveness is enforced when the nondeterministic supervisor selects control sets for which at least one transition is closed-loop enabled. Extensions of the approach of [114, 72] to partial observability and decentralized control appear in [217] and [32], respectively. A related approach has been proposed for controlled Petri nets with a state machine structure in [26]. The specification corresponds to the requirement that the marking μ should satisfy at least one set of constraints $L_i\mu \leq b_i$, $i = 1 \ldots p$, where the matrices L_i have nonnegative elements. Further, an extension of [72, 115] to controlled Petri nets having an ordinary Petri net structure appears in [71]. In [71], the specification is given in terms of a set $\mathcal{F}$ of subsets of places, by defining the set of forbidden states as $\mathcal{M}_F = \{\mu : \exists F \in \mathcal{F} \ \forall p \in F, \mu(p) \geq 1\}$. Compared to [72, 115], since the Petri nets are no longer assumed to be safe cyclic marked graphs, the specification is no longer able to capture all possible sets of forbidden markings.

In [124], the theory of [159] is extended to Petri nets, with an emphasis on state specifications. The work of [124] is done under the no concurrency assumption. The extension to the concurrency assumption appears in [126]. The Petri net models are assumed to be pure, that is, without self-loops. Each transition of the Petri net corresponds to an unique event. As in [159], the supervisor is allowed to disable events from a set of controllable events. However, here partial observability consists of the partial observation of the state (marking) of the Petri net. Note that the Petri nets are assumed to have a fixed initial marking. This observation is important because not all results can be directly applied to the case when the initial marking is viewed as a variable. Two kinds of problems are considered: static state-feedback control and dynamic state-feedback control. Static state-feedback control refers to specifications described by predicates in terms of the state (marking) of the system. For instance, enforcing $L\mu \leq b$ would be seen as a static state-feedback control in [124]. On the other hand, dynamic state-feedback control corresponds to the enforcement of specifications that refer to the history of the event occurrences in the plant. As an example, denoting by v the Parikh vector and C an integer matrix, enforcing $Cv \leq b$ is seen as a dynamic state-feedback in [124]. In the context of static state-feedback control, controllability and observability are defined for predicates, following the approach of the supervisory control of automata, in which controllability and observability are defined for languages [159]. Then, it is shown that a predicate can be exactly enforced if and only if it is controllable and observable. (A supervisor exactly enforcing a predicate ensures that the set of reachable states equals the set of states satisfying the predicate.) The dynamic state-feedback control is reduced to the static case by extending the plant with a memory. The memory is a DES that accepts the language of the specification. Thus, the memory is finite if its DES representation is finite. Note that [124] considers also the specifications $Cv \leq b$. The memory extension of [124] for the enforcement of $Cv \leq b$ corresponds to the C-transformation of the plant in chapter 4.

Note that constraints closely related to the form $Cv \leq b$ have arisen prior to the work of [124] in the context of the synchrony theory of Petri nets; see [50, pp. 132–134]. The synchrony theory [50, 61, 153] deals with the dependence between the number of firings of two or more transitions in a Petri net. In this context, constraints $Cv \leq b$ can arise from specifications requiring upper bounds on synchronic distances.

In [126], the work of [124] is extended for the concurrency assumption. However, [126] does not address partial observability. The authors show first that the controllability and concurrent well-posedness of a state predicate is necessary and sufficient for the existence of a deterministic supervisor that exactly enforces the predicate. Controllability is defined the same way as under the no concurrency assumption, while concurrent well-posedness is a technical condition defined in the paper. Further, the authors show that controllability alone is necessary and sufficient for the existence of a nondeterministic supervisor exactly enforcing a state predicate. Finally, the authors show how to construct the most permissive nondeterministic supervisor.

Of particular interest in this book are the approaches for the enforcement of constraints $L\mu \leq b$. These constraints have been proposed for a variety of applications: a constrained optimal control problem of chemical processes [213], the coordination of AGVs [115], manufacturing constraints [141], and mutual exclusion in batch processing [193]. Moreover, by considering also classes of constraints that can be reduced to the form $L\mu \leq b$ on transformed Petri nets, specifically the generalized linear constraints of chapter 4, two other applications can be mentioned here as well: the supervisory control of railway networks [58] and fairness enforcement, such as bounding the difference between the number of occurrences of two events, in protocols [50] and manufacturing [124]. Other types of constraints that can be reduced to the form $L\mu \leq b$ on transformed Petri nets include language constraints and disjunctions $[L_1\mu \leq b_1] \vee [L_2\mu \leq b_2] \vee \ldots [L_p\mu \leq b_p]$, the latter under certain boundedness assumptions. These are discussed in chapter 4. An interesting property of the constraints $L\mu \leq b$ is that they can describe any set of markings of a safe Petri net [213, 57]. Further, the derivation of the form $L\mu \leq b$ from a Boolean expression can be carried out rather easily [213, 214].

In the supervisory control literature, numerous researchers have considered the constraints $L\mu \leq b$. Historically, the first published papers on the topic are [213, 115, 57, 125]. In [213] a constrained optimal control problem is considered. The objective is to reach a final state from a given initial state. The constraints on the inputs (the controllable transitions) and the state (marking) are described by inequalities of the form $L\mu \leq b$ and $L\mu + Hq \leq b$, where the latter type of constraints is discussed in chapter 4. In [115], a particular form of constraints $L\mu \leq b$ has been used in order to simplify the specification of forbidden states. Later, the enforcement of the constraints $L\mu \leq b$ was considered in [57]. Subsequently, a computationally efficient method for the enforcement of $L\mu \leq b$ in partially controllable Petri nets appeared in [125].

The reference [57] considers the enforcement of constraints $L\mu \leq b$ when all elements of L and b are nonnegative. Note that the results can easily extend to the general case, in which L and b have integer elements [214, 141, 142]. The paper deals with the redundancy, equivalence, and modeling power of the specifications $L\mu \leq b$ and the enforcement of $L\mu \leq b$ for fully controllable Petri nets. The paper shows also that in the case of partially controllable Petri nets the enforcement problem is more difficult, as the optimal solution requires solving a general forbidden state problem.

The reference [214] presents the construction of the supervisors of $L\mu \leq b$ and emphasizes the qualities of this solution. Thus, the supervisor consists only of control places, such that the closed-loop system is still a Petri net. Further, the number of control places equals the number of inequalities in $L\mu \leq b$ and the design of the supervisor only involves matrix multiplications. It is also shown that for safe Petri nets any logic constraints on the state can be written as constraints $L\mu \leq b$. Moreover, a method to transform constraints of the form $L\mu + Hq \leq b$ into constraints $L\mu \leq b$ is presented. (The constraints $L\mu + Hq \leq b$, involving both the marking and the firing vector q, are considered in chapter 4.) The presentation of [214] focuses on the fully controllable and observable case, as all transitions that need to be controlled/observed are assumed to be controllable/observable. The extension of the results to the partially controllable and observable case appears in [141, 142].

Among the methods dealing with the enforcement of constraints $L\mu \leq b$ on partially controllable and observable Petri nets, of special interest are the structural methods that rely on the conditions for admissibility (3.4)–(3.5) defined in this chapter. This class of methods was discussed already in section 3.5. They are especially interesting because, as shown in section 3.5, they can be rather easily extended to the more general settings of labeled Petri nets and double-labeled Petri nets.

The references [141, 142] propose methods for the design of supervisors enforcing $L\mu \le b$ under the partial controllability and observability setting of individually controllable and observable transitions. The approach involving the parameterization (3.10)–(3.11) and the admissibility conditions (3.4)–(3.5) was discussed already in section 3.5. While optimal for constraints satisfying (3.4)–(3.5), the approach is suboptimal for inadmissible constraints. As discussed in section 3.5, approaches providing more permissive solutions are possible under certain assumptions [31, 181].

An important class of methods for the enforcement of the constraints $L\mu \le b$ are developed under the partial controllability setting of individually controllable transitions and assume full observability. Further, they aim to calculate the *maximal controlled invariant set* [115, 158] defined as

$$\mathcal{A}_F = \{\mu : \mathcal{R}(\mathcal{N}_u, \mu) \cap \mathcal{M}_F = \emptyset\}, \tag{3.20}$$

where $\mathcal{M}_F$ is the set of forbidden markings, T_{uc} is the set of uncontrollable transitions, and $\mathcal{N}_u = (P, T_{uc}, D^-(\cdot, T_{uc}), D^+(\cdot, T_{uc}))$ is the *uncontrolled subnet* of the Petri net. For the constraints $L\mu \le b$, $\mathcal{M}_F = \{\mu : L\mu \not\le b\}$. Once we know $\mathcal{A}_F$, the optimal solution to the enforcement of $L\mu \le b$ is to disable the firings of controllable transitions that lead to markings outside $\mathcal{A}_F$. An especially interesting case is the situation when $\mathcal{A}_F$ can be expressed by a conjunction of linear constraints: $\mathcal{A}_F = \{\mu : L_a\mu \le b_a\}$. When this is the case, the optimal solution can be implemented by the SBPI with the specification $L_a\mu \le b_a$ (which is admissible!). It is known that $\mathcal{A}_F$ cannot always be represented by a conjunction of inequalities $L_a\mu \le b_a$ [57]. Thus, in general a solution of the form $L_a\mu \le b_a$ is not optimal, that is, $\mathcal{A}_F \supseteq \{\mu : L_a\mu \le b_a\}$.

In [125], $\mathcal{A}_F$ is calculated under certain assumptions. The assumptions involve subnets associated with uncontrollable transitions that affect the term $L\mu$. Under these assumptions, the set $\mathcal{A}_F$ is calculated without resorting to the traditional integer programming methods. In the most general case, the solution is represented by a disjunction $[L_1\mu \le b_1] \vee [L_2\mu \le b_2] \vee \ldots [L_p\mu \le b_p]$.

In [30], given a single constraint $l\mu \le c$ with $l \in \mathbb{N}^m$ and $c \in \mathbb{N}$, the influential subnet $\mathcal{N}_u^l$ is defined. $\mathcal{N}_u^l$ is the subnet of $\mathcal{N}_u$ containing the places p with $l(p) \neq 0$ and the directed paths of $\mathcal{N}_u$ to these places. The main result of the paper shows how to express $\mathcal{A}_F$ as the set of markings satisfying a disjunction of linear marking inequalities. This result relies on two conditions, as follows. First, $\mathcal{N}_u^l$ should be a marked graph. Second, for all reachable markings of $(\mathcal{N}, \mu_0)$, every directed circuit of $\mathcal{N}_u^l$ should have at least one token. A similar result was obtained in [34] for the case in which $\mathcal{N}_u^l$ is a state machine, instead of a marked graph. For this case, it is shown that $\mathcal{A}_F$ has the form $\mathcal{A}_F = \{\mu : l_a\mu \le c\}$, where l_a can be easily computed. Thus, the control place enforcing $l_a\mu \le c$ is the least restrictive supervisor.

The reference [182] considers the enforcement of k-safeness on state machines, where k-safeness is expressed by the constraints $\mu(p) \le k \ \forall p \in P$.

It is shown that the set $\mathcal{A}_F$ can be expressed by a particular form of the constraints $L\mu \leq b$. An algorithm is developed to minimize the number of control places that implement the specification. The results are obtained under the transition-bag concurrency setting. For the supervision problem of general Petri nets with arbitrary forbidden set specifications and the same concurrency setting, [183] shows that the calculation of $\mathcal{A}_F$ can be done on a subnet $\mathcal{N}_A$ of the uncontrolled subnet $\mathcal{N}_u$. This result is applied in [185] to the calculation of $\mathcal{A}_F$ for specifications $L\mu \leq b$. The set $\mathcal{A}_F$ is obtained in the form $L_a\mu \leq b_a$ under three hypotheses: $\mathcal{N}_A$ is acyclic, the transitions t of $\mathcal{N}_A$ satisfy $|\bullet t| \leq 1$ as well as a condition which, in particular, is satisfied when the input arcs of t have the weight 1. The computation of $\mathcal{A}_F$ has low polynomial complexity. The observation that the computation of $\mathcal{A}_F$ is easier when the uncontrolled subnet $\mathcal{N}_u$ is acyclic, was made also in [32].

A reachability approach to the supervision of partially controllable Petri nets appears in [51], in the setting of individually controllable transitions. Full observability is assumed. Here, the supervisor is designed as a set of control places acting upon the Petri net plant. First, a subset of the reachability graph is obtained, such that from any of the markings of the subgraph, forbidden states and blocking states cannot be reached by firing uncontrollable transitions. This subgraph becomes the desired reachability graph that is to be achieved by the closed loop. Then, the authors deal with the design of supervisors that ensure the closed loop has the specified reachability graph. Given a set Ω containing the pairs (μ, t) such that t should be disabled at the marking μ, control places are designed, such that each control place deals with at least one of the pairs (μ, t) of Ω. The connections between a control place and the plant are determined by finding an integer solution to a system of inequalities. Due to the particular form of the inequalities, the solution can be found using linear programming.

We have already mentioned results on the supervision of marked graphs in the setting of controlled Petri nets. There are also results for marked graphs in the partial controllability and observability setting of individually controllable and observable transitions. In [52], a solution involving the online computation of the supervisory policy is proposed. The computation involves solving linear programs. Particular cases in which the computation is simpler are also identified. The paper assumes Petri nets with a marked graph structure, full observability, and the no concurrency setting. The Petri nets are not required to be safe. The results of [52] are further extended in [1] for partial observability. As in [52], at every reachable state, the controllable transitions that may be enabled are found online by means of linear programming. The paper assumes all controllable transitions are observable.

Important results on the control of live marked graphs appear in [39], under the setting of individually controllable and observable transitions. The specification considered in the paper has the form $av \leq c$, where v is the Parikh vector, $a \in \mathbb{Z}^{1 \times |T|}$ and $c \in \mathbb{Z}$. The set of transitions T is partitioned into three disjoint subsets, $T = T_c \cup T_f \cup T_i$, where T_c is the set of controllable transitions, and $T_o = T_c \cup T_f$ the set of observable transitions. (Thus, all controllable transitions are assumed to be observable.) The approach of the paper is as follows. *Suspect* vectors are defined as Parikh vectors v such that $v|_{T_o} = v'|_{T_o}$ for some v' with the property that there is a $v'' \geq v'$ such that v'' is forbidden (i.e., $av'' > c$) and all nonzero entries of $v'' - v'$ correspond to uncontrollable transitions. The paper shows that any deterministic supervisor has to avoid reaching the set of suspect vectors and that the projections of these vectors on T_o are the integral points of a convex set. The paper shows also how to compute this set. Since the complement of the set of suspect vectors may not correspond to the integral points of a convex set, it follows that the least restrictive supervisor may not be implementable by control places. Even when control places can be used, the paper shows that their number may be exponential. Another observation of the authors is that the number of linear constraints defining the set of suspect vectors may depend exponentially on the size of $D(\cdot, T_{uc})$. The alternative to the computation of this set is to solve online at every state and for all $t \in T_c$ a linear program, in order to decide whether t should be enabled. Since linear (not integer linear) programming is used, the computation has polynomial complexity.

So far, we have focused on specifications related to the form $L\mu \leq b$. There has also been work on different types of specifications. In [185], a method is given to calculate the maximal controlled-invariant set $\mathcal{A}_F$ for specifications given in the form of disjunctions $\bigvee_i [l_i \mu \leq c_i]$, where $l_i \in \mathbb{N}^{1 \times m}$ and $c_i \in \mathbb{N}$. The result is obtained in the form of another disjunction of linear inequalities. Several assumptions are made in terms of a certain uncontrolled subnet $\mathcal{N}_A$. Two of the assumptions are that $\mathcal{N}_A$ is acyclic and that the transitions t of $\mathcal{N}_A$ satisfy $D^-(p, t) = 1 \ \forall p \in \bullet t$. The solution assumes full observability and is obtained under the no concurrency assumption and the setting of individually controllable transitions.

A detailed discussion of a number of results on the supervision of labeled Petri nets appears in the survey [70]. A feature of the labeled Petri net results is that typically they are obtained under the no concurrency assumption. Further, typically they assume that all events are observable. A Petri net plant and a Petri net specification can be composed as shown in [54] to represent the closed-loop system, assuming no controllability and observability issues arise. The case in which there are controllability issues was addressed in [17]. It was also addressed in [116] for the case in which both the plant and the specification are deterministic labeled Petri nets. (A labeled Petri net $(\mathcal{N}, \rho, \mu_0)$ is deterministic when all firing sequences σ enabled at μ_0 generate distinct words $\rho(\sigma)$.) In this context, [116] shows that the computation of the least restrictive supervisor can be reduced to a forbidden marking problem.

The methods presented in this book address the following problems. The enforcement of specifications consisting of disjunctions $\bigvee_i [L_i \mu \leq b_i]$ and P-type languages is studied in chapter 4. A more general type of constraints $\bigvee_i [L_i \mu + H_i q + C_i v \leq b_i]$ is also considered in chapter 4, where q and v are the firing vector and the Parikh vector, respectively. The significance of the sets of constraints $L\mu + Hq + Cv \leq b$ is that they can describe the operation of any free-labeled Petri net. Chapter 4 presents the results in the general settings of double-labeled Petri nets and labeled Petri nets. The concurrency setting is the transition-bag assumption. While some of the results of chapter 4 could be used in future work for the design of optimal (least restrictive) supervisors, the main application of the chapter consists of suboptimal structural approaches to the enforcement of the aforementioned specifications. A common feature of these methods is that by means of direct and inverse Petri net transformations, they reduce the supervisor design problem to problems that can be addressed by the methods of section 3.5.

The remaining part of the book focuses on the setting of individually controllable and observable transitions. This simplifies the presentation of the results, often without much loss of generality. The decentralized control of Petri nets is considered in chapter 5 for specifications $L\mu \leq b$. A literature review of decentralized control in the context of DESs is included in the same chapter. The problem of liveness enforcement is then considered in chapters 6–8. Typically, the Petri net literature has considered separately the problem of enforcing a forbidden state or language specification and the problem of liveness enforcement. Thus, for the vast majority of the literature results on forbidden state and language specifications, additional methods are needed to ensure the closed loop is live. A structural approach for liveness enforcement is presented in chapters 7–8. The literature on liveness enforcement is overviewed in chapter 7.

3.7 Concluding Remarks

This chapter has presented an introduction to the supervision of Petri nets. Of particular interest has been the method known as the supervision based on place invariants (SBPI), as it is used or referred to in several chapters of the book. Even though the method has been presented under the assumption of no concurrency, it applies also under the more general concurrency settings [181]. Further, section 3.5 shows that the SBPI can be applied to various settings of partial controllability and observability. Among these, the setting of double-labeled Petri nets is used in the next chapter and the setting of individually controllable and observable transitions in the rest of this book. Further, except for the next chapter, the setting of no concurrency is assumed. Moreover, throughout the book, deterministic supervision is assumed. It should be noticed that these simplifying assumptions come often without significant loss of generality, while facilitating the presentation of the material.

4

Enforcing General Specifications

4.1 Introduction

The previous chapter introduced the SBPI as a simple structural approach for the supervision of Petri nets. This chapter shows that the method can be extended for more general specifications. We consider first a linear type of specification that involves not only the marking, but also the firing vector and the Parikh vector. Then we consider language specifications. Finally, we consider the enforcement of disjunctive constraints.

First, we consider constraints of the form

$$L\mu + Hq + Cv \leq b, \qquad (4.1)$$

where q and v are the firing vector and the Parikh vector, respectively. This class of constraints effectively extends the constraints $L\mu \leq b$ considered in the previous chapter, as they can represent control places arbitrarily connected to the transitions of a Petri net. As shown in section 4.2.2, this type of specification corresponds to the class of P-type languages of free-labeled Petri nets.

The second type of specification is described by the P-type languages of labeled Petri nets. Such a specification is given in the form of a labeled Petri net, and the supervisor is used to ensure the plant generates a sublanguage of the specification language.

The third type of specification considered here is described by disjunctive constraints. They are described by expressions of the form

$$\bigvee_i L_i\mu \leq b_i. \qquad (4.2)$$

This type of specification requires that for any reachable marking μ, there is one set of constraints $L_i\mu \leq b_i$ that is satisfied. For this type of specification our approach requires certain boundedness assumptions, as will be seen later.

4.2 Enforcing Generalized Linear Constraints

4.2.1 Introduction

Recall that in the previous chapter we considered the SBPI for constraints of the form

$$L\mu \le b, \tag{4.3}$$

where $L \in \mathbb{Z}^{n_c \times m}$, $b \in \mathbb{Z}^{n_c}$, $\mathbb{Z}$ is the set of integers, m is the number of places, and n_c the number of constraints. Constraints of the form (4.3) can describe (generalized) mutual exclusion, deadlock prevention constraints, and others [141]. This section will extend (4.3) to include also the firing and the Parikh vector. The analysis will typically be done in the transition-bag concurrency setting, allowing multiple transition firings at the same time. In practice, this could correspond to the situation in which the supervisor is not as fast as the plant. Thus, when the supervisor issues control decisions allowing multiple firings, the plant is not slowed down to fire transitions at the same pace the supervisor issues the control decisions.

The constraints (4.3) have been extended in [141, 214] to the form

$$L\mu + Hq \le b, \tag{4.4}$$

which adds a firing vector term, where $H \in \mathbb{N}^{n_c \times n}$, $\mathbb{N}$ is the set of nonnegative integers, and n is the number of transitions. (Without loss of generality, H has been assumed to have nonnegative elements.) In such constraints, an element q_i of the firing vector q is set to 1 if the transition t_i is to be fired next from μ; else $q_i = 0$. Alternatively, if multiple firings are allowed at the same time, the elements q_i of q represent how many times the transition t_i is fired at the next firing instance. The constraint is interpreted as follows. In the no concurrency case, a supervisor enforcing (4.4) ensures that (i) all markings μ must satisfy (4.3); (ii) if $\mu \xrightarrow{t_i} \mu'$, t_i is allowed to fire only if $L\mu + Hq \le b$ and $L\mu' \le b$. The concurrency interpretation of (4.4) is more involved. While a supervisor may perceive the firing of q as the concurrent firing of transitions, physically, the firing of q may consist of sequential firings $q_1, q_2, \ldots, q_k$, where $q_1 + q_2 + \cdots q_k = q$. Thus, it is important to guarantee that at all possible intermediary states that could be reached while firing q, (4.4) is satisfied. Formally, a supervisor enforcing (4.3) ensures that (i) all markings μ satisfy (4.3); (ii) q is allowed to fire only if $\forall q', q'' \ge 0$, if $q' + q'' \le q$ then $L\mu' + Hq'' \le b$, where $\mu \xrightarrow{q'} \mu'$.[1] The requirement (ii) is the one that ensures that (4.4) is satisfied at all intermediary states reached while firing q. (Note that (i) is implied by (ii) for $q' = q'' = 0$.) The form (4.4) describes constraints on the enabling of transitions (as opposed to the constraints on

---

[1]This interpretation of the constraints (4.4) is applicable to the concurrency settings in which the domain Δ of the firing vectors is left-closed, meaning that for all $q \in \Delta$, $0 \le q' \le q \Rightarrow q' \in \Delta$.

the state, naturally described by (4.3)). Several applications can be found in [141, 58]. For instance, in Fig. 4.1, the constraint that a railway track should only contain trains going in the same direction can be expressed by $kq_3 + \mu_1 \le k$ and $kq_2 + \mu_2 \le k$, where k is the maximum number of trains on the track and μ_1 (μ_2) is the number of trains on the track that come from the left (right).

In this chapter we consider constraints which add to (4.4) a Parikh vector term:

$$L\mu + Hq + Cv \le b, \tag{4.5}$$

where $C \in \mathbb{Z}^{n_c \times n}$. In (4.5) v is the Parikh vector; that is, v_i counts how many times the transition t_i has fired. A supervisor enforcing (4.5) ensures that (i) all states (μ, v) satisfy $L\mu + Cv \le b$; (ii) if $\mu \xrightarrow{t_i} \mu'$ and $v' = v + q$, then $L\mu + Hq + Cv \le b$ and $L\mu' + Cv' \le b$. The concurrency interpretation of (4.5) is similar to that of (4.4). The Parikh vector term can be viewed as a marking term in a Petri net extended with sink places on transitions. Regardless of the viewpoint, whether we look at the constraints as involving Parikh vector terms or marking terms involving additional sink places, it is apparent that such constraints need to be considered, as they effectively increase the expressiveness power of the constraints (4.4). In fact, we will show that (4.5) can represent any supervisor implemented as additional places (*control places*) connected to the transitions of a plant Petri net. This means that the operation of any Petri net can be entirely described by constraints (4.5), with a one-to-one correspondence between each place and each inequality of (4.5). We also show that (4.5) are as expressive as the constraints of the form

$$Hq + Cv \le b. \tag{4.6}$$

While the marking term in (4.5) does not make (4.5) more expressive, in practice it may be more intuitive to write constraints that also involve the marking. This is one reason we consider constraints of the form (4.5) instead of just (4.6). Finally, note that Parikh vector terms can be used to describe fairness requirements, such as the constraint that the difference between the number of firings of two transitions is limited by one.

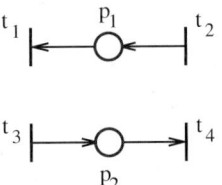

Fig. 4.1. Illustration for the constraints (4.4).

The material is organized as follows. In section 4.2.2 we show that any place of a Petri net can be seen as a supervisor place enforcing a constraint of the form (4.6). This property was noticed first for constraints of the form $Cv \leq b$ and Petri nets without self-loops [125]. Then we show how to obtain supervisors enforcing constraints (4.5) in Petri nets. We first give the solution for the case of fully controllable and observable Petri nets in section 4.2.3. Then, the problem is addressed for partially controllable and observable Petri nets. Section 4.2.4 defines the concepts of admissibility and feasibility. Admissibility is defined for specifications (4.5) and feasibility for arbitrary specifications. Feasibility and admissibility have been introduced already for specifications (4.3) in section 3.4. Next, section 4.2.5 shows that the design problem can be reduced without loss of permissiveness to a design problem involving a transformed Petri net and a set of constraints (4.3). The results are presented in more general terms, showing that the problem of enforcing a disjunction $\bigvee_{i=1}^{k} L_i\mu + H_i q + C_i v \leq b_i$ can be reduced to the problem of enforcing a disjunction $\bigvee_{i=1}^{k} L_{HCi}\mu \leq b_i$ in a transformed Petri net. Thus, based on a feasible solution $\bigvee_{i=1}^{r} L_{HCi,a}\mu \leq b_{i,a}$ to the latter problem, a feasible solution $\bigvee_{i=1}^{r} L_{i,a}\mu + H_{i,a}q + C_i v \leq b_{i,a}$ to the former problem can be easily derived. The approach of section 4.2.5 is specialized for structural admissibility constraints in section 4.2.7. This structural approach allows enforcing constraints (4.5) based on the literature approaches for constraints (4.3) discussed in section 3.5. As the approach of section 4.2.7 is structural, it promises computational benefits and can be carried out independently of the initial marking. However, unlike the approach of section 4.2.5, it is suboptimal, though optimal when compared to the methods based on the proposed structural conditions for admissibility. Note that the supervision problem is formulated in the general setting of double-labeled Petri nets, and thus the results apply also to the less general settings of labeled Petri nets and individually controllable and observable transitions. Finally, an example illustrating an application of the constraints (4.5) is given in section 4.2.6.

4.2.2 On the Significance of the Constraints

In this section it is shown that any place of a Petri net can be seen as a control place enforcing a constraint of the form (4.6). Let D denote the incidence matrix, and D^+ and D^- the input and output matrices. The common algebraic Petri net representation is via the state equation

$$\mu = \mu_0 + Dv, \tag{4.7}$$

where μ_0 is the initial marking. The operation of a Petri net can also be described through inequalities of the form (4.6). Indeed, from (4.7) we derive

$$(-D)v \leq \mu_0. \tag{4.8}$$

Let $C = -D$. The inequality $Cv \leq \mu_0$ determines the operation of a Petri net only if the net has no self-loops and we are in the no concurrency framework.

To deal with self-loops and concurrency, an additional term is introduced:

$$Hq + Cv \leq \mu_0, \tag{4.9}$$

where $H = D^-$. Note that $H_{i,j} \geq 0$ for all indices i and j. The constraints (4.9) completely describe the operation of a Petri net, regardless of whether it has self-loops or not. Indeed, after we fire from μ_0 a sequence σ of Parikh vector v, a firing vector q is enabled if and only if $Hq+Cv \leq \mu_0$ and $C(v+q) \leq \mu_0$. (Note that as $H = D^-$ and $C = -D$, we have that $Hq+Cv \leq \mu_0 \Rightarrow C(v+q) \leq \mu_0$.)

While (4.9) can describe any Petri net, the simpler constraints $Cv \leq \mu_0$ can describe any Petri net without self-loops under the no concurrency assumption [125]. In this context, if q is the firing vector associated with the firing of a transition t, $-D(v+q) \leq \mu_0$ and $-Dv \leq \mu_0$ imply $D^-q - Dv \leq \mu_0$. Thus, the Hq term can be omitted and the Petri net described by $Cv \leq \mu_0$ for $C = -D$. However, an Hq term is necessary under more general concurrency assumptions or when the Petri net has self-loops, as seen from the following example.

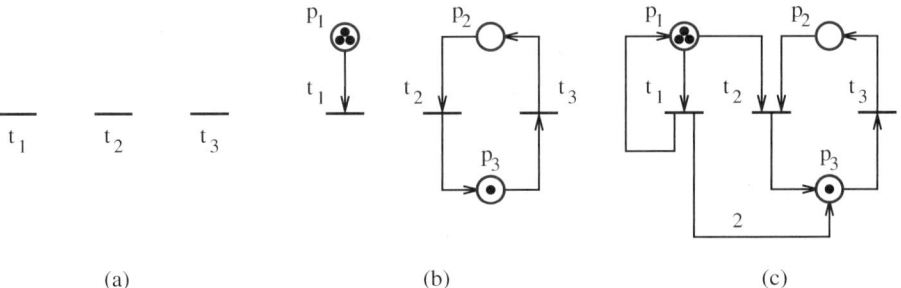

Fig. 4.2. Petri nets for Example 4.1.

EXAMPLE 4.1

Consider the Petri nets of Fig. 4.2 under the no concurrency assumption. The Petri net (a) is not restricted: the firings of t_1, t_2, and t_3 are free. Thus, it is described by (4.9) when H and C are empty matrices. By adding the places p_1, p_2, and p_3, we can obtain the Petri net (b). Assuming that the figure shows the Petri net at the initial marking, we can describe the Petri net by the following inequalities $Cv \leq \mu_0$, where $v_i = v(t_i)$:

$$v_1 \leq 3, \tag{4.10}$$

$$v_2 - v_3 \leq 0, \tag{4.11}$$

$$-v_2 + v_3 \leq 1. \tag{4.12}$$

The inequalities (4.10)–(4.12) correspond, in this order, to p_1, p_2, and p_3. For instance, p_1 ensures that t_1 may fire only $\mu_0(p_1)$ times, and p_2 that the number of firings of t_2 do not exceed the number of firings of t_3 plus $\mu_0(p_2)$. Similarly, the Petri net (c) can be described by the following equations (4.9):

$$q_1 + v_2 \leq 3, \tag{4.13}$$

$$v_2 - v_3 \leq 0, \tag{4.14}$$

$$-2v_1 - v_2 + v_3 \leq 1. \tag{4.15}$$

In the first equation, note that the term q_1 is necessary in order to describe that p_1 only controls whether t_1 is enabled or not, as p_1 does not control how many times t_1 may fire.

Now, assuming we are no longer under the no concurrency assumption, let's note that both (4.10)–(4.12) and (4.13)–(4.15) allow the simultaneous firing of t_2 and t_3 at the initial marking, while neither of the Petri nets (b) and (c) allows this firing. This problem does not appear if we model the Petri nets by (4.9) with $H = D^-$ and $C = -D$.

To see that the form (4.5) is more expressive than (4.4), note that in the Petri net of Fig. 4.3 there is no place invariant involving the control place C. Therefore, C cannot be described by (4.4). However, the following constraint in the form (4.5) describes it:

$$-v_1 + v_2 + v_3 \leq 1.$$

In fact, every place of a Petri net can be seen as a control place restricting the firings of the net transitions. This result is stated next.

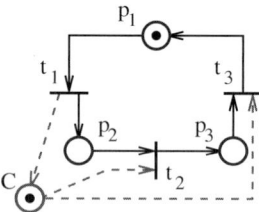

Fig. 4.3. A control place that is not in a place invariant.

Proposition 4.1. *Every place of a Petri net can be seen as a control place enforcing a single inequality of the form (4.5).*

Proof. The proof follows immediately from (4.9): the constraint of each place p_i is $hq + cv \leq \mu_{0i}$, where h and c are the ith rows of D^- and D^+. $\qquad\square$

An immediate consequence of Proposition 4.1 is that we can identify redundant places in a Petri net by finding the redundant constraints of a set of constraints (4.5) describing the Petri net. This gives an interesting interpretation to the literature approaches for the identification of redundant places, such as that of [171].

An important consequence of Proposition 4.1 is that the P-type languages of free-labeled Petri nets correspond to the constraints (4.5). A P-type language specification constrains a plant to generate a P-type language that is a subset of the P-type language given in the specification. Thus, the specification is given in terms of a Petri net $(\mathcal{N}_s, \mu_{0s})$ having the desired P-type language. In the fully observable and controllable case, $(\mathcal{N}_s, \mu_{0s})$ is also the supervisor, and the parallel composition of $(\mathcal{N}_s, \mu_{0s})$ and the plant $(\mathcal{N}, \mu_0)$ results in the closed loop. Assuming that $\mathcal{N}_s$ is free-labeled, the parallel composition simply enhances $\mathcal{N}$ with control places (which are the places of $\mathcal{N}_s$). However, in view of Proposition 4.1, any control places can be described by constraints (4.5). It follows that the constraints (4.5) can describe any P-type languages of free-labeled Petri nets. On the other hand, as will be shown in subsection 4.2.3, enforcing (4.5) results in control places. Provided $\mathcal{N}$ is free-labeled, by deleting from the closed loop the places of the plant, we obtain the Petri net generating a P-type language specification that is equivalent to (4.5).

Note that both μ and v can describe the state of a Petri net. We choose to denote by $\mathcal{R}(\mathcal{N}, \mu_0)$ all pairs (μ, v) such that $\mu_0 \xrightarrow{\sigma} \mu$, and the Parikh vector of the firing sequence σ is v.

4.2.3 Design in the Fully Controllable and Observable Case

A least restrictive supervisor can be implemented as follows. Let

$$D_{lc}^+ = \max(0, -LD - C), \tag{4.16}$$
$$D_{lc}^- = \max(0, LD + C), \tag{4.17}$$

$$D_c^+ = D_{lc}^+ + \max(0, H - D_{lc}^-), \tag{4.18}$$
$$D_c^- = \max(D_{lc}^-, H). \tag{4.19}$$

In the equations (4.16)–(4.19), the operator max is taken element by element. That is, $Y = \max(0, X)$ means $Y_{ij} = \max(0, X_{ij})$ and $Z = \max(X, Y)$ means $Z_{ij} = \max(X_{ij}, Y_{ij})$. The matrices D_c^+ and D_c^- describe a Petri net structure with the same transitions as the plant. This Petri net structure represents the Petri net implementation of the supervisor. The initial marking μ_{c0} of the supervisor depends on the initial marking μ_0 of the plant as follows:

$$\mu_{c0} = b - L\mu_0 - Cv_0. \tag{4.20}$$

Note that $b - L\mu_0 - Cv_0 \geq 0$ is a consequence of the fact that (4.5) is satisfied at the initialization of the plant.

By definition, the constraints $L\mu + Hq + Cv \leq b$ are interpreted as requiring that $\forall q', q'' \geq 0$, $q' + q'' \leq q \Rightarrow L\mu' + Hq'' + Cv' \leq b$, where $\mu \xrightarrow{q'} \mu'$ and $v' = v + q'$. It is important to notice that this interpretation of (4.5) can be simply expressed by the inequality

$$L\mu + D_c^- q + Cv \leq b. \tag{4.21}$$

This is proved in the following lemma.

Lemma 4.2. μ, $q \geq 0$, and v satisfy (4.21) if and only if $\forall q', q'' \geq 0$, $q' + q'' \leq q \Rightarrow L\mu' + Hq'' + Cv' \leq b$, where $\mu \xrightarrow{q'} \mu'$ and $v' = v + q'$.

Proof. First, let's note that $L\mu' + Hq'' + Cv' \leq b$ can be written as

$$L\mu + Cv + (LD + C)q' + Hq'' \leq b. \tag{4.22}$$

"$\Rightarrow$" By (4.17) and (4.19),

$$D_c^- = \max(0, LD + C, H). \tag{4.23}$$

The conclusion follows based on the observation that $D_c^- q \geq D_c^-(q' + q'') \geq (LD + C)q' + Hq''$.

"$\Leftarrow$" Let l, c, h, and e denote the kth row of L, C, H, and b. We prove that if $\forall q', q'' \geq 0$, $q' + q'' \leq q \Rightarrow l\mu + cv + (lD + c)q' + hq'' \leq e$, then $l\mu + d_c^- q + cv \leq e$, where $d_c^- = \max(0, lD + c, h)$. We prove it by showing that the maximum of $[(lD + c)q' + hq'']$ subject to $q', q'' \geq 0$ and $q' + q'' \leq q$, equals $d_c^- q$.

Let $q = [q_1, q_2, \ldots, q_n]^T$, $q' = [q_1', q_2', \ldots, q_n']^T$ and $q'' = [q_1'', q_2'', \ldots, q_n'']^T$. Note that $\max[(lD + c)q' + hq''] = \max[\sum_i ((lD + c)_i q_i' + h_i q_i'')]$, where $(lD + c)_i$ and h_i are the ith components of $lD + c$ and h. Since, $\max[(lD + c)_i q_i' + h_i q_i''] = q_i \max(0, (lD + c)_i, h_i)$, we obtain $\max[(lD + c)q' + hq''] = \sum_i q_i \max(0, (lD + c)_i, h_i) = d_c^- q$, which completes our proof. $\square$

An interesting consequence of Lemma 4.2 is that the negative elements of H do not count.

Theorem 4.3. *The supervisor defined by (4.18), (4.19), and (4.20) enforces (4.5) and is least restrictive.*

Proof. By Lemma 4.2, a supervisor enforces (4.5) exactly if and only if a firing vector q is enabled when

$$D_c^- q \leq b - L\mu - Cv. \tag{4.24}$$

Based on (4.16)–(4.19), it can easily be checked that

$$D_c^+ - D_c^- = -LD - C. \tag{4.25}$$

Further, (4.20) and (4.25) imply that the supervisor marking μ_c satisfies

$$\mu_c = b - Cv - L\mu \qquad (4.26)$$

at all reachable states (μ, v). Next, note that the supervisor enables a firing vector q if and only if $\mu_c \geq D_c^- q$. However, in view of (4.26), $\mu_c \geq D_c^- q$ is the same as (4.24)! This proves that the supervisor enforces (4.5) and that it is least restrictive. $\qquad\square$

The proof of the theorem has shown that the supervisor marking μ_c always satisfies (4.26). This implies that the supervisors we build for (4.5) may not create a place invariant. Note that if we substitute (4.20) and $\mu = \mu_0 + Dv$ in (4.26), we have that

$$\mu_c = \mu_{c0} - (C + LD)v. \qquad (4.27)$$

Finally, the fact that negative entries in H do not increase the expressiveness of (4.5) or (4.4) can also be seen from (4.18) and (4.19). Indeed, due to the max operators, the value of a negative entry in H has no effect on D_c^- and D_c^+.

4.2.4 Admissibility and Feasibility

The design of a supervisor in the case of fully controllable and observable Petri nets is straightforward, as it only involves the simple matrix operations of (4.16)–(4.20). Sometimes, this simple construction can also be used for partially controllable and observable Petri nets. *Admissibility*, which is defined with respect to a given plant, identifies a class of constraints (4.5) with the property that a valid supervisor optimally enforcing them can easily be derived based on the construction (4.16)–(4.20).

A specification is said to be **enforced** by a supervisor Ξ of a plant $(\mathcal{N}, \mu_0)$ if the closed loop $(\mathcal{N}, \mu_0, \Xi)$ allows only firing sequences that satisfy the specification. A specification is said to be **optimally enforced** if the closed loop $(\mathcal{N}, \mu_0, \Xi)$ disables only the firing sequences of the plant that do not satisfy the specification. In other words, a supervisor Ξ that optimally enforces the specification has the permissiveness of a least restrictive supervisor designed in the setting of fully controllable and observable Petri nets.

Before defining admissibility, the stronger concept of feasibility is introduced. Note that a supervisor enforcing a specification on a Petri net may not be feasible, as it may require control and observation abilities that are not provided by the plant. *Feasibility* identifies the specifications for which there is a valid supervisor optimally enforcing them. Admissibility is a stronger concept because it is intended only for the supervisors obtained based on the construction (4.16)–(4.20).

In the context of double-labeled Petri nets, two labeling functions are defined. The events used for control are mapped by $\rho : T \rightarrow \mathcal{K}$ and the events used for observation by $o : T \rightarrow \mathcal{O}$. In particular, for labeled Petri nets

$\rho(t) = o(t) \ \forall t \in T$ and $\mathcal{K} = \mathcal{O} = \Sigma$, where Σ is the set of events. Further, for Petri nets with individually controllable and observable transitions, $\rho(t) = o(t) = \{t\} \ \forall t \in T$ and $\mathcal{K} = \mathcal{O} = T$. In order to define formally feasibility, the following notation is introduced.

1. Let $\mathcal{K}_c \subseteq \mathcal{K}$ denote the set of controllable events. Given a firing vector q, $\rho^*(q)$ denotes a vector $z \in \mathbb{N}^{|\mathcal{K}_c|}$ indexed by the events of $\mathcal{K}_c$, such that $\forall e \in \mathcal{K}_c, \ z(e) = \sum_{t \in \rho^{-1}(e)} q(t)$.

2. Let $\mathcal{O}_o \subseteq \mathcal{O}$ denote the set of observable events. Given a firing vector q, $o^*(q)$ denotes a vector $z \in \mathbb{N}^{|\mathcal{O}_o|}$ indexed by the events of $\mathcal{O}_o$, such that $\forall e \in \mathcal{O}_o, \ z(e) = \sum_{t \in o^{-1}(e)} q(t)$.

3. Given a firing sequence $\sigma = q_1 q_2 q_3 \ldots$ let $o^*(\sigma)$ denote the sequence of observation vectors $o^*(q_1) o^*(q_2) o^*(q_3) \ldots$.

Definition 4.4. *A specification on a Petri net* $(\mathcal{N}, \mu_0)$ *is* **feasible** *if a supervisor optimally enforcing the specification ensures the following:*

1. *If q and q' are two plant-enabled firing vectors and $\rho^*(q) = \rho^*(q')$, then the closed loop enables either both q and q' or none of them.*
2. *If σ_1 and σ_2 are two firing sequences closed loop enabled at the initial state, $o^*(\sigma_1) = o^*(\sigma_2)$, and $q \neq 0$ is a firing vector such that both $\sigma_1 q$ and $\sigma_2 q$ are plant enabled at the initial state, then either both $\sigma_1 q$ and $\sigma_2 q$ or none of them are closed-loop enabled at the initial state.*

In Definition 4.4, note that the firing vectors q and q' are not necessarily nonzero, and the sequences σ_1 and σ_2 are not necessarily nonempty. In our convention, a firing vector $q = 0$ and an empty firing sequence σ are always enabled. Note also that feasible supervisors have been defined in section 3.3. A feasible specification has the property that there is a feasible supervisor that optimally enforces it.

The following is a formal definition of admissibility with respect to the method of subsection 4.2.3. The definition is with respect to three control and observation settings: individually controllable and observable transitions, labeled Petri nets, and double-labeled Petri nets.

Definition 4.5. *Given a set of constraints (4.5) on a Petri net* $(\mathcal{N}, \mu_0)$, *consider the construction of section 4.2.3. The set of constraints (4.5) is* **admissible** *if for all reachable states* (μ, v) *of the closed-loop net it is true that:*

1. *For any firing vector q such that $q(t) = 0$ for all controllable transitions t, if $\mu|_\mathcal{N}$ enables[2] $q' = q + x$ in $\mathcal{N}$ for some $x \in \mathbb{N}^n$ and either $x = 0$ or μ enables x in the closed-loop net, then μ enables q' in the closed-loop net.*
2. *If t is unobservable and μ enables t, then $D_c^+(\cdot, t) = D_c^-(\cdot, t)$.*

In addition, if transitions are observed through a map $o : T \to \mathcal{O}$:

---

[2]We denote by $\mu|_\mathcal{N}$ the restriction of μ to the places of $\mathcal{N}$.

3. *If $o(t_1)$ is an observable event, $o(t_1) = o(t_2)$ and none of t_1 and t_2 is dead at the initial state of the closed loop, then $D_c(\cdot, t_1) = D_c(\cdot, t_2)$.*

In addition, if transitions are disabled by means of control events $\mathcal{K}$, mapped to transitions by $\rho : T \to \mathcal{K}$:

4. *For all reachable states, if q and q' are two firing vectors enabled in the plant $\mathcal{N}$ at the marking $\mu|_{\mathcal{N}}$ and $\rho^*(q) = \rho^*(q')$, then the closed loop enables either both q and q' or none of them.*

The conditions 1 and 2 of the definition are required for the simplest PN models, in which transitions are individually controllable/observable. The condition 2 ensures that the marking of the supervisor is not varied by the firings of unobservable transitions. The definition accounts also for the more complex Petri net models. The condition 3 is added to account for the case in which transitions are labeled by events (labeled PNs). It ensures that the events with the same label affect the supervisor in the same way. The condition 4 is added to account for PNs in which we have separate events for observation and control (double-labeled PNs).

In the general case, reachability analysis may be required to check feasibility and admissibility. Checking these properties is especially difficult when we deal with unbounded Petri nets. A suboptimal alternative is to use sufficient conditions for admissibility that are based on the structure of the Petri net. Structural conditions for admissibility have been presented already in section 3.5 for specifications (4.3). For specifications (4.5), the conditions (3.16)–(3.19) given for double-labeled Petri nets and (4.3) can be rewritten as follows:

$$\forall t \in T, \; \rho(t) \in \mathcal{K}_{uc} \cup \{\lambda\} \Rightarrow D_c^-(\cdot, t) = 0, \tag{4.28}$$

$$\forall t \in T, \; o(t) \in \mathcal{O}_{uo} \cup \{\lambda\} \Rightarrow D_c(\cdot, t) = 0, \tag{4.29}$$

$$\forall t_1, t_2 \in T, \; o(t_1) = o(t_2) \Rightarrow D_c(\cdot, t_1) = D_c(\cdot, t_2), \tag{4.30}$$

$$\forall i = 1 \ldots n_c, \; \forall t_1, t_2 \in T, \; \forall \alpha \in \mathcal{K}, \; \rho(t_1) = \rho(t_2) = \alpha \Rightarrow \\ D_c^-(i, t_1) = D_c^-(i, t_2) \vee [D_c^-(i, t_1) = 0 \wedge D_c^-(i, t_2) = 0]. \tag{4.31}$$

These conditions ensure that the supervisor (4.18)–(4.20) is correct with respect to the controllability and observability of the plant. Thus, the condition (4.28) ensures that the supervisor does not disable an uncontrollable event. The condition (4.29) ensures that the occurrence of an unobservable event does not affect the supervisor. The condition (4.30) requires that the firing of a transition should have the same effect on the supervisor as the firing of any other transition with the same label. The condition (4.31) is sufficient to guarantee that any two transitions with the same label are either both disabled or both enabled in the closed loop.

Proposition 4.6. *Let D_c^+ and D_c^- be defined by (4.16)–(4.19). Then (4.5) is admissible if (4.28)–(4.31) are satisfied.*

The structural conditions (4.28)–(4.31) are very useful due to their simplicity. Further, it appears that the interesting result stated in Theorem 3.3 could be generalized to the setting of double-labeled Petri nets, the admissibility constraints (4.28)–(4.31), and the specifications (4.5). However, it should be noticed that it is possible to use structural admissibility conditions that are less restrictive than (4.28)–(4.31). For instance, constraints (4.9) describing the closed loop may identify events that never occur, such as by means of integer linear programming. Such events need not be considered in (4.28)–(4.31). Further, a similar approach could be used to identify other control events that need not be considered in (4.31).

4.2.5 Design in the Partially Controllable and Observable Case

Here we consider the enforcement of constraints (4.5) by means of methods for the enforcement of the constraints (4.3), and direct and inverse transformations of Petri nets and constraints. This approach is convenient because there are already numerous methods for the enforcement of the constraints (4.3), as seen in section 3.6. Further, the transformations proposed here have low polynomial complexity. This ensures that the constraints (4.5) can be enforced with the same efficiency as (4.3).

The approach described here applies not only for conjunctions (4.5), but also for disjunctions $\bigvee_{i=1}^{k} L_i\mu + H_iq + C_iv \leq b_i$. Therefore, it is presented in terms of the more general specifications involving disjunctions of constraints (4.5). Given a disjunction of constraints (4.5) and a Petri net, first a transformation is applied, resulting in another Petri net and a disjunction of constraints (4.3). The solution is found by applying the inverse transformation of the solution to the problem of enforcing the disjunction of constraints (4.3). This approach is optimal, in the sense that if the solution found for the enforcement of the disjunction of constraints (4.3) is optimal with respect to permissiveness, than so is the solution for the enforcement of the disjunction of constraints (4.5). This approach is specialized for the methods based on structural conditions for admissibility in subsection 4.2.7. We begin by defining the transformation for conjunctions of constraints (4.5).

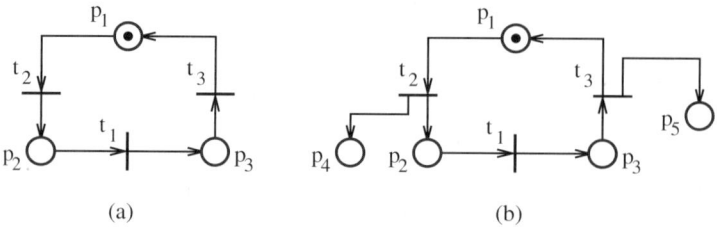

(a)   (b)

Fig. 4.4. Illustration of the C-transformation.

The C-transformation removes the Cv term from the constraints (4.5). The idea of the C-transformation is illustrated in an example. Consider the Petri net of Fig. 4.4(a), and assume that we desire to enforce the following constraint:

$$\mu_1 + q_1 + v_2 - v_3 \leq 3. \tag{4.32}$$

The idea is to transform the net such that the Cv term is transformed into a marking term. Thus, by transforming the net as in Fig. 4.4(b), (4.32) can be written without referring to v:

$$\mu_1 + q_1 + \mu_4 - \mu_5 \leq 3. \tag{4.33}$$

The Petri net of Fig. 4.4(b) and the constraint (4.33) are the C-transformation of the Petri net of Fig. 4.4(a) and of (4.32). The extension of the Petri net through the C-transformation corresponds to the one-dimensional memory of [124] that is used for the enforcement of linear dynamic specifications $Cv \leq b$.

The inverse C-transformation is also possible. Given the constraint

$$\mu_1 - 3\mu_4 + 2\mu_5 + q_1 \leq 5 \tag{4.34}$$

on the Petri net of Fig. 4.4(b), we can map it to

$$\mu_1 + q_1 - 3v_4 + 2v_5 \leq 5 \tag{4.35}$$

in the original Petri net. We proceed next to formally define the direct and inverse transformations.

The C-transformation

Input: The Petri net $\mathcal{N}$, the constraints $L\mu + Hq + Cv \leq b$, and optionally the initial marking μ_0, the initial Parikh vector v_0, and a set $T_{s,C} \subseteq T$ (by default, $T_{s,C} = \emptyset$).

Output: The C-transformed Petri net $\mathcal{N}_C = (P_C, T, D_C^-, D_C^+)$, the C-transformed constraints $L_C\mu_C + Hq \leq b$, and the initial marking μ_{0C} of $\mathcal{N}_C$.

1. Initialize $\mathcal{N}_C$ to be identical to $\mathcal{N}$, with the same controllability and observability attributes. Initialize L_C to L and μ_{0C} to μ_0.

2. For all $t \in T$ such that either $C(\cdot, t)$ is not zero or $t \in T_{s,C}$:

   (a) Add a new place p to $\mathcal{N}_C$ such that $p\bullet = \emptyset$ and $\bullet p = \{t\}$.

   (b) Set $L_C(\cdot, p) = C(\cdot, t)$ and $\mu_{0C}(p) = v_0(t)$.

The C^{-1}-transformation

Input: The Petri net $\mathcal{N} = (P, T, D^-, D^+)$, the C-transformed net $\mathcal{N}_C = (P_C, T, D_C^-, D_C^+)$, and a set of constraints $L_C\mu_C + Hq \leq b$ on $\mathcal{N}_C$.

Output: The C^{-1}-transformed constraints $L\mu + Hq + Cv \leq b$.

1. Set $L(\cdot, p) = L_C(\cdot, p)\ \forall p \in P$ and C to the null matrix.

2. For all $p \in P_C \setminus P$:

    (a) Let t be the transition such that $\bullet p = \{t\}$.

    (b) Set $C(\cdot, t) = L_C(\cdot, p)$.

Note that $L_C D_C = LD + C$. Thus, the supervisor (4.16)–(4.20) enforcing $L_C \mu_C + Hq \leq b$ in $(\mathcal{N}_C, \mu_{0C})$ is identical to the supervisor (4.16)–(4.20) enforcing (4.5) on $(\mathcal{N}, \mu_0)$.

The C-transformation is very straightforward. From a supervisor designed in the C-transformed Petri net, we can immediately obtain an equivalent supervisor on the original Petri net by a simple change of notation, by substituting the marking of the places $P_C \setminus P$ with the corresponding components of the Parikh vector. Thus, from now on we will ignore the Parikh vector term and focus on constraints $L\mu + Hq \leq b$.

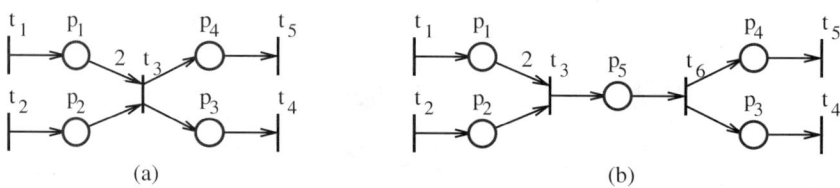

Fig. 4.5. Example for the H-transformation.

The H-transformation is a modification of the indirect method for enforcing firing vector constraints in [142]. The idea of the transformation is illustrated in the following example. Consider the Petri net of Fig. 4.5(a). Assume that we desire to enforce

$$\mu_1 + \mu_2 + 2\mu_3 + q_3 \leq 5. \qquad (4.36)$$

Then we can transform the Petri net as shown in Fig. 4.5(b). The transformation adds a place and a transition which correspond to the factor q_3. Then

$$\mu_1 + \mu_2 + 2\mu_3 + 4\mu_5 \leq 5 \qquad (4.37)$$

is the transformed constraint, where the term $4\mu_5$ is obtained as follows. Consider firing t_3 in the transformed net. If $\mu \xrightarrow{t_3} \mu'$ and a is the coefficient of μ_5, we desire

$$a + \mu_1' + \mu_2' + 2\mu_3' = 1 + \mu_1 + \mu_2 + 2\mu_3,$$

where the factor 1 is the coefficient of q_3 in (4.36). Thus we obtain $a = 4$.

Next, the H-transformation is formally defined. The transformation is formulated in the most general case, that of double-labeled Petri nets. Note that labeled Petri nets correspond to $\rho(t) = o(t)$ for all transitions t, and Petri nets with individually controllable and observable transitions correspond to $\rho(t) = o(t) = \{t\}$ for all transitions t.

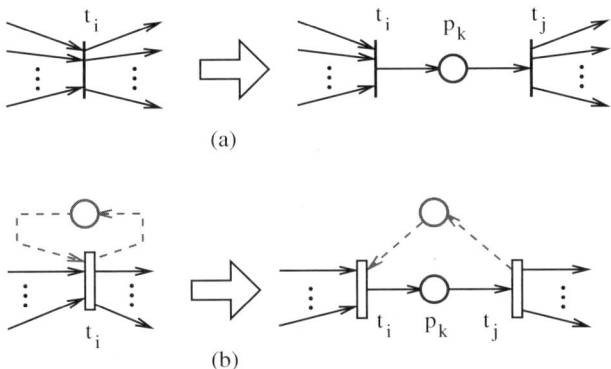

(a)

(b)

Fig. 4.6. Illustration of the transition split operation.

The H-transformation

Input: The Petri net structure $\mathcal{N} = (P, T, D^-, D^+)$, the constraints $L\mu + Hq \leq b$, and optionally the initial marking μ_0 and a set $T_{s,H} \subseteq T$ (by default, $T_{s,H} = \emptyset$).

Output: The H-transformed structure $\mathcal{N}_H = (P_H, T_H, D_H^-, D_H^+)$, the H-transformed constraints $L_H \mu_H \leq b$, and the initial marking μ_{H0} of $\mathcal{N}_H$.

1. Let $H_d = \max(LD, H, 0)$, $T^1 = T_{s,H} \cup \{t \in T : H_d(\cdot, t) \neq 0\}$, and $T_s = \{t \in T : \rho(t) = \rho(t') \text{ for some } t' \in T^1\}$. (Thus $T_s \supseteq T^1$.)

2. Initialize $\mathcal{N}_H$ to be identical to $\mathcal{N}$, with the same controllability and observability attributes. Also initialize L_H to L and μ_{H0} to μ_0.

3. For all $t \in T_s$:

   (a) Add a new place p_k and a new transition t_j to $\mathcal{N}_H$ as in Fig. 4.6(a).

   (b) Set $L_H(\cdot, p_k) = H_d(\cdot, t_i) + LD^-(\cdot, t_i)$ and $\mu_{H0}(p_k) = 0$.

4. For all $t \in T_s$, the controllability and observability of the transitions t_j is defined as follows:

   (a) $o(t \bullet \bullet) = o(t)$.

   (b) The set of control events is extended such that $\rho(t \bullet \bullet) \notin \{\rho(t) : t \in T\}$.

(c) $\rho(t \bullet \bullet)$ is controllable if and only if $\rho(t)$ is controllable.

(d) For $t, t' \in T_s$, $\rho(t \bullet \bullet) = \rho(t' \bullet \bullet)$ if and only if $\rho(t) = \rho(t')$.

The H^{-1}-transformation

Input: The Petri net $\mathcal{N} = (P, T, D^-, D^+)$, the H-transformed net $\mathcal{N}_H = (P_H, T_H, D_H^-, D_H^+)$, and a set of constraints $L_H \mu_H \leq b$ on $\mathcal{N}_H$.

Output: The H^{-1}-transformed constraints $L\mu + Hq \leq b$.

1. Set $L(\cdot, p) = L_H(\cdot, p) \ \forall p \in P$ and H to the null matrix.

2. For all $p_k \in P_H \setminus P$:

    (a) Let t_i be the transition such that $\{t_i\} = \bullet p_k$.

    (b) Set $H(\cdot, t_i) = L_H(\cdot, p_k) - L_H D_H^-(\cdot, t_i)$.

Note several properties of the H- and H^{-1}-transformations. To simplify our notation, assume single constraints $l\mu + hq \leq b$ and $l_H \mu_H \leq b$. Then, note that if $l_H \mu_H \leq b$ is the H-transformation of $l\mu + hq \leq b$:

$$l_H(p) = \begin{cases} l(p) & \text{if } p \in P, \\ h_d(\bullet p) + l D^-(\cdot, \bullet p) & \text{if } p \in P_H \setminus P. \end{cases} \tag{4.38}$$

In addition, the relation between $\mathcal{N}_H$ and $\mathcal{N}$ is such that

$$\forall t \in T \setminus \bullet(P_H \setminus P) : D_H^-(p, t) = \begin{cases} D^-(p, t) & \text{for } p \in P, \\ 0 & \text{for } p \in P_H \setminus P, \end{cases} \tag{4.39}$$

$$D_H^+(p, t) = \begin{cases} D^+(p, t) & \text{for } p \in P, \\ 0 & \text{for } p \in P_H \setminus P, \end{cases} \tag{4.40}$$

$$\forall t \in T \cap \bullet(P_H \setminus P) : D_H^-(p, t) = \begin{cases} D^-(p, t) & \text{for } p \in P, \\ 0 & \text{for } p \notin P_H \setminus P, \end{cases} \tag{4.41}$$

$$D_H^+(p, t) = \begin{cases} 0 & \text{for } p \notin (P_H \setminus P) \cap t\bullet, \\ 1 & \text{for } p = (P_H \setminus P) \cap t\bullet, \end{cases} \tag{4.42}$$

$$\forall t \in T_H \setminus T : D_H^-(p, t) = \begin{cases} 0 & \text{for } p \neq \bullet t, \\ 1 & \text{for } p = \bullet t, \end{cases} \tag{4.43}$$

$$D_H^+(p, t) = \begin{cases} D^+(p, \bullet \bullet t) & \text{for } p \in P, \\ 0 & \text{for } p \notin P. \end{cases} \tag{4.44}$$

Furthermore, if $l\mu + hq \leq b$ is the H^{-1}-transformation of $l_H \mu_H \leq b$

$$l(p) = l_H(p) \ \forall p \in P, \tag{4.45}$$

$$h(t) = \begin{cases} l_H(p) - l_H D_H^-(\cdot, t) & \text{if } t \bullet \cap(P_H \setminus P) = p, \\ 0 & \text{if } t \bullet \cap(P_H \setminus P) = \emptyset. \end{cases} \tag{4.46}$$

The following relation can be easily verified based on (4.38)–(4.44). The relation will prove very useful in the further developments.

$$L_H D_H(\cdot, t) = \begin{cases} LD(\cdot, t) & \text{for } t \in T \setminus \bullet(P_H \setminus P), \\ H_d(\cdot, t) & \text{for } t \in T \cap \bullet(P_H \setminus P), \\ LD(\cdot, \bullet\bullet t) - H_d(\cdot, \bullet\bullet t) & \text{for } t \in T_H \setminus T. \end{cases} \quad (4.47)$$

Let D_c, D_c^-, and D_c^+ denote the incidence, output, and input matrices of the supervisor enforcing $L\mu + Hq \leq b$. Similarly, let's define $D_{c,H}$, $D_{c,H}^-$, and $D_{c,H}^+$ for the supervisor enforcing $L_H \mu_H \leq b$ in $\mathcal{N}_H$. Note that $D_{c,H} = -L_H D_H$ and $D_{c,H}^- = \max(0, L_H D_H)$. Thus, based on (4.47), the following is obtained:

$$D_{c,H}^-(\cdot, t) = \begin{cases} D_c^-(\cdot, t) & \text{for } t \in T, \\ 0 & \text{for } t \in T_H \setminus T. \end{cases} \quad (4.48)$$

Further, since $D_{c,H}^+ = \max(0, -L_H D_H)$

$$D_{c,H}^+(\cdot, t) = \begin{cases} D_c^+(\cdot, t) & \text{for } t \in T \setminus \bullet(P_H \setminus P), \\ 0 & \text{for } t \in T \cap \bullet(P_H \setminus P), \\ D_c^+(\cdot, \bullet\bullet t) & \text{for } t \in T_H \setminus T. \end{cases} \quad (4.49)$$

We introduce the following notation. If a transition t_i is split in the H-transformation as in Fig. 4.6, let $\sigma_H(t_i)$ be the firing sequence $t_i t_j$. If a transition t_i is not split, let $\sigma_H(t_i)$ equal t_i. Further, we also use σ_H for firing vectors: $\sigma_H(q) = q_H q'_H$, where $q_H(t_i) = q'_H(t_j) = q(t_i)$ for a transition t_i split in t_i and t_j, $q_H(t_i) = q(t_i)$ for a transition t_i that is not split, $q'_H(t_i) = 0$ $\forall t_i \in T$ and $q_H(t_j) = 0$ $\forall t_j \in T_H \setminus T$. If $\sigma = q_1 q_2 \ldots$ is a firing sequence in $\mathcal{N}$, let $\sigma_H(\sigma) = \sigma_H(q_1)\sigma_H(q_2)\ldots$. Further, let m_H map the markings of $\mathcal{N}$ into markings of $\mathcal{N}_H$ as follows:

$$\mu_H = m_H(\mu) \Rightarrow \mu_H(p) = \begin{cases} \mu(p) & \text{for } p \in P, \\ 0 & \text{for } p \in P_H \setminus P. \end{cases} \quad (4.50)$$

Proposition 4.7. *Given $(\mathcal{N}, \mu_0)$ and $(\mathcal{N}_H, m_H(\mu_0))$, let q be a firing vector in $\mathcal{N}$ and $\sigma_H(q) = q_H q'_H$.*

(a) *At all reachable markings, q_H is enabled if and only if $\sigma_H(q)$ is enabled.*
(b) *q is enabled at the marking μ_1 if and only if $\sigma_H(q)$ is enabled at the marking $m_H(\mu_1)$.*

Proof. (a) By (4.42) and (4.43), q_H is enabled if and only if $q_H q'_H$ is enabled.
    (b) Let $\mu_{H1} = m_H(\mu_1)$. Note that $\mu_1 \geq D^- q \Leftrightarrow \mu_{H1} \geq D_H^- q_H$, by (4.39) and (4.41). Therefore, q is enabled if and only if q_H is enabled, which concludes our proof by part (a). □

Proposition 4.8. *Consider $(\mathcal{N}, \mu_0)$ in closed loop with a supervisor Ξ optimally enforcing $L\mu + Hq \leq b$, and $(\mathcal{N}_H, m_H(\mu_0))$ in closed loop with a supervisor Ξ_H optimally enforcing $L_H \mu_H \leq b$. Let q be a firing vector in $\mathcal{N}$ and $\sigma_H(q) = q_H q'_H$.*

(a) *At all reachable markings, q_H is closed-loop enabled if and only if $\sigma_H(q)$ is closed-loop enabled.*

(b) *μ is reachable and q is closed-loop enabled at μ if and only if $\mu_H = m_H(\mu_1)$ is reachable and $\sigma_H(q)$ is closed-loop enabled at μ_H.*

Proof. (a) By (4.48), $D^-_{c,H}q'_H = 0$. Thus, Ξ_H never restricts the firing of q'_H. Therefore, in view of Proposition 4.7(a), q_H is closed-loop enabled if and only if $q_H q'_H$ is closed-loop enabled.

(b) We show that a sequence $\mu_0 \xrightarrow{q_1} \mu_1 \xrightarrow{q_2} \mu_2 \ldots \xrightarrow{q_k} \mu_k$ is possible in the closed loop of $\mathcal{N}$ if and only if $\mu_{H0} \xrightarrow{\sigma_H(q_1)} \mu_{H1} \xrightarrow{\sigma_H(q_2)} \mu_{H2} \ldots \xrightarrow{\sigma_H(q_k)} \mu_{Hk}$ is possible in the closed loop of $\mathcal{N}_H$. Note that $q_1 q_2 \ldots q_k$ is plant enabled if and only if $\sigma_H(q_1 q_2 \ldots q_k)$ is plant enabled, based on Propositions 4.7(b) and 4.9(a), where Proposition 4.9(a) shows that $\mu_{Hi} = m_H(\mu_i)$ for $i = 0, 1, \ldots, k$. Thus, we only need to prove that if $q_1 q_2 \ldots q_i$ and $\sigma_H(q_1 q_2 \ldots q_i)$ are closed-loop enabled, then $q_1 q_2 \ldots q_i q_{i+1}$ is supervisor enabled if and only if $\sigma_H(q_1 q_2 \ldots q_i q_{i+1})$ is supervisor enabled.

Note that the constraints $L_H \mu_H \le b$ are not violated by firing q_H when $L_H \mu_H + D^-_{c,H} q_H \le b$. Further, the constraints $L\mu + Hq \le b$ are not violated by firing q when $L\mu + H_d q \le b$. By definition, $H_d = D^-_c$. Further, by (4.48) $D^-_{c,H} q_H = H_d q$, and by (4.38) and $\mu_H = m_H(\mu)$, $L_H \mu_H = L\mu$. It follows that $L_H \mu_{Hi} + D^-_{c,H} q_{Hi+1} \le b \Leftrightarrow L\mu_i + H_d q_{i+1} \le b$ (where q_{Hi+1} is the first term of $\sigma_H(q_{i+1}) = q_{Hi+1} q'_{Hi+1}$). Therefore, q_{i+1} is supervisor enabled if and only if q_{Hi+1} is supervisor enabled. By part (a), this concludes the proof. $\square$

Given a firing sequence σ of $\mathcal{N}$, we have already defined $\sigma_H(\sigma)$ to denote the equivalent firing sequence σ_H of $\mathcal{N}_H$. In the following developments, we will need also the converse operation $\sigma(\sigma_H)$, associating a firing sequence σ of $\mathcal{N}$ to each firing sequence σ_H of $\mathcal{N}_H$. Assume μ_0 and $\mu_{H0} = m_H(\mu_0)$ are the initial markings of $\mathcal{N}$ and $\mathcal{N}_H$. Given a firing sequence σ_H of $\mathcal{N}_H$, let $\overline{\sigma}_H$ be the firing count vector and $\nu_H(\sigma_H)$ be the largest integer vector v_H such that $v_H \le \overline{\sigma}_H$ and $\forall t \in \bullet(P_H \setminus P)$, $v_H(t) = v_H(t \bullet \bullet)$. Further, let $\chi_H(\sigma_H) = \overline{\sigma}_H - \nu_H(\sigma_H)$. Thus, if $q_H = \chi_H(\sigma_H)$, then $\forall t \in T_H \setminus \bullet(P_H \setminus P)$, $q_H(t) = 0$. Let $\nu(\sigma_H)$ and $\chi(\sigma_H)$ be the restrictions of $\nu_H(\sigma_H)$ and $\chi_H(\sigma_H)$ to the transitions in T. If $\sigma_H = q_{H1} q_{H2} \ldots q_{Hx}$, let σ_{H0} be an empty sequence, $\sigma_{H1} = q_{H1}$, $\sigma_{H2} = q_{H1} q_{H2}$, $\ldots$, $\sigma_{Hx} = q_{H1} q_{H2} \ldots q_{Hx}$, and $q_i = \nu(\sigma_{Hi}) - \nu(\sigma_{Hi-1})$ for $i = 1 \ldots x$. We define $\sigma(\sigma_H)$ as the sequence $q_1 q_2 \ldots q_x$.

Proposition 4.9. *Consider $(\mathcal{N}, \mu_0)$, the set of constraints $L\mu + Hq \le b$, and their H-transformation $(\mathcal{N}_H, \mu_{H0})$ and $L_H \mu_H \le b$, where $\mu_{H0} = m_H(\mu_0)$.*

(a) *If $\sigma_H(q) = q_H q'_H$, $\mu_1 \xrightarrow{q} \mu_2$, $\mu_{H1} \xrightarrow{q_H} \mu'_{H1} \xrightarrow{q'_H} \mu_{H2}$, and $\mu_{H1} = m_H(\mu_1)$, then $\mu_{H2} = m_H(\mu_2)$ and $L_H \mu'_{H1} = L\mu_1 + H_d q$.*

(b) *If $\mu_{H0} \xrightarrow{\sigma_H} \mu_H$, then $\sigma(\sigma_H)$ is enabled at μ_0 and firing it results in $\mu = \mu_0 + D\nu(\sigma_H)$. Further, $q = \chi(\sigma_H)$ is enabled at μ and $L_H \mu_H = L\mu + H_d q$.*

(c) *Given σ_H and q_H, if $\sigma_H q_H$ is enabled at μ_{H0} and x is the restriction of q_H to T, then $\sigma(\sigma_H)q$ is enabled at μ_0, where $q = \chi(\sigma_H) + x$.*

(d) *Let Ξ be a supervisor optimally enforcing $L\mu + Hq \leq b$ in $(\mathcal{N}, \mu_0)$ and Ξ_H a supervisor optimally enforcing $L_H \mu_H \leq b$ in $(\mathcal{N}_H, \mu_{H0})$. If σ_H is closed-loop enabled at μ_{H0}, then $\sigma(\sigma_H)$ is closed-loop enabled at μ_0.*

Proof. (a) $L_H \mu'_{H1} = L\mu_1 + H_d q$ follows from (4.38)–(4.42) and $\mu_{H2} = m_H(\mu_2)$ from (4.39)–(4.44).

(b) Let μ_{H1}, μ_{H2}, ..., μ_{Hx} be markings such that $\mu_{H0} \xrightarrow{q_{H1}} \mu_{H1} \xrightarrow{q_{H2}} \mu_{H2} \ldots \xrightarrow{q_{Hx}} \mu_{Hx}$. Let σ_{H0} be an empty sequence, $\sigma_{H1} = q_{H1}$, $\sigma_{H2} = q_{H1}q_{H2}$, ..., $\sigma_{Hx} = q_{H1}q_{H2} \ldots q_{Hx}$. Further, let $\mu_i = \mu_{i-1} + D(\nu(\sigma_{Hi}) - \nu(\sigma_{Hi-1}))$ and $u_i = \chi(\sigma_{Hi})$ for $i = 1 \ldots x$. We show by induction that μ_i is reachable from μ_{i-1} by firing $q_i = \nu(\sigma_{Hi}) - \nu(\sigma_{Hi-1})$, where $\mu_i = \mu_{i-1}$ if $q_i = 0$, and that $L_H \mu_{Hi} = L\mu_i + H_d u_i$ for $i = 1 \ldots x$. For $i = 1$, note that $\nu(\sigma_{H1}) = 0$ and $\mu_1 = \mu_0$. Further, $L_H \mu_{H1} = L\mu_1 + H_d u_1$ is satisfied by part (a). Now, assume the induction hypothesis satisfied at step i. Let $q_{i+1} = \nu(\sigma_{Hi+1}) - \nu(\sigma_{Hi})$. Let's show first that if $q_{i+1} \neq 0$ then q_{i+1} is plant enabled, that is, $\mu_i \geq D^- q_{i+1}$. Note that $\mu_{Hi} = \mu_{H0} + D_H \nu_H(\sigma_{Hi}) + D_H \chi_H(\sigma_{Hi})$, and so $\mu_{Hi} = m_H(\mu_i) + D_H \chi_H(\sigma_{Hi})$. Since q_{Hi+1} is enabled, $\mu_{Hi} \geq D_H^- q_{Hi+1}$. Then, by (4.39)–(4.42), $\mu_i \geq D^-(u_i + x)$, where x is the restriction of q_{Hi+1} to T. Note that since $\mu_{H0} = m_H(\mu_0)$, any firing of a transition $t \in T_H \setminus T$ must be preceded by a firing of the transition $\bullet \bullet t$. Thus, $\forall t \in T_H \setminus T$: $q_{Hi+1}(t) \leq u_i(\bullet \bullet t)$ and $q_{Hi+1}(t) = q_{i+1}(\bullet \bullet t)$. Further, $\forall t \in T \setminus \bullet(P_H \setminus P)$: $x(t) = q_{i+1}(t)$. Therefore, we can conclude that $q_{i+1} \leq u_i + x$ and so $\mu_i \geq D^- q_{i+1}$. Next we show that $L_H \mu_{Hi+1} = L\mu_{i+1} + H_d u_{i+1}$. Now, $L_H \mu_{Hi+1} = L_H \mu_{Hi} - D_{c,H} q_{Hi+1}$. Let's decompose q_{Hi+1} as $q_{Hi+1} = \alpha_H + \beta_H + \gamma_H$, where $\alpha_H(t) = q_{Hi+1}(t)$ for $t \in T_H \setminus T$ and $\alpha_H(t) = 0$ otherwise, $\beta_H(t) = q_{Hi+1}(t)$ if $t \in T \setminus \bullet(P_H \setminus P)$ and $\beta_H(t) = 0$ otherwise, and $\gamma_H(t) = q_{Hi+1}(t)$ if $t \in T \cap \bullet(P_H \setminus P)$ and $\gamma_H(t) = 0$ otherwise. By (4.39)–(4.44), $D_{c,H} q_{Hi+1} = D_c^+ \alpha + D_c \beta - D_c^- \gamma$, where β and γ are the restrictions of β_H and γ_H to T, and $\alpha(t) = \alpha_H(t \bullet \bullet)$ for $t \in T \cap \bullet(P_H \setminus P)$ and $\alpha(t) = 0$ otherwise. Thus, from $L_H \mu_{Hi+1} = L_i \mu_i + H_d u_i - D_{c,H} q_{Hi+1}$ and $D_c^- = H_d$ we obtain $L_H \mu_{Hi+1} = L_i \mu_i - D_c(\alpha + \beta) + H_d(u_i - \alpha + \gamma)$. Note that $q_{i+1} = \alpha + \beta$ and $u_{i+1} = u_i - \alpha + \gamma$, so $L_H \mu_{Hi+1} = L_{i+1} \mu_{i+1} + H_d u_{i+1}$, which concludes our induction proof.

It only remains to show that $q = \chi(\sigma_H)$ is enabled at μ. Let q'_H be defined as $q'_H(t) = \mu_H(\bullet t)$ $\forall t \in T_H \setminus T$ and $q'_H(t) = 0$ otherwise. Thus, μ_H enables q'_H. Therefore, by the first part of the proof, μ enables $q_z = \nu(\sigma_H q'_H) - \nu(\sigma_H)$. Note that $q_z = q$. Therefore, μ enables q.

(c) Let $x_H = q_H(t)$ $\forall t \in T$ and $x_H(t) = 0$ otherwise. Let q'_H be defined as $q'_H(t) = \mu_H(\bullet t)$ $\forall t \in T_H \setminus T$ and $q'_H(t) = 0$ otherwise, where $\mu_H \xleftarrow{\sigma_H} \mu_{H0}$. Note that $\sigma_H x_H q'_H$ is enabled. Further, let x^*_H and q^*_H be defined as $x^*_H(t) = x_H(t)$ $\forall t \in T \cap \bullet(P_H \setminus P)$, $x^*_H(t) = 0$ otherwise, $q^*_H(t) = x_H(t)$ $\forall t \in T \setminus \bullet(P_H \setminus P)$, $q^*_H(t) = q'_H(t)$ $\forall t \in T_H \setminus T$, and $q^*_H(t) = 0$ otherwise. (So $x^*_H + q^*_H = x_H + q'_H$.) In view of (4.39)–(4.44), since $\sigma_H x_H q'_H$ is enabled, $\sigma_H x^*_H q^*_H$ is too. Note that $\sigma(\sigma_H x^*_H q^*_H) = \sigma(\sigma_H)q$ for $q = \chi(\sigma_H) + x$. Then, $\sigma(\sigma_H)q$ is enabled by part (b).

(d) The induction proof of part (b) can be used, once we show that q_{i+1} is supervisor enabled at the marking μ_i, that is, $L\mu_i + H_d q_{i+1} \leq b$. Since q_{Hi+1} is closed-loop enabled, $L_H \mu_{Hi} + D_{c,H}^- q_{Hi+1} \leq b$. By (4.48), $D_{c,H}^- q_{Hi+1} = H_d x$. By $L_H \mu_{Hi} = L\mu_i + H_d u_i$, $L\mu_i + H_d q_{i+1} + H_d(x + u_i - q_{i+1}) \leq b$. Since $x + u_i \geq q_{i+1}$, $L\mu_i + H_d q_{i+1} \leq b$, and so q_{i+1} is supervisor enabled. $\square$

Next, a relaxed concept of feasibility is introduced for specifications on $\mathcal{N}_H$. Compared to Definition 4.4, the second requirement is relaxed to constrain only the firing sequences σ_H of $\mathcal{N}_H$ that have the form $\sigma_H = \sigma_H(\sigma)$, where σ is a sequence of $\mathcal{N}$.

Definition 4.10. *A specification on $(\mathcal{N}_H, \mu_{H0})$ is* **h-feasible** *if a supervisor optimally enforcing the specification ensures that*

1. *If q_H and q_H' are two plant-enabled firing vectors and $\rho^*(q_H) = \rho^*(q_H')$, then the closed loop enables either both q_H and q_H' or neither of them.*
2. *Let $q \neq 0$ be a firing vector of $\mathcal{N}$ and σ_1 and σ_2 be two sequences of firing vectors of $\mathcal{N}$. If $\sigma_H(\sigma_1)$ and $\sigma_H(\sigma_2)$ are enabled by the closed loop at the initial state, $o^*(\sigma_H(\sigma_1)) = o^*(\sigma_H(\sigma_2))$, and both $\sigma_H(\sigma_1 q)$ and $\sigma_H(\sigma_2 q)$ are plant enabled at the initial state, then either both $\sigma_H(\sigma_1 q)$ and $\sigma_H(\sigma_2 q)$ or neither of them are closed-loop enabled at the initial state.*

There are several literature methods that result in designs corresponding to the enforcement of disjunctions of constraints (4.3). Here, a disjunction of constraints has the form

$$\bigvee_{i=1}^{n_d} [L_i \mu + H_i q + C_i v \leq b_i] \tag{4.51}$$

requiring that all reachable states satisfy

$$\bigvee_{i=1}^{n_d} [L_i \mu + C_i v \leq b_i] \tag{4.52}$$

and that a firing vector q should be enabled only if μ, q, and v satisfy

$$\bigvee_{i=1}^{n_d} [L_i \mu + H_{d,i} q + C_i v \leq b_i], \tag{4.53}$$

where $H_{d,i} = \max(L_i D, H_i, 0)$. $H_{d,i}$ is the H_d matrix defined in the H-transformation, which is also the same as $D_{c,i}^-$ calculated by (4.19).

Note that this interpretation of a disjunction (4.51) is not the most general. Recall that the constraints (4.5) were defined to require the inequality $L\mu + Hq + Cv \leq b$ be satisfied for all possible intermediary states reached during the firing of q; that is, $\forall q', q'' \geq 0$, if $q' + q'' \leq q$ then $L\mu' + Hq'' + Cv' \leq b$, where $\mu \xrightarrow{q'} \mu'$ and $v' = v + q'$. Thus, it was shown in Lemma 4.2 that the constraints

(4.5) enable a firing vector q if and only if the inequality $L\mu + H_d q + Cv \le b$ is satisfied. On the other hand, the requirement that $\forall q', q'' \ge 0$, if $q' + q'' \le q$ then $\bigvee_i L_i \mu' + H_i q'' + C_i v' \le b_i$, is weaker than the requirement that μ, q, and v satisfy (4.53). However, (4.53) is easier to check online and allows us to easily extend our results from conjunctions of constraints to disjunctions of constraints. In the particular case of no concurrency and $H_i = 0$ for all i, these two interpretations of (4.51) are equivalent.

The C- and H-transformations have been defined for conjunctions (4.5). The transformations can easily be extended to disjunctions (4.51) as follows. The C-transformation of (4.51) results in a transformed Petri net $\mathcal{N}_C$ and a disjunction

$$\bigvee_{i=1}^{n_d} [L_{C,i}\mu_C + H_i q \le b_i] \tag{4.54}$$

according to the following algorithm.

The C-transformation for constraints (4.51)

1. Let $T_{s,C} = T_{s,C} \cup \bigcup_{i=1}^{n_d} \{t \in T : C_i(\cdot, t) \ne 0\}$.

2. For all $i = 1 \dots n_d$, apply the C-transformation to the constraints $L_i \mu + H_i q + C_i v \le b_i$ with the argument $T_{s,C}$ calculated at step 1. Let $L_{C,i}\mu_C + H_i q \le b_i$ be the transformed constraints.

3. The result of the C-transformation consists of the disjunction (4.54), the Petri net $\mathcal{N}_C$, and the initial marking μ_{0C}, where $\mathcal{N}_C$ and μ_{0C} are obtained from any of the C-transformations of step 2.

Note that the choice of the set $T_{s,C}$ guarantees that the same Petri net $\mathcal{N}_C$ is obtained by all C-transformations of step 2. The C^{-1}-transformation of a disjunction (4.54) results in a disjunction (4.51), obtained by taking the disjunction of the C^{-1}-transformations of the constraints $L_{C,i}\mu_C + H_i q \le b_i$.

The C^{-1}-transformation for constraints (4.54)

1. For all $i = 1 \dots n_d$, apply the C^{-1}-transformation to the constraints $L_{C,i}\mu_C + H_i q \le b_i$. Let $L_i \mu + H_i q + C_i v \le b_i$ be the transformed constraints.

2. The result of the C^{-1}-transformation is the disjunction (4.51).

Similarly, the H-transformation of a disjunction

$$\bigvee_{i=1}^{n_d} [L_i \mu + H_i q \le b_i] \tag{4.55}$$

results in a disjunction

$$\bigvee_{i=1}^{n_d} [L_{H,i}\mu_H \leq b_i], \tag{4.56}$$

where the constraints $L_{H,i}\mu_H \leq b_i$ are obtained by taking the H-transformation of each $L_i\mu + H_iq \leq b_i$.

The H-transformation for constraints (4.55)

1. Let $H_{d,i} = \max(L_iD, H_i, 0)$ and $T_{s,H} = T_{s,H} \cup \bigcup_{i=1}^{n_d}\{t \in T : H_{d,i}(\cdot, t) \neq 0\}$.

2. For all $i = 1 \ldots n_d$, apply the H-transformation to the constraints $L_i\mu + H_iq \leq b_i$ with the argument $T_{s,H}$ calculated at step 1. Let $L_{H,i}\mu_H \leq b_i$ be the transformed constraints.

3. The result of the H-transformation consists of the disjunction (4.56), the Petri net $\mathcal{N}_H$, and the initial marking μ_{H0}, where $\mathcal{N}_H$ and μ_{H0} are obtained from any of the H-transformations of step 2.

Note that the choice of the set $T_{s,H}$ guarantees that the same Petri net $\mathcal{N}_H$ is obtained by all H-transformations of step 2. The H^{-1}-transformation of a disjunction (4.56) results in a disjunction (4.55), obtained by taking the disjunction of the H^{-1}-transformations of the constraints $L_{H,i}\mu_H \leq b_i$.

The H^{-1}-transformation for constraints (4.56)

1. For all $i = 1 \ldots n_d$, apply the H^{-1}-transformation to the constraints $L_{H,i}\mu_H \leq b_i$. Let $L_i\mu + H_iq < b_i$ be the transformed constraints.

2. The result of the H^{-1}-transformation is the disjunction (4.55).

As previously mentioned, the C-transformation is straightforward, as we can obtain an equivalent supervisor Ξ_C on $\mathcal{N}_C$ from a supervisor Ξ on $\mathcal{N}$ by simply substituting the Parikh vectors v with a vector of sink place markings. The H-transformation is more involved, so we will focus on deriving its properties.

The next result shows that Proposition 4.8 can be extended to the case of disjunctions of constraints.

Proposition 4.11. *Consider $(\mathcal{N}, \mu_0)$ in closed loop with a supervisor Ξ optimally enforcing (4.55), and $(\mathcal{N}_H, m_H(\mu_0))$ in closed loop with a supervisor Ξ_H optimally enforcing (4.56). Let q be a firing vector in $\mathcal{N}$ and $\sigma_H(q) = q_Hq'_H$.*

(a) *At all reachable markings, q_H is closed-loop enabled if and only if $\sigma_H(q)$ is closed-loop enabled.*

(b) *μ is reachable and q is closed-loop enabled at μ if and only if $\mu_H = m_H(\mu_1)$ is reachable and $\sigma_H(q)$ is closed-loop enabled at μ_H.*

Proof. (a) By Proposition 4.7(a), q_H is plant enabled if and only if $\sigma_H(q)$ is plant enabled. By (4.48), $D^-_{c,H,i}q'_H = 0$ for all $i = 1 \ldots n_d$, and so firing q'_H cannot violate any of the constraints $L_{Hi}\mu_H \leq b_i$ that are satisfied. The conclusion follows.

(b) The proof is the same as in Proposition 4.8(b), once we substitute $L\mu + Hq \leq b$ ($L_H\mu_H \leq b$) by the constraints $L_j\mu + H_jq \leq b_j$ ($L_{Hj}\mu_H \leq b_j$), $j \in \{1, 2, \ldots, n_d\}$, that are satisfied when q_{i+1} (q_{Hi+1}) is fired at μ_i (μ_{Hi}). □

Part (a) of the next result shows that the parts (c) and (d) of Proposition 4.9 can also be extended to the case of disjunctions of constraints.

Proposition 4.12. *Let Ξ be a supervisor optimally enforcing* (4.55) *in* $(\mathcal{N}, \mu_0)$ *and Ξ_H a supervisor optimally enforcing* (4.56) *in* $(\mathcal{N}_H, \mu_{H0})$, *where $\mu_{H0} = m_H(\mu_0)$.*

(a) *If σ_H is closed-loop enabled at μ_{H0}, then $\sigma(\sigma_H)$ is closed-loop enabled at μ_0.*

(b) *Assume that σ_H is closed-loop enabled at μ_{H0} and σ_Hq_H is plant enabled at μ_{H0}. Then σ_Hq_H is closed-loop enabled at μ_{H0} if and only if $q = \chi(\sigma_H) + x$ is closed-loop enabled at $\mu = \mu_0 + D\nu(\sigma_H)$, where x is the restriction of q_H to T.*

Proof. (a) The proof of Proposition 4.9(d) can be adapted here based on the following observation. For any closed-loop enabled sequence $\mu_{H0} \xrightarrow{q_{H1}} \mu_{H1} \xrightarrow{q_{H2}} \mu_{H2} \ldots \xrightarrow{q_{Hx}} \mu_{Hx}$, there is a sequence of indices $k_0, k_1, \ldots, k_{x-1} \in \{1, 2, \ldots, n_d\}$ such that $L_{H,k_i}\mu_{Hi} + D^-_{c,H,k_i}q_{Hi+1} \leq b_{k_i}$, for all $i = 0, 1, \ldots, x - 1$. Thus, the proof of Proposition 4.9(d) can be used to show that $L_{H,k_i}\mu_{Hi} + D^-_{c,H,k_i}q_{Hi+1} \leq b_{k_i} \Rightarrow L_{k_i}\mu_i + H^-_{d,k_i}q_{i+1} \leq b_{k_i}$, where $q_1q_2 \ldots q_x$ denotes the sequence $\sigma(\sigma_H)$.

(b) By part (a) and Proposition 4.9(b), μ is reachable in the closed loop by firing $\sigma(\sigma_H)$. Let $\mu_H \xleftarrow{\sigma_H} \mu_{H0}$. For all $i = 1, 2, \ldots, n_d$, $D^-_{c,H,i}q_H = D^-_{c,i}x$ by (4.48), and $L_{H,i}\mu_H = L_i\mu + H_{d,i}\chi(\sigma_H)$ by Proposition 4.9(b). Thus, $L_{H,i}\mu_H + D^-_{c,H,i}q_H = L_i\mu + H_{d,i}q$. If q is closed-loop enabled, then there is $i \in \{1, 2, \ldots, n_d\}$ such that $L_i\mu + H_{d,i}q \leq b_i$. Thus, $L_{H,i}\mu_H + D^-_{c,H,i}q_H \leq b_i$, which shows that q_H is supervisor enabled at μ_H. On the other hand, if q_H is closed-loop enabled at μ_H, there is $i \in \{1, 2, \ldots, n_d\}$ such that $L_{H,i}\mu_H + D^-_{c,H,i}q_H \leq b_i$, so $L_i\mu + H_{d,i}q \leq b_i$. Thus, q is supervisor enabled at μ. Therefore, in view of Proposition 4.9(c), q is closed-loop enabled at μ. □

Theorem 4.13. *Let* (4.56) *denote the H-transformation of* (4.55), μ_0 *the initial marking of $\mathcal{N}$, and $\mu_{H0} = m_H(\mu_0)$ the initial marking of $\mathcal{N}_H$. Then* (4.56) *is h-feasible if and only if* (4.55) *is feasible.*

Proof. The proof shows that each of the two requirements of Definition 4.4 implies its corresponding requirement in Definition 4.10 and vice versa. The proof for the first requirement is by contradiction.

Case 1a: The first requirement is satisfied in Definition 4.10 but not in Definition 4.4. Thus, there is a reachable marking μ of $\mathcal{N}$ such that two plant-enabled firing vectors q_1 and q_2 satisfy that $\rho^*(q_1) = \rho^*(q_2)$ and that the closed loop enables q_1 but disables q_2. Since q_1 is supervisor enabled and q_2 is supervisor disabled, there is $k \in \{1, 2, \ldots, n_d\}$ such that $L_k\mu + H_{d,k}q_1 \leq b_k$, and $L_i\mu + H_{d,i}q_2 \not\leq b_i \; \forall i = 1 \ldots n_d$. Let $\sigma_H(q_1) = q_{H1}q'_{H1}$ and $\sigma_H(q_2) = q_{H2}q'_{H2}$. By Propositions 4.11(b) and 4.7(b), $\mu_H = m_H(\mu)$ is reachable in the closed loop, q_{H1} is closed-loop enabled, and q_{H2} is only plant enabled. However, this contradicts the first requirement of Definition 4.10, since $\rho^*(q_1) = \rho^*(q_2) \Rightarrow \rho^*(q_{H1}) = \rho^*(q_{H2})$.

Case 1b: The first requirement is satisfied in Definition 4.4 but not in Definition 4.10. Thus, there is a reachable marking μ_H of $\mathcal{N}_H$ such that two firing vectors q_{H1} and q_{H2} satisfy that $\rho^*(q_{H1}) = \rho^*(q_{H2})$ and that q_{H1} is closed-loop enabled and q_{H2} is only plant enabled. For $i = 1, 2$, let x_{Hi} be defined as $x_{Hi}(t) = q_{Hi}(t) \; \forall t \in T$ and $x_{Hi}(t) = 0$ otherwise. By (4.48), x_{H1} is closed-loop enabled and x_{H2} is only plant enabled. Let σ_H be a firing sequence such that $\mu_{H0} \xrightarrow{\sigma_H} \mu_H$ and let x_1 and x_2 be the restrictions of x_{H1} and x_{H2} to T, $q_1 = \chi(\sigma_H) + x_1$, and $q_2 = \chi(\sigma_H) + x_2$. By Propositions 4.12 and 4.9(b)–(c), $\mu_0 \xrightarrow{\sigma(\sigma_H)} \mu$, q_1 is closed-loop enabled at μ and q_2 is only plant enabled at μ. This contradicts the first requirement of Definition 4.4, since $\rho^*(q_{H1}) = \rho^*(q_{H2}) \Rightarrow \rho^*(q_1) = \rho^*(q_2)$.

Case 2: We show that the second requirement in Definition 4.10 is not satisfied if and only if the second requirement in Definition 4.4 is not satisfied. The second requirement of Definition 4.4 is not satisfied if and only if there are two sequences σ_1 and σ_2 and a firing vector q such that $\sigma_1 q$ and σ_2 are closed-loop enabled, $\sigma_2 q$ is only plant enabled, and $o^*(\sigma_1) = o^*(\sigma_2)$. Further, $\sigma_1 q$ and σ_2 are closed-loop enabled and $\sigma_2 q$ is only plant enabled if and only if $\sigma_H(\sigma_1 q)$ and $\sigma_H(\sigma_2)$ are closed-loop enabled and $\sigma_H(\sigma_2 q)$ is only plant enabled, by Propositions 4.11(b) and 4.7(b). Since $o^*(\sigma_1) = o^*(\sigma_2) \Leftrightarrow o^*(\sigma_H(\sigma_1)) = o^*(\sigma_H(\sigma_2))$, the conclusion follows.   □

Given $(\mathcal{N}, \mu_0)$, we say that a supervisor Ξ_1 is at least as restrictive as a supervisor Ξ_2, which we write $\Xi_1 \preceq \Xi_2$, if any sequence σ closed-loop enabled at the initial state of $(\mathcal{N}, \mu_0, \Xi_1)$ is also closed-loop enabled at the initial state of $(\mathcal{N}, \mu_0, \Xi_2)$. Further, Ξ_1 is more restrictive than Ξ_2, which we write $\Xi_1 \prec \Xi_2$, if $\Xi_1 \preceq \Xi_2$ and there is a sequence σ closed-loop enabled at the initial state of $(\mathcal{N}, \mu_0, \Xi_2)$ that is not closed-loop enabled at the initial state of $(\mathcal{N}, \mu_0, \Xi_1)$. Let $\mathcal{S}$ denote a set of constraints $\bigvee_{i=1}^{n_d}[L_i\mu + H_iq \leq b_i]$ and $\mathcal{S}'$ denote $\bigvee_{i=1}^{n'_d}[L'_i\mu + H'_iq \leq b_i]$. Let $\mathcal{S}_H$ denote $\bigvee_{i=1}^{n_d}[L_{Hi}\mu_H \leq b_i]$, the H-transformation of $\mathcal{S}$, and $\mathcal{S}'_H$ denote $\bigvee_{i=1}^{n'_d}[L'_{Hi}\mu_H \leq b'_i]$, the H-transformation of $\mathcal{S}'$. In order to ensure that the H-transformations of $\mathcal{S}$ and $\mathcal{S}'$ result in the same Petri net $\mathcal{N}_H$, we define the **joint H-transformation** of $\mathcal{S}$ and $\mathcal{S}'$ to consist of an H-transformation of $\mathcal{S}$ and an H-transformation of $\mathcal{S}'$ that use the same parameter $T_{s,H} \supseteq \bigcup_{i=1}^{n_d}\{t \in T : H_{d,i}(\cdot, t) \neq 0\} \cup \bigcup_{i=1}^{n'_d}\{t \in T :$

$H'_{d,i}(\cdot,t) \neq 0\}$, where $H_{d,i} = \max(L_i D, H_i, 0)$ and $H'_{d,i} = \max(L'_i D, H'_i, 0)$. The **joint C-transformation** of two sets of constraints $\mathcal{S}$ and $\mathcal{S}'$ of the form (4.51) is similarly defined, consisting of a C-transformation of $\mathcal{S}$ and a C-transformation of $\mathcal{S}'$ that use the same parameter $T_{s,C} \supseteq \bigcup_{i=1}^{n_d} \{t \in T : C_i(\cdot,t) \neq 0\} \cup \bigcup_{i=1}^{n'_d} \{t \in T : C'_i(\cdot,t) \neq 0\}$.

Theorem 4.14. *Let $\mathcal{S}$ and $\mathcal{S}'$ be two sets of constraints (4.55), and $\mathcal{S}_H$ and $\mathcal{S}'_H$ their joint H-transformation. Let Ξ, Ξ', Ξ_H, and Ξ'_H be supervisors optimally enforcing $\mathcal{S}$, $\mathcal{S}'$, $\mathcal{S}_H$, and $\mathcal{S}'_H$, respectively, in $(\mathcal{N},\mu_0)$ and $(\mathcal{N}_H,\mu_{H0})$, where $\mu_{H0} = m_H(\mu_0)$. $\Xi \preceq \Xi'$ ($\Xi \prec \Xi'$) if and only if $\Xi_H \preceq \Xi'_H$ ($\Xi_H \prec \Xi'_H$).*

Proof. The proof is by contradiction. First, we prove $\Xi_H \preceq \Xi'_H \Rightarrow \Xi \preceq \Xi'$. Assume σ enabled at μ_0 in $(\mathcal{N},\mu_0,\Xi)$ and not in $(\mathcal{N},\mu_0,\Xi')$. Then, $\sigma_H(\sigma)$ is enabled at μ_{H0} in $(\mathcal{N}_H,\mu_{H0},\Xi_H)$ but not in $(\mathcal{N}_H,\mu_{H0},\Xi'_H)$, by Proposition 4.11(b). This contradicts $\Xi_H \preceq \Xi'_H$. Next we prove that $\Xi \preceq \Xi' \Rightarrow \Xi_H \preceq \Xi'_H$. Assume σ_H enabled at μ_{H0} in $(\mathcal{N}_H,\mu_{H0},\Xi_H)$ and $(\mathcal{N}_H,\mu_{H0},\Xi'_H)$, but $\sigma_H q_H$ enabled only in $(\mathcal{N}_H,\mu_{H0},\Xi_H)$. Let q be defined as in Proposition 4.12(b). Then, $\sigma(\sigma_H)q$ is enabled at μ_0 in $(\mathcal{N},\mu_0,\Xi)$ and not in $(\mathcal{N},\mu_0,\Xi')$, by Proposition 4.12(b). This contradicts $\Xi \preceq \Xi'$. Now, we prove $\Xi \prec \Xi' \Rightarrow \Xi_H \prec \Xi'_H$. Assume $\Xi_H \not\prec \Xi'_H$. Since $\Xi \prec \Xi' \Rightarrow \Xi \preceq \Xi' \Rightarrow \Xi_H \preceq \Xi'_H$, it must be that Ξ_H and Ξ'_H are equally permissive. Thus, $\Xi_H \succeq \Xi'_H$. Then, $\Xi \succeq \Xi'$, which contradicts $\Xi \prec \Xi'$. The proof of $\Xi_H \prec \Xi'_H \Rightarrow \Xi \prec \Xi'$ is similar. $\qquad\square$

In the following developments, it will be useful to guarantee that the successive application of the H^{-1}- and H-transformations to a set of constraints (4.56) produces exactly the same set of constraints. To this end, each component $L_H \mu_H \leq b$ of a disjunction (4.56) will be constrained to satisfy

$$\forall p \in P_H \setminus P : \begin{cases} L_H(\cdot,p) \geq L_H D_H^+(\cdot,p\bullet), \\ L_H(\cdot,p) \geq L_H D_H^-(\cdot,\bullet p), \end{cases} \tag{4.57}$$

$$\forall t \in T \setminus \bullet(P_H \setminus P) : L_H D_H(\cdot,t) \leq 0. \tag{4.58}$$

The following theorem summarizes the properties of (4.57)–(4.58).

Theorem 4.15. (a) *The H-transformation of any set of constraints $L\mu + Hq \leq b$ satisfies (4.57)–(4.58).*

(b) *Given an H-transformed net $\mathcal{N}_H$ and a set of constraints $L_H \mu_H \leq b$, let $L\mu + Hq \leq b$ denote the H^{-1}-transformation of $L_H \mu_H \leq b$ and let $L'_H \mu'_H \leq b$ and $\mathcal{N}'_H$ denote the H-transformation of (4.4). If L_H satisfies (4.57)–(4.58) and the H-transformation generating $L'_H \mu'_H \leq b$ has the parameter $T_{s,H} = \bullet(P_H \setminus P)$, then $\mathcal{N}_H$ and $\mathcal{N}'_H$ are identical, and $L'_H = L_H$.*

Proof. (a) By definition, $H_d(\cdot,\bullet p) = \max(0, LD(\cdot,\bullet p), H(\cdot,\bullet p)) \; \forall p \in P_H \setminus P$. Further, by (4.41) and (4.44), $LD(\cdot,\bullet p) = L_H D_H^+(\cdot,p\bullet) - L_H D_H^-(\cdot,\bullet p)$ and $LD^-(\cdot,\bullet p) = L_H D_H^-(\cdot,\bullet p)$. Then, (4.57) is obtained by substituting LD in

H_d, then H_d and LD^- in $\forall p \in P_H \setminus P$: $L_H(\cdot, p) = H_d(\cdot, \bullet p) + LD^-(\cdot, \bullet p)$, where this expression is true by (4.38). According to the H-transformation, all transitions t for which $H_d(\cdot, t) \neq 0$ are split. Therefore, $\forall t \in T \setminus \bullet(P_H \setminus P)$, $H_d(\cdot, t) = 0$, and so $LD(\cdot, t) \leq 0$. By (4.47), this proves (4.58).

(b) By definition, $H_d = \max(LD, H, 0)$. For $t \in T \cap \bullet(P_H \setminus P)$ we have $H_d(\cdot, t) = H(\cdot, t)$, in view of (4.57), $LD(\cdot, t) = L_H D_H^+(\cdot, t\bullet\bullet) - L_H D_H^-(\cdot, t)$, and $H(\cdot, t) = L_H(\cdot, t\bullet) - L_H D_H^-(\cdot, t)$ (by (4.46)). For $t \in T \setminus \bullet(P_H \setminus P)$, $H_d(\cdot, t) = H(\cdot, t) = 0$, in view of (4.46), (4.47), and (4.58). This shows that $H_d = H$. Then, by (4.38), (4.41), and (4.46) we get $L_H'(\cdot, p) = L_H(\cdot, p) \; \forall p \in P_H'$. Note that $H_d = H \Rightarrow P_H' \subseteq P_H$; $P_H' = P_H$ is guaranteed by $T_{s,H} = \bullet(P_H \setminus P)$. $\square$

Let $\mathcal{S}$ denote the specification (4.51) on $(\mathcal{N}, \mu_0)$. Based on the results obtained so far, the following procedure could be used to find a feasible specification $\mathcal{S}_a$ that is at least as restrictive as $\mathcal{S}$. The procedure could be used whenever $\mathcal{S}$ is not feasible or its feasibility is not known.

PROCEDURE 4.1

1. Apply the C-transformation and then the H-transformation. Let $\mathcal{S}_{HC}$ and $(\mathcal{N}_{HC}, \mu_{HC0})$ be the transformed constraints and Petri net.

2. Find h-feasible constraints $\mathcal{S}_{HCa}$ that satisfy (4.57)–(4.58) such that $\Xi_{HCa} \preceq \Xi_{HC}$, where Ξ_{HCa} and Ξ_{HC} are supervisors optimally enforcing $\mathcal{S}_{HCa}$ and $\mathcal{S}_{HC}$, respectively. If no solution is found, declare failure and exit.

3. Apply to $\mathcal{S}_{HCa}$ the H^{-1}-transformation and then the C^{-1}-transformation. Let $\mathcal{S}_a$ be the result. Enforce $\mathcal{S}_a$ in $(\mathcal{N}, \mu_0)$.

The set of constraints obtained by this procedure has interesting properties when the H-transformation splits all transitions and the C-transformation adds sink places to all transitions. Therefore, let's define the **total H-transformation** as the H-transformation with parameter $T_{s,H} = T$, and the **total C-transformation** as the C-transformation with parameter $T_{s,C} = T$. Let $\mathcal{X}$ be the set of all supervisors optimally enforcing feasible constraints of the form (4.51). Let $\mathcal{X}_{HC}$ be the set of all supervisors optimally enforcing h-feasible constraints of the form (4.55) that satisfy (4.57)–(4.58).

Theorem 4.16. *Assume the notation of Procedure 4.1. Let Ξ and Ξ_a be supervisors optimally enforcing $\mathcal{S}$ and $\mathcal{S}_a$, respectively.*

(a) *$\mathcal{S}_a$ is feasible and $\Xi_a \preceq \Xi$.*

Assume that the total C- and H-transformations are applied at the first step of the procedure.

(b) *Ξ_a is least restrictive among the supervisors of $\mathcal{X}$ enforcing $\mathcal{S}$ if and only if Ξ_{HCa} is least restrictive among the supervisors of $\mathcal{X}_{HC}$ enforcing $\mathcal{S}_{HC}$.*

(c) *There is no supervisor $\Xi^* \succ \Xi_a$ of $\mathcal{X}$ that enforces $\mathcal{S}$ if there is no supervisor $\Xi^*_{HC} \succ \Xi_{HCa}$ of $\mathcal{X}_{HC}$ that enforces $\mathcal{S}_{HC}$.*

Proof. (a) Let P_C and P_{HC} be the set of places of the Petri nets obtained by the C- and H-transformation of $\mathcal{S}$. In view of Theorem 4.15(b), the same Petri net $\mathcal{N}_{HC}$ is obtained by the C- and H-transformations of $\mathcal{S}_a$, when the transformations use the parameters $T_{s,C} = \bullet(P_C \backslash P)$ and $T_{s,H} = \bullet(P_{HC} \backslash P_C)$. Further, $\mathcal{S}_{HCa}$ is the C- and H-transformation of $\mathcal{S}_a$. Therefore, $\mathcal{S}_a$ is feasible by Theorem 4.13 and $\Xi_a \preceq \Xi$ in view of $\Xi_{HCa} \preceq \Xi_{HC}$ and Theorem 4.14.

(b) Note that the total C- and H-transformation of any set of constraints (4.51) results in the same Petri net $\mathcal{N}_{HC}$. By Theorem 4.15(b), the total C- and H-transformation of $\mathcal{S}_a$ is $\mathcal{S}_{HCa}$. The proof is by contradiction. Assume there is another supervisor $\Xi' \in \mathcal{X}$ enforcing $\mathcal{S}$ such that $\Xi' \npreceq \Xi_a$. Since $\Xi' \in \mathcal{X}$, Ξ' optimally enforces a feasible set of constraints $\mathcal{S}'$ of the form (4.51). By Theorems 4.13 and 4.15(a), $\Xi'_{HC} \in \mathcal{X}_{HC}$, where Ξ'_{HC} is a supervisor optimally enforcing the $\mathcal{S}'_{HC}$, the total C- and H-transformation of $\mathcal{S}'$. By Theorem 4.14, $\Xi'_{HC} \preceq \Xi_{HC}$. Therefore, $\Xi'_{HC} \preceq \Xi_{HCa}$, since Ξ_{HCa} is least restrictive. By Theorem 4.14, $\Xi' \preceq \Xi_a$, which contradicts the original assumption.

(c) The proof is similar to that of part (b). $\square$

Theorem 4.16 shows that the problem of enforcing constraints (4.51) can be solved in terms of the simpler constraints (4.56) in a transformed Petri net, without loss of permissiveness. Since our results were derived under the transition-bag concurrency setting, there is a loss of permissiveness when the Procedure 4.1 is used for other concurrency settings. Indeed, a feasible least restrictive supervisor enforcing (4.51) may be too restrictive for the other concurrency settings, though it would still enforce (4.51). This suggests that the design at step 2 of Procedure 4.1 needs additional constraints besides (4.57)–(4.58), depending on the concurrency setting, to ensure an optimal supervisor can be obtained. Finally, no specific method has been referenced for step 2 of the procedure. A structural solution will be given in section 4.2.7, based on the concept of admissibility. It should be noted that the solution of section 4.2.7 is suboptimal, though in a certain sense optimal with respect to structural admissibility conditions, as will be shown there.

4.2.6 Example

Consider the plant Petri net of Fig. 4.7. Here we assume the no concurrency framework (i.e., no transitions fire at the same time). The example corresponds to a region of a factory cell in which autonomous vehicles (AVs) access a shared area (SA). The number of AVs which may be in the SA at the same time is limited. The AVs enter the SA from two directions: left and right; AVs coming from the left side enter via t_4 or t_{13}, and AVs coming from the right side via t_5 or t_{14}. The AVs exit the restricted area via t_9 or t_{10}. The total marking of p_1, p_2, and p_7 corresponds to the number of left AVs waiting in line to enter

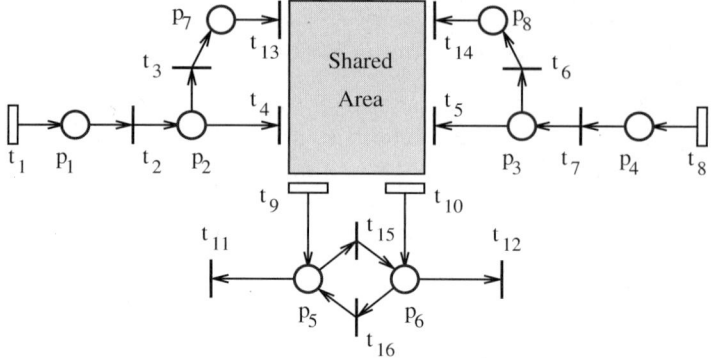

Fig. 4.7. Plant Petri net in the example.

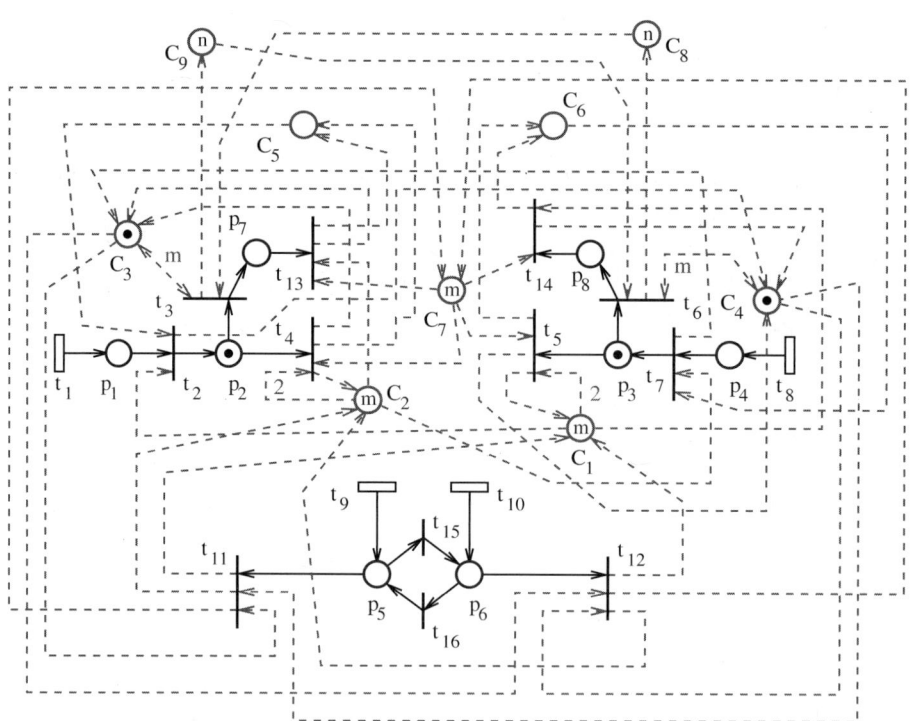

Fig. 4.8. Closed-loop Petri net.

the SA; only one AV should be in the states p_2 and p_7, that is, $\mu_2 + \mu_7 \leq 1$. The marking of p_3, p_4, and p_8 has a similar meaning.

Let m be the maximum number of AVs which can be in the SA at the same time; note that the number of AVs in the SA is $v_{13} + v_{14} + v_4 + v_5 - v_9 - v_{10}$. When the number of vehicles in the restricted area is $m - 1$ and both a left and a right AV attempt to enter the restricted area (i.e., both $\mu_2 + \mu_7 = 1$ and $\mu_3 + \mu_8 = 1$), arbitration is required. When an AV is in p_2 and no arbitration is required, it can enter the SA without stopping. When arbitration is required, it stops (enters the state p_7) and awaits the arbitration result. The same applies to p_3 and p_8. We desire the following. When an AV enters the SA, if an arbitration was required to decide that it may enter, the AV should enter via t_{13} or t_{14}; if no arbitration was required, it should enter via t_4 or t_5. These constraints can be written as follows:

$$2q_5 + \mu_2 + \mu_7 \leq m - (v_{13} + v_{14} + v_4 + v_5 - v_9 - v_{10}) + 1, \quad (4.59)$$
$$2q_4 + \mu_3 + \mu_8 \leq m - (v_{13} + v_{14} + v_4 + v_5 - v_9 - v_{10}) + 1, \quad (4.60)$$
$$mq_3 \leq \mu_3 + \mu_8 + v_{13} + v_{14} + v_4 + v_5 - v_9 - v_{10}, \quad (4.61)$$
$$mq_6 \leq \mu_2 + \mu_7 + v_{13} + v_{14} + v_4 + v_5 - v_9 - v_{10}. \quad (4.62)$$

In addition we have the requirements that

$$\mu_2 + \mu_7 \leq 1, \quad (4.63)$$
$$\mu_3 + \mu_8 \leq 1. \quad (4.64)$$

The requirement on the maximum number of AVs in the SA is

$$v_{13} + v_{14} + v_4 + v_5 - v_9 - v_{10} \leq m. \quad (4.65)$$

We add the fairness constraints

$$v_3 - v_6 \leq n, \quad (4.66)$$
$$-v_3 + v_6 \leq n. \quad (4.67)$$

As t_1, t_8, t_9, t_{10} are uncontrollable and t_9, t_{10} unobservable, the constraints (4.59)–(4.62) and (4.65) are inadmissible. However, they can be transformed to the admissible constraints

$$2q_5 + \mu_2 + \mu_5 + \mu_6 + \mu_7 + v_{13} + v_{14} + v_4 + v_5 - v_9 - v_{10} \leq m + 1, \quad (4.68)$$
$$2q_4 + \mu_3 + \mu_5 + \mu_6 + \mu_8 + v_{13} + v_{14} + v_4 + v_5 - v_9 - v_{10} \leq m + 1, \quad (4.69)$$
$$mq_3 - \mu_3 - \mu_8 - \mu_5 - \mu_6 - (v_{13} + v_{14} + v_4 + v_5 - v_9 - v_{10}) \leq 0, \quad (4.70)$$
$$mq_6 - \mu_2 - \mu_7 - \mu_5 - \mu_6 - (v_{13} + v_{14} + v_4 + v_5 - v_9 - v_{10}) \leq 0, \quad (4.71)$$
$$v_{13} + v_{14} + v_4 + v_5 - v_9 - v_{10} + \mu_5 + \mu_6 \leq m. \quad (4.72)$$

Note that the constraints (4.61) and (4.62) cannot be transformed by any method to admissible constraints that are at least as restrictive as (4.61) and

(4.62). This is due to the fact that the Petri net model does not relate in any way the firings of the transitions t_9 and t_{10} with the firings of the transitions t_4, t_5, t_{13}, and t_{14}. Thus (4.70) and (4.71) are admissible relaxations of the constraints (4.61) and (4.62).

The closed-loop Petri net is shown next to the plant in Fig. 4.8, where the control places $C_1 \ldots C_9$ correspond to the constraints (4.68), (4.69), (4.70), (4.71), (4.63), (4.64), (4.72), (4.66), and (4.67), in this order.

4.2.7 The Structural Approach

Here, Procedure 4.1 is specialized for the case in which the structural admissibility conditions (4.28)–(4.31) are used for the design of the supervisor. The conditions (4.28)–(4.31) can be used for both conjunctions of constraints (4.5) and disjunctions (4.51), where in the latter case each component $L_i\mu + H_iq + C_iv \leq b_i$ of the disjunction is required to satisfy (4.28)–(4.31). A different set of conditions will be required for the components $L_{HC,i}\mu_{HC} \leq b_i$ of the disjunction $\mathcal{S}_{HC}$. First, given a set of constraints $L\mu + Hq + Cv \leq b$, let's try to express D_c in terms of L_{HC} and D_{HC}, the incidence matrix of $\mathcal{N}_{HC}$. Then, by substituting D_c in (4.29)–(4.30) we can obtain relaxed admissibility conditions on the constraints $L_{HC}\mu_{HC} \leq b$. Indeed, we are only interested in the properties of $L_{HC}\mu_{HC} \leq b$ that ensure that the H^{-1}- and C^{-1}-transformed constraints are admissible. Note that for constraints $L\mu + Hq + Cv \leq b$, $D_c = -LD - C$. Let

$$\Delta(t) = \begin{cases} D_{c,HC}(\cdot, t) & \text{for } t \in T \setminus \bullet(P_{HC} \setminus P_C), \\ D_{c,HC}^+(\cdot, t \bullet \bullet) - D_{c,HC}^-(\cdot, t) & \text{for } t \in T \cap \bullet(P_{HC} \setminus P_C), \end{cases} \tag{4.73}$$

where $D_{c,HC}$, $D_{c,HC}^+$, and $D_{c,HC}^-$ denote the incidence, input, and output matrices of the supervisor (4.18)–(4.20) enforcing $L_{HC}\mu_{HC} \leq b$ in $\mathcal{N}_{HC}$. The significance of $\Delta(t)$ is that $\Delta(t) = D_c(\cdot, t)$, as shown next.

Proposition 4.17. *For all $t \in T$, $\Delta(t) = -LD(\cdot, t) - C(\cdot, t)$.*

Proof. The result can be verified based on (4.47) and the following equations: $L_C D_C = LD + C$, $D_{c,HC} = -L_{HC}D_{HC}$, $D_{c,HC}^+ = \max(0, D_{c,HC})$, and $D_{c,HC}^- = \max(0, -D_{c,HC})$. $\qquad\square$

Given $L_{HC}\mu_{HC} \leq b$, the following relaxed admissibility conditions will be used, instead of (3.16)–(3.19):

$$\forall t \in T, \ \rho(t) \in \mathcal{K}_{uc} \cup \{\lambda\} \Rightarrow D_{c,HC}(\cdot, t) \leq 0, \tag{4.74}$$

$$\forall t \in T, \ o(t) \in \mathcal{O}_{uo} \cup \{\lambda\} \Rightarrow \Delta(\cdot, t) = 0, \tag{4.75}$$

$$\forall t_1, t_2 \in T, \ o(t_1) = o(t_2) \Rightarrow \Delta(\cdot, t_1) = \Delta(\cdot, t_2), \tag{4.76}$$

$$\forall i = 1 \ldots n_c, \ \forall t_1, t_2 \in T, \ \forall \alpha \in \mathcal{K}, \ \rho(t_1) = \rho(t_2) = \alpha \Rightarrow$$
$$[D_{c,HC}(i, t_1) = D_{c,HC}(i, t_2)] \vee [D_{c,HC}(i, t_1) \leq 0 \wedge D_{c,HC}(i, t_2) \leq 0]. \tag{4.77}$$

Assuming labeled Petri nets, the admissibility conditions have the form

$$L_{HC}A \leq 0 \text{ and } L_{HC}B = 0, \tag{4.78}$$

where A and B are matrices of integers. However, in the case of double-labeled Petri nets, due to the condition (4.77), the admissibility conditions have the form

$$\bigvee_{i=1}^{w} [L_{HC}A_i \leq 0] \text{ and } L_{HC}B = 0. \tag{4.79}$$

Note that various methods from the literature could be used to find admissible constraints $L_{HCa}\mu_{HC} \leq b_a$ subject to (4.78) or even (4.79), as mentioned in section 3.5. Compared to the constraints (4.28)–(4.31), which have been intended for generalized constraints (4.5), note that (4.75)–(4.76) are a relaxed version of (4.29)–(4.30). Graphically, the fact that this relaxation is possible can be seen in Fig. 4.6(b). For instance, assuming t_i is unobservable but controllable, the control place can control the transition in $\mathcal{N}$ without observing it, if connected to t_i by arcs of equal weight. However, in $\mathcal{N}_H$ this corresponds to the requirement that the arc of the control place to t_i has the same weight as the one from t_j. This requirement is expressed by (4.75).

As in section 4.2.5, we will focus on the properties of the H-transformation, as the C-transformation results in trivial changes in the supervisory problem.

Theorem 4.18. *Let $L_H\mu_H \leq b$ be the H-transformation of $L\mu + Hq \leq b$. The constraints $L\mu + Hq \leq b$ satisfy (4.28)–(4.31) if and only if $L_H\mu_H \leq b$ satisfy (4.74)–(4.77).*

Proof. Based on (4.47), $H_d = D_c^- = \max(LD, H, 0)$, and $H_d(\cdot, t) = 0 \ \forall t \in T \setminus \bullet(P_H \setminus P)$, note that (4.28) $\Leftrightarrow$ (4.74). Further, (4.29) $\Leftrightarrow$ (4.75) and (4.30) $\Leftrightarrow$ (4.76) are immediate consequences of Proposition 4.17. Finally, by (4.48), (4.31) $\Leftrightarrow$ (4.77). $\qquad\square$

Based on this result, a possibility would be to simply replace the h-feasibility requirement in the second step of the Procedure 4.1 by (4.74)–(4.77). Still another change can be made, namely replacing the requirement $\Xi_{HCa} \preceq \Xi_{HC}$ by $\mathcal{S}_{HCa} \preceq \mathcal{S}_{HC}$, where the relation $\preceq$ is defined for sets of constraints next. This change is motivated by the fact that some of the structural methods in the literature are well suited for requirements $\mathcal{S}_{HCa} \preceq \mathcal{S}_{HC}$.

Let $\mathcal{S}$ denote a set of constraints $\bigvee_{i=1}^{n_d}[L_i\mu + H_iq + C_iv \leq b_i]$ and $\mathcal{S}'$ denote $\bigvee_{i=1}^{n_d'}[L_i'\mu + H_i'q + C_i'q \leq b_i']$. Given a Petri net structure $\mathcal{N}$, we say that $\mathcal{S}$ is at least as restrictive as $\mathcal{S}'$, which we write $\mathcal{S} \preceq \mathcal{S}'$, if for all $\mu \in \mathbb{N}^{|P|}$, $q \in \mathbb{N}^{|T|}$, and $v \in \mathbb{N}^{|T|}$ that satisfy $\mu \geq D^-q$:

$$\bigvee_{i=1}^{n_d} [L_i\mu + H_{d,i}q + C_iv \leq b_i] \implies \bigvee_{i=1}^{n_d'} [L_i'\mu + H_{d,i}'q + C_i'q \leq b_i']. \tag{4.80}$$

Further, we say that S is more restrictive than S', which we write $S \prec S'$, if $S \preceq S'$ and there are $\mu \in \mathbb{N}^{|P|}$, $q \in \mathbb{N}^{|T|}$, and $v \in \mathbb{N}^{|T|}$ satisfying $\mu \geq D^- q$ such that $L_i \mu + H_{d,i} q + C_i v \not\leq b_i \ \forall i \in \{1, 2, \ldots, n_d\}$ and $\bigvee_{i=1}^{n_d'} [L_i' \mu + H_{d,i}' q + C_i' q \leq b_i']$. Compared to the $\preceq$ relation between supervisors, the relation between the sets of constraints is stronger: $S \preceq S' \Rightarrow \Xi \preceq \Xi'$, where Ξ and Ξ' denote supervisors optimally enforcing S and S', respectively, in a Petri net $(\mathcal{N}, \mu_0)$.

Note that $H_{d,i}$ ($H_{d,i}'$) and not H_i (H_i') appear in (4.80), due to Lemma 4.2 and our interpretation (4.53) of the disjunctions of constraints. Further, note that this definition is most appropriate for the situations in which nonzero values are allowed for the initial value of v. This is due to the fact that given a set of constraints (4.5), $L\mu + H_d q + Cv \leq b$ may not imply there is a firing sequence from some initial state $(\mu_0, 0)$ to (μ_n, v_n) such that $\mu_n = \mu$, $v_n = v$, and $L\mu_i + H_d q_i + Cv_i \leq b$ at every intermediary step (μ_i, q_i, v_i), $i = 0 \ldots n-1$. Note also that the $H_{d,i}$ and $H_{d,i}'$ terms appear in (4.80) even when $H_i = 0$ and $H_i' = 0$. However, it can be shown that they can be omitted in the no concurrency setting when $H_i = 0$ and $H_i' = 0$.

Theorem 4.19. *Let S and S' denote the sets of constraints $L\mu + Hq \leq b$ and $L'\mu + H'q \leq b'$, and let S_H and S_H' be the constraints obtained after a joint H-transformation. Then $S \preceq S'$ if and only if $S_H \preceq S_H'$. Furthermore, $S \prec S'$ if and only if $S_H \prec S_H'$.*

Proof. The proof is by contradiction. First, we show that $S \preceq S'$ and $S_H \not\preceq S_H'$ is not possible. If $S_H \not\preceq S_H'$, there are μ_H and q_H such that $\mu_H \geq D_H^- q_H$, $L_{H,i} \mu_H + H_{d,H,i} q_H \leq b$ for some $i \in \{1, 2, \ldots, n_d\}$ and $L_{H,j}' \mu_H + H_{d,H,j}' q_H \not\leq b'$ for all $j = 1, 2, \ldots, n_d'$. Then, in view of (4.41)–(4.42) and (4.48), there is a possibly empty sequence σ_H from a marking μ_H^* such that $\mu_H^* = m_H(\mu^*)$, μ_H^* satisfies S_H, and firing σ_H satisfies S_H. Therefore, Proposition 4.12(b) applies for $\mu_0 = \mu^*$ and $\mu_{H0} = \mu_H^*$. Thus, $\mu = \mu_0 + D\nu(\sigma_H)$ and $q = \chi(\sigma_H) + x$ satisfy S and $\mu \geq D^- q$, where x is the restriction of q_H to T. Moreover, by Proposition 4.9(b), $L_i \mu + H_{d,i} \chi(\sigma_H) = L_{Hi} \mu_H$, and by (4.48) $H_{d,i} x = H_{d,H,i} q_H$. The same relations apply for L_j', $H_{d,j}'$, L_{Hj}', and $H_{d,H,j}'$. Thus, we have found q and $\mu \geq D^- q$ such that $L_i \mu + H_{d,i} q = L_{H,i} \mu_H + H_{d,H,i} q_H \leq b$ and $L_j' \mu + H_{d,j}' q = L_{H,j}' \mu_H + H_{d,H,j}' q_H \not\leq b'$ for all $j = 1, 2, \ldots, n_d'$. However, this contradicts $S \preceq S'$.

Now, assume $S \not\preceq S'$ and $S_H \preceq S_H'$. Since $S \not\preceq S'$, there are $\mu \geq D^- q$ and q such that $L_i \mu + H_{d,i} q \leq b$ for some $i \in \{1, 2, \ldots, n_d\}$ and $L_j' \mu + H_{d,j}' q \not\leq b'$ for all $j = 1, 2, \ldots, n_d'$. Let $\mu_H = m_H(\mu)$ and q_H be the first component of $\sigma_H(q)$: $\sigma_H(q) = q_H q_H'$. By Proposition 4.7(a), μ_H enables $\sigma_H(q)$. Let $\mu_H' \xlongequal{q_H} \mu_H$. By Proposition 4.9(a), $L_i \mu + H_{d,i} q = L_{Hi} \mu_H'$ and $L_j' \mu + H_{d,j}' q = L_{Hj}' \mu_H'$ for all $j = 1, 2, \ldots, n_d'$. Thus, at the marking μ_H', S_H is satisfied and S_H' is not. This contradicts $S_H \preceq S_H'$.

To prove that $S \prec S' \Rightarrow S_H \prec S_H'$, note that $S \prec S' \Rightarrow S_H \preceq S_H'$, by the first part of the proof. Therefore, if $S_H \not\prec S_H'$, it must be that S_H' and S_H are equally permissive, and so $S_H \succeq S_H'$. Then, $S \succeq S'$, which contradicts $S \prec S'$. Therefore, $S_H \prec S_H'$. The proof of $S_H \prec S_H' \Rightarrow S \prec S'$ is similar. $\square$

Procedure 4.1 refers to no method for the design problem at its second step. An optimal method for step 2 could be quite involved, as the design is constrained by rather complicated h-feasibility conditions and a plant that is (in the most general case) a double-labeled Petri net. Here, based on the structural admissibility conditions (4.74)–(4.77), the procedure is modified to provide a rather simple way to find a solution at step 2 of Procedure 4.1.

PROCEDURE 4.2

1. *Apply the C-transformation and then the H-transformation. Let $\mathcal{S}_{HC}$ and $\mathcal{N}_{HC}$ be the transformed constraints and Petri net.*

2. *Find constraints $\mathcal{S}_{HCa}$ that satisfy (4.57)–(4.58) and (4.74)–(4.77) such that $\mathcal{S}_{HCa} \preceq \mathcal{S}_{HC}$. If no solution is found, declare failure and exit.*

3. *Apply to $\mathcal{S}_{HCa}$ the H^{-1}-transformation and then the C^{-1}-transformation. Let $\mathcal{S}_a$ be the result. Enforce $\mathcal{S}_a$ in $\mathcal{N}$.*

Compared to Procedure 4.1, Procedure 4.2 is suboptimal, as it relies on the structural admissibility conditions (4.74)–(4.77). However, the problem of finding a feasible set of constraints $\mathcal{S}_a$ is reduced to a problem involving the enforcement of the simpler constraints (4.56) in a transformed Petri net $\mathcal{N}_{HC}$. Further, the design of Procedure 4.1 requires the solution of step 2 to satisfy h-feasibility constraints and (4.57)–(4.58) in a double-labeled Petri net $(\mathcal{N}_{HC}, \mu_{HC0})$. On the other hand, Procedure 4.2 represents the requirement of feasibility in a double-labeled Petri net by algebraic constraints (4.74)–(4.77). This simplifies significantly the problem of finding $\mathcal{S}_{HCa}$. In fact, there are already literature methods that could design $\mathcal{S}_{HCa}$ subject to (4.74)–(4.76) and (4.57)–(4.58), as discussed in section 3.5. Note also that in Procedure 4.2, as long as we impose $\mathcal{S}_{HCa} \preceq \mathcal{S}_{HC}$ and not $\Xi_{HCa} \preceq \Xi_{HC}$ (where Ξ_{HCa} and Ξ_{HC} are the supervisors optimally enforcing $\mathcal{S}_{HCa}$ and $\mathcal{S}_{HC}$ with respect to an initial marking), the result can be calculated independently of the initial marking. The drawback of Procedure 4.2 is that it is suboptimal: a supervisor designed according to the procedure may not be least restrictive, and a feasible solution $\mathcal{S}_a$ may not be found even when one exists. With respect to our structural conditions of admissibility, the Procedure 4.2 is optimal, as shown in the next result.

Let $\mathcal{Y}$ be the set of all sets of constraints $\mathcal{S}_x$ of the form (4.51) such that $\mathcal{S}_x \preceq \mathcal{S}$ and $\mathcal{S}_x$ satisfies (4.28)–(4.31). Let $\mathcal{Y}_{HC}$ be the set of all sets of constraints $\mathcal{S}_{HCx}$ of the form (4.55) that satisfy (4.57)–(4.58), (4.74)–(4.77), and $\mathcal{S}_{HCx} \preceq \mathcal{S}_{HC}$.

Theorem 4.20. *Assume the notation of Procedure 4.2.*

(a) $\mathcal{S}_a$ *is feasible and* $\mathcal{S}_a \preceq \mathcal{S}$.

Assume that the total C- and H-transformations are applied at the first step of the procedure.

(b) $\mathcal{S}_a$ *is least restrictive among the sets of constraints in* $\mathcal{Y}$ *if and only if* $\mathcal{S}_{HCa}$ *is least restrictive among the sets of constraints of* $\mathcal{Y}_{HC}$.

(c) *There is no* $\mathcal{S}^* \in \mathcal{Y}$ *such that* $\mathcal{S}^* \succ \mathcal{S}_a$ *if there is no* $\mathcal{S}_{HC}^* \in \mathcal{Y}_{HC}$ *such that* $\mathcal{S}_{HC}^* \succ \mathcal{S}_{HCa}$.

Proof. The proof is similar to that of Theorem 4.16, where Theorems 4.18 and 4.19 are used in the place of Theorems 4.13 and 4.14.   □

4.3 Language Constraints

The previous section has dealt with the enforcement of constraints (4.5) on Petri nets. As shown in subsection 4.2.2, the constraints (4.5) can describe the P-type language of any free-labeled Petri net. This section deals with the enforcement of P-type languages of Petri nets that are not necessarily free-labeled.

As an example, consider the Petri net and the specification shown in Fig. 4.9. In this example, the specification is described by a Petri net labeled by the events a and b. To simplify the notation, it is assumed that all events of the plant that do not appear in the specification are always enabled in the specification. The closed loop in our example can be computed immediately by a parallel composition of the plant and specification, and is shown in Fig. 4.10(a). Note that in the closed loop, the transition t_1 of the plant appears in the form of t_1^1 and t_1^2, corresponding to the synchronization of t_1 with the transitions t^1 and t^2 of the supervisor. Similarly, t_2^3 and t_2^4 correspond to the synchronization of t_2 with t_3 and t^4. A formal description of the parallel composition of Petri nets is given in Algorithm 2.2.

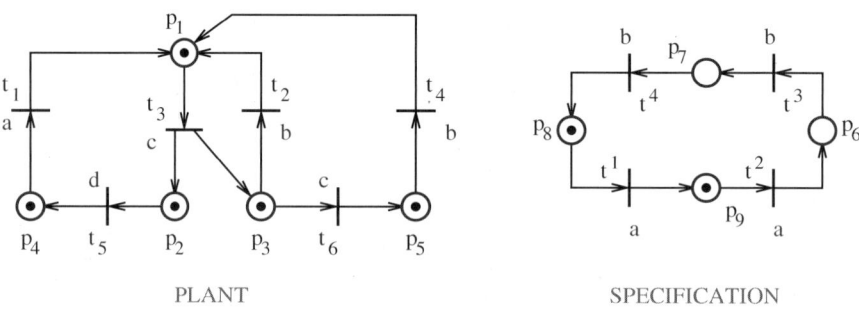

PLANT   SPECIFICATION

Fig. 4.9.

The supervision is interpreted as follows. The plant and the supervisor each have a distinct set of transitions, T_p and T_s, respectively. The supervisor cannot observe/control the plant transitions directly, but it can observe/control

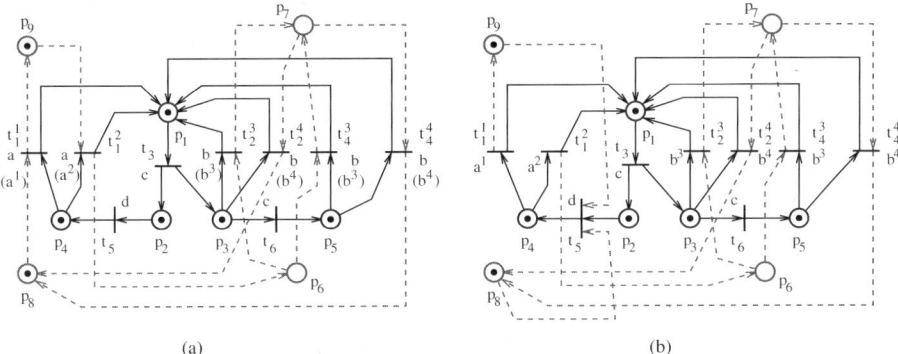

Fig. 4.10.

events generated by the plant. When the plant generates the event a, the supervisor picks one of its own enabled transitions $t \in T_s$ that is labeled by a, and fires it. Note that the supervisor is free to choose which of its enabled transitions labeled by a fires. For instance, in Fig. 4.9, when the plant generates a, the supervisor can select either of t^1 or t^2, since both are enabled and labeled by a. So we can relabel the closed loop, to indicate that the supervisor can distinguish between its own transitions that have the same label. Thus, in Fig. 4.10 we have the following new labels: a^1 for t_1^1, a^2 for t_1^2, b^3 for t_2^3 and t_4^3, and b^4 for t_2^4 and t_4^4.

As mentioned in the previous section, in the closed loop, every place of the supervisor corresponds to a specification in terms of constraints (4.5). For instance, p_9 enforces $v_1^2 - v_1^1 \leq 1$ and p_8 enforces $v_1^1 - v_2^4 - v_4^4 \leq 1$. This gives us a readily available approach for supervisor design in the case of partial controllability and partial observability:

1. Compose the Petri net plant and the Petri net specification (supervisor).
2. Relabel the closed loop to take into account that the supervisor can distinguish between its own transitions.
3. Find the constraints (4.5) corresponding to the constraints enforced by the monitors of the closed loop.
4. Transform these constraints (4.5) to a feasible form, which is at least as restrictive.

A practical way to implement the last step is to use the structural approach of subsection 4.2.7.

As an illustration, assume that in our example t_1 (the event a) is uncontrollable but the other transitions are controllable. Assume all other events are observable. Notice that in Fig. 4.10(a) p_8 and p_9 may attempt disabling t_1. So, the specification is inadmissible. However, the constraints enforced by p_8 and p_9, namely $v_1^1 - v_2^4 - v_4^4 \leq 1$ and $v_1^2 - v_1^1 \leq 1$, can be transformed to the admissible form $v_1^1 - v_2^4 - v_4^4 + \mu_4 \leq 1$ and $v_1^2 - v_1^1 + \mu_4 \leq 1$. The resulting closed loop and supervisor are shown in Fig. 4.10(b) and Fig. 4.11, respectively. The supervision is admissible, while ensuring the plant generates only words that satisfy the original specification of Fig. 4.9.

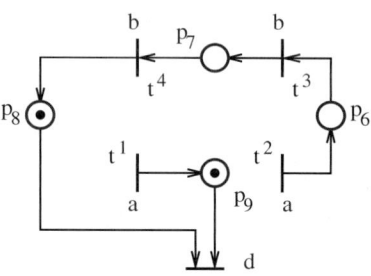

Fig. 4.11.

It is known that the supremal controllable sublanguage of a P-type Petri net language may not be a P-type Petri net language [55]. This is an indication that the approach presented here is suboptimal, in the sense that it may not lead to the least restrictive supervisor. Note that in the literature it has been shown that the computation of the least restrictive supervisor can be reduced to a forbidden marking problem, provided both the plant and specification generate deterministic languages [116]. (Given a labeled Petri net $(\mathcal{N}, \rho, \mu_0)$, the P-language it generates is deterministic if for any of its strings w, there is a unique transition sequence σ enabled by μ_0 that generates w: $\rho(\sigma) = w$.) In the setting of [116], the Petri nets are assumed to be partially controllable and fully observable (i.e., all events are observable).

4.4 Disjunctions of Constraints

This section shows that under certain assumptions, disjunctions of constraints can be enforced by Petri net supervisors, without loss of permissiveness. The disjunctions of constraints considered here have the form

$$\bigvee_i L_i \mu \leq b_i, \tag{4.81}$$

where $L_i \in \mathbb{Z}^{k_i \times n}$ and $b_i \in \mathbb{Z}^{k_i}$. The approach is presented under the no concurrency setting and assuming full controllability and observability. The feasibility of the solution under partial controllability and observability is addressed at the end of the section.

Since each set of constraints (4.81) is a conjunction of constraints $l\mu \leq c$ (where $l \in \mathbb{Z}^{1 \times n}$ and $c \in \mathbb{Z}$), (4.81) can also be written as

$$\bigwedge_j \bigvee_{i \in A_j} l_i \mu \leq c_i, \tag{4.82}$$

where $l_i \in \mathbb{Z}^{1 \times n}$, $c_i \in \mathbb{Z}$, and A_j is a set of integers. The main idea of our approach is to include additional binary variables δ_i for each constraint $l_i \mu \leq c_i$ such that

$$[l_i \mu \leq c_i] \leftrightarrow [\delta_i = 1]. \tag{4.83}$$

Then, the disjunction (4.81) can be replaced by

$$\sum_{i \in A_j} \delta_i \geq 1 \tag{4.84}$$

for all indices j. If we know that $l_i \mu$ is between the bounds m_i and M_i, (4.83) is equivalent to the following system of inequalities:

$$l_i \mu + (M_i - c_i)\delta_i \leq M_i, \tag{4.85}$$
$$l_i \mu + (c_i + 1 - m_i)\delta_i \geq c_i + 1. \tag{4.86}$$

Note that this technique of adding auxiliary variables has been used to solve propositional logic via integer programming in [205, 206]. This technique has also been applied to hybrid systems in [22]. Note also the assumptions that were made:

1. $l_i \mu$ is bounded for all i and all plant markings μ that are reachable (in the closed loop).
2. Some lower bound m_i and upper bound M_i are known for all $l_i \mu$.

The first assumption is reasonable for the specifications that can be implemented in practice. Further, the second assumption appears to be satisfied often in practice.

So far we have shown that enforcing (4.81) is equivalent to enforcing constraints (4.84)–(4.86), which can be done using the SBPI. However, note that the SBPI approach cannot be applied directly, as the constraints contain variables δ_i that do not correspond to the markings of any of the plant places. Thus, supervisor places d_i are created first, to represent the places of marking δ_i. Then the SBPI is applied. Each place d_i corresponds to a constraint $l_i \mu \leq c_i$ and is added to the Petri net according to the following algorithm:

1. Let $T_i^+ = \{t \in T : l_i D(\cdot, t) < 0\}$ and $T_i^- = \{t \in T : l_i D(\cdot, t) > 0\}$.
2. Add a place d_i, a copy t_j^+ of each transition $t_j \in T_i^+$, and a copy t_j^- of each transition $t_j \in T_i^-$. (We say that t' is a copy of t if $D^-(\cdot, t') = D^-(\cdot, t)$ and $D^+(\cdot, t') = D^+(\cdot, t)$.)
3. Connect d_i to the transitions t_j^+ by input arcs (t_j^+, d_i) of weight 1 and to the transitions t_j^- by output arcs (d_i, t_j^-) of weight 1.

Once the places d_i have been added to the Petri net, the constraints (4.84)–(4.86) are enforced, with δ_i denoting the marking of d_i. The role of the transitions t^- and t^+ is to reset (set) δ_i whenever there is a transition from (to) a marking satisfying $l_i\mu \leq c_i$ to (from) μ' with $l_i\mu' \not\leq c_i$. The result of this construction can be seen as the closed loop of the plant with a Petri net supervisor enforcing (4.81).

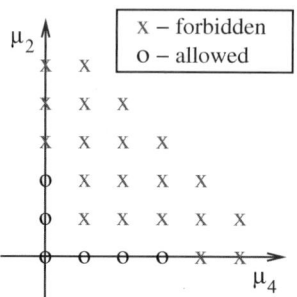

Fig. 4.12.

The algorithm is illustrated on the following example. Assume we desire to enforce

$$[\mu_2 \leq 0] \vee [\mu_4 \leq 0] \tag{4.87}$$

on the Petri net of Fig. 4.13(a). Assume also the following bounds are known: $\mu_2 \leq 2$ and $\mu_4 \leq 3$. Note that (4.87) cannot be represented by conjunctions of inequalities that use only the variables μ_2 and μ_4 (Fig. 4.12). For $\mu_2 \leq 2$, the relations (4.85)–(4.86) become (for $c_i = 0$, $m_i = 0$ and $M_i = 2$)

$$\mu_2 + 2\delta_1 \leq 2, \tag{4.88}$$
$$\mu_2 + \delta_1 \geq 1. \tag{4.89}$$

Similarly, for $\mu_4 \leq 3$ we have

$$\mu_4 + 3\delta_2 \leq 3, \tag{4.90}$$
$$\mu_4 + \delta_2 \geq 1. \tag{4.91}$$

The places d_1 and d_2 are shown in Fig. 4.13(b). Fig. 4.13(c) also shows the monitors a_1, e_1, a_2, and e_2, which correspond to (4.88)–(4.91), in this order. Finally, our disjunction (4.87) can be implemented by enforcing $\delta_1 + \delta_2 \geq 1$ (Fig. 4.13(d)). The supervisor can be represented as in Fig. 4.14, where the Petri net of Fig. 4.13(d) can be seen as the composition of the plant in Fig. 4.13(a) and the supervisor. In Fig. 4.14, α_i denotes the label of t_i, for $i = 2, 3, 4, 5$. The operation of the supervisor can be described as in Fig. 4.15, for the transitions t_2 and t_3. Similar operations are performed for t_4 and t_5.

A careful examination of the construction presented so far shows that the disjunction is not implemented in a least restrictive fashion. In our preliminary construction, if firing a transition t of the plant would change the value of more

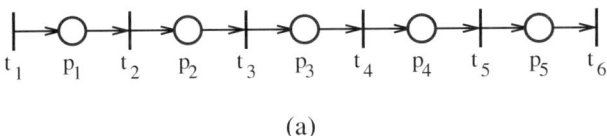

(a)

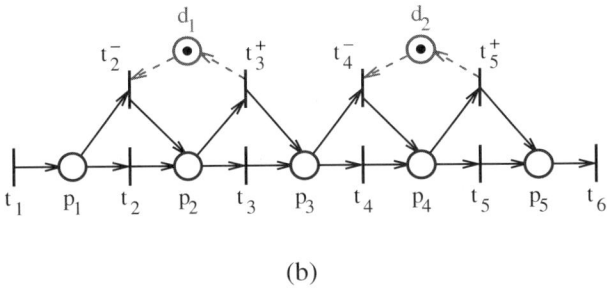

(b)

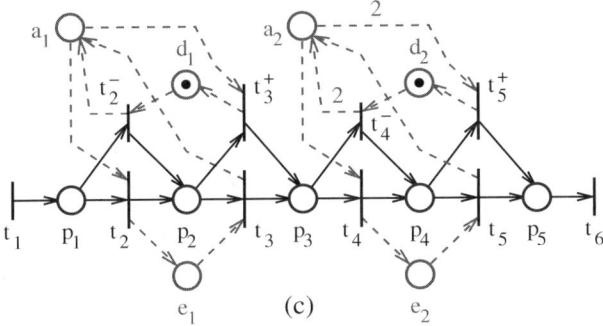

(c)

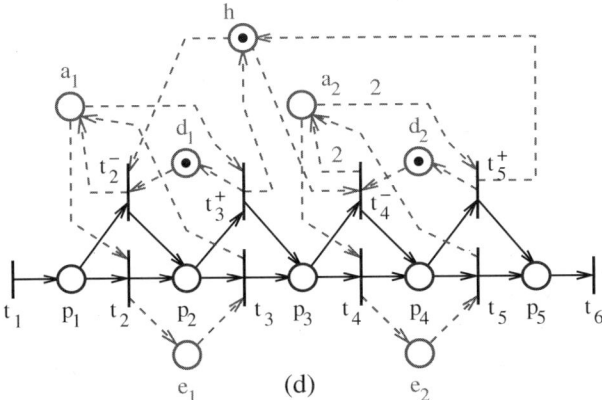

(d)

Fig. 4.13.

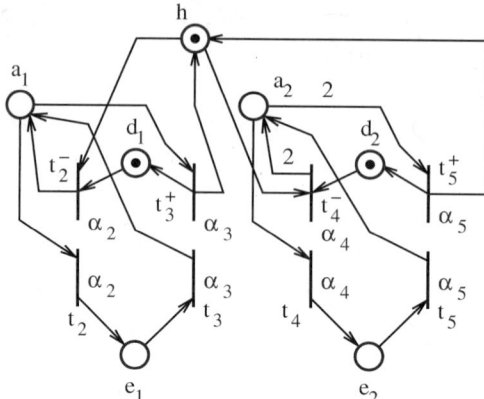

Fig. 4.14.

than one of the variables δ_i, the transition t would be disabled. This problem can be fixed with the following algorithm. Given a disjunction $\bigvee_{i \in A_j} l_i \mu \leq c_i$:

1. Let $f^-(t) = \{i \in A_j : l_i D(\cdot, t) > 0\}$, $f^+(t) = \{i \in A_j : l_i D(\cdot, t) < 0\}$, and $f(t) = f^-(t) \cup f^+(t)$.
2. For all $i \in A_j$ add a place d_i; for all $t_k \in T$ add the copies t_k^σ of t_k for all $\sigma \in 2^{f(t_k)} \setminus \{\emptyset\}$.
3. Connect the places d_i to the transitions t_k^σ by input arcs (t_k^σ, d_i) of weight 1, if $i \in f^+(t_k) \cap \sigma$, and by output arcs (d_i, t_k^σ) of weight 1, if $i \in f^-(t_k) \cap \sigma$.

The method introduced here for the enforcement of disjunctions of constraints has been presented under the assumption of full controllability and observability. The issues arising in the presence of partial controllability and observability are a matter of further investigation. It can be noticed, though, that in the setting of individually controllable and observable transitions, if the specification is feasible, this construction will not break any of the controllability constraints. Further, it is apparent that in the setting of double-labeled Petri nets, if each constraint $l_i \mu \leq c_i$ satisfies (3.16)–(3.19), the construction presented here respects the controllability and observability constraints of the plant.

A related topic is the study of [184] on the enforcement of specifications that require the marking to stay within a union of legal sets $\mathcal{M}_1 \cup \mathcal{M}_2$. Structural conditions are given under which the least restrictive supervisor enforcing a union of legal sets $\mathcal{M}_1 \cup \mathcal{M}_2$ can be implemented by combining the least restrictive supervisor enforcing $\mathcal{M}_1$ with the one enforcing $\mathcal{M}_2$. Further, in [185] a method is given to calculate the maximal controlled-invariant set for specifications $\bigvee_i [l_i \mu \leq c_i]$. The result is obtained in the form of another disjunction of linear inequalities. Several assumptions are made in terms of a certain uncontrolled subnet $\mathcal{N}_A$. Two of the assumptions are that $\mathcal{N}_A$ is

```
enable t₂ in plant if a₁ ∨ (d₁ ∧ h)

if t₂ fires in plant then
    if a₁      /* t₂ fires in closed loop */
        a₁ → a₁ − 1, e₁ → e₁ + 1
    else      /* t₂⁻ fires in closed loop */
        a₁ → a₁ + 1, d₁ → d₁ − 1, and h → h − 1
    end
end
```

```
enable t₃ in plant if a₁ ∨ e₁

if t₃ fires in plant then
    if a₁      /* t₃⁺ fires in closed loop */
        a₁ → a₁ − 1, d₁ → d₁ + 1, and h → h + 1
    else      /* t₃ fires in closed loop */
        a₁ → a₁ + 1, e₁ → e₁ − 1
    end
end
```

Fig. 4.15. Description of the operation of the supervisor. For simplicity of notation, a_1, e_1, ... denote $\mu(a_1)$, $\mu(e_1)$, ...

acyclic and that the transitions t of $\mathcal{N}_{\mathcal{A}}$ satisfy $D^-(p,t) = 1$ $\forall p \in \bullet t$. The result is obtained under the no concurrency assumption.

5

Decentralized Supervision of Petri Nets

5.1 Introduction

The decentralized control of DESs has received considerable attention in the recent years [164]. The current research effort has been focused on the automata setting and has considered both versions of decentralized control, with communication and with no communication. This chapter considers the decentralized control of Petri nets by means of the SBPI [57, 142, 214]. Note that the SBPI has been introduced in chapter 3 in section 3.2. Compared to the supervision settings defined in section 3.3, the presentation of this chapter assumes the no concurrency setting. Further, as can be seen later in this chapter, the specific type of partial controllability and observability of our decentralized setting assumes that the controllable (observable) transitions can be individually controlled (observed). However, it should be noticed that some of these assumptions are not critical and can be easily relaxed.

Compared to the automata models for decentralized control, the Petri net models have the major benefit that they can be used to model concurrent systems more concisely, as they do not represent explicitly the state space of the system. Therefore, Petri net methods relying on the structure of the net rather than the state space are of special interest, as the size of the state space, when finite, can be exponentially related to the size of the net. Among the structural methods, the SBPI offers an efficient technique for the design of supervisors enforcing the class of state predicates described by linear marking inequalities. Recalling the main benefits of the SBPI, let's note that the SBPI specifications have been shown to be relevant to several types of applications, as mentioned in section 3.6. Further, as shown in chapter 4, the SBPI can be extended to more general specifications, including language specifications and disjunctive constraints. Chapter 4 also shows that it is possible to extend the SBPI to various centralized supervision settings. This chapter considers the extension of the SBPI to the decentralized supervision setting.

The decentralized setting of this chapter follows that of the previous works in the field (e.g., [167, 164]). Thus, a global plant model is given, together with a specification that is to be enforced by the joint operation of n supervisors $\mathcal{S}_i$, $i = 1 \ldots n$. Each supervisor $\mathcal{S}_i$ may control/observe a subset of the plant events $T_{c,i}/T_{o,i}$. Both versions of the decentralized supervision problem are considered here, with and without communication. Here, communication allows a local supervisor $\mathcal{S}_i$ to observe events that are (directly) observed by other supervisors and to send to other supervisors requests to disable events that are not locally accessible (controllable).

Admissibility is a key concept in the SBPI of Petri nets that are only partially controllable and observable. When dealing with such Petri nets, the SBPI approach classifies the specifications as admissible and inadmissible, where the former can be directly enforced and the latter are first transformed to an admissible form and then enforced. As shown in section 3.4, admissibility is a special type of feasibility, defined with respect to a given supervision method and a given type of specifications. For instance, the supervision method could be the SBPI for fully controllable and observable Petri nets, and the specifications could be of the form $L\mu \leq b$ defined in (4.3). In this case, if a set of constraints (4.3) is admissible, the SBPI method for fully controllable and observable Petri nets provides a feasible supervisor. In contrast, if (4.3) is feasible, there is some feasible supervisor enforcing the specification, even though the SBPI for fully controllable and observable Petri nets may not provide a feasible supervisor.

The main contributions of this chapter are as follows. First, we define *d-admissibility* (decentralized admissibility) as an extension of admissibility to the decentralized setting. D-admissibility extends the centralized admissibility while allowing the supervisors enforcing d-admissible constraints to be designed with very low computational complexity (the complexity of matrix multiplication), just as in the SBPI. Thus, d-admissibility identifies a class of constraints for which the supervisors can be *easily* computed, rather than the class of feasible constraints (the class of constraints for which supervisors can be computed). Thus, d-admissibility does not parallel controllability and co-observability in the automata setting [167]. An algorithm for the design of supervisors enforcing *d-admissible* constraints and an algorithm testing d-admissibility are provided. Unlike the former, the algorithm testing d-admissibility is computationally more complex, as it may require some reachability analysis. To overcome this difficulty, a simplification is proposed, similar to the structural admissibility test in the SBPI [142], again allowing very low computational effort. The trade-off is that only a subclass of d-admissible constraints is identified, called globally d-admissible constraints. Results concerning this class of constraints are also presented. Next, to deal with constraints that are not d-admissible, we provide two supervisor design approaches. The first one involves two stages. The first stage solves the problem in a centralized setting by assuming all locally observable and controllable transitions as centrally available. The solution is then distributed in the second stage to the local supervisors, by means of communication. To design the communication between the supervisors such that a communication cost is minimized, an integer programming approach is proposed. The second approach dealing with constraints that are not d-admissible uses constraint transformations. The constraint transformations replace the specification, which is not d-admissible, with a (more restrictive) d-admissible specification. The solution here is obtained via integer programming. This approach can handle design both with and without communication. Further, the design process can incorporate communication constraints, such as limits on the average network traffic, and minimize a communication cost function.

Comparing the two supervisor design approaches, note the following. The first approach is expected to generate more permissive solutions than the second. However, the first is intended for problems with unrestricted communication, while the second for problems with restricted or no communication. Decentralization is useful in the first case, as it may result in a solution with a better communication cost. In the second case a decentralized solution is unavoidable, due to the absence or restriction of the means of communication. With regard to our use of integer programming, note that while the development of alternative methods that are less computationally intensive are a direction for future research, in the automata setting it was shown that a decentralized solution cannot be found with polynomial complexity [164]. Note also that the size of the integer program depends on the size of the Petri net structure, and not on the size of its state space (i.e., the size of its equivalent automaton), which may not be finite.

Apart from our work, which is presented in this chapter, the decentralized supervision of Petri nets has been considered also in [32, 63]. In [63], the distributed supervision of controlled Petri nets is considered based on results for centralized supervision from [71]. The solution involves communication between local supervisors. In [32], distributed supervisors and a central coordinator are designed, where the distributed supervisors implement the distributed part of the specification and the coordinator implements the rest of the specification. In the automata setting, much of the work on decentralized control is presented in [164] and the references therein. In particular, we mention [167] for the decentralized control with no communication and [16, 165] for decentralized control with communication. Our communication setting differs from that of most papers in that the communication involves events rather than state estimates [16] or sequences of events [200]. Further, related to our approach is also [200], which considers the problem of finding a decentralized solution with the same performance as a centralized solution, when communication is available. The vast majority of the decentralized control papers assume language specifications. However, in this chapter we will focus our attention on the particular class of state predicate specifications supported by the SBPI. In the automata setting, the existence of a decentralized solution enforcing state predicates is studied in [189].

The chapter is organized as follows. A detailed literature review is given in section 5.2. Section 5.3 describes the notation and outlines the SBPI. Section 5.4 describes the decentralized setting of our approach. D-admissibility is defined in section 5.5. There, the properties of d-admissibility and the related algorithms are presented. The first approach for the enforcement of constraints that are not d-admissible is presented in section 5.6. This approach assumes communication is available and does not incorporate communication constraints. The second approach, intended for the settings with no communication or restricted communication, is presented in section 5.7. Finally, a manufacturing example adapted from [129] is presented in section 5.8.

5.2 Related Work

A survey of some of the current results on the decentralized control of DES can be found in [164]. The current work, at least for the most part, has been done in the supervisory control framework of Ramadge and Wonham [159]. In this framework, a plant G is to be controlled by using local supervisors $\mathcal{S}_1 \ldots \mathcal{S}_n$, where the supervisor $\mathcal{S}_i$ observes the subset of events $\Sigma_{o,i}$ of G and controls the subset of events $\Sigma_{c,i}$ of G. Based on the observation that the results for two supervisors can usually be extended to results on n supervisors, many of the problems addressed in the literature are formulated in terms of only two local supervisors $\mathcal{S}_1$ and $\mathcal{S}_2$. In the following, $\mathcal{L}(G)$ will denote the language generated by G, $\mathcal{L}_m(G)$ the marked language of G, and $\overline{E}$ the prefix closure of a language E. The general setting of decentralized control does not require a system to be *physically* divided into subsystems. Thus, the supervisors $\mathcal{S}_i$ do not necessarily control distinct physical entities. However, in the following it is convenient to denote by *subsystem i* the tuple of G, $\Sigma_{c,i}$, and $\Sigma_{o,i}$.

The case in which the specification is already given in a decomposed form for each supervisor $\mathcal{S}_i$ to implement is studied in [129]. When additional requirements, not included in the decomposed specification, are to be enforced, the authors propose to use a central coordinator. Depending on the application, all or some of the locally controllable/observable events can be included in the set of controllable/observable events of the coordinator. It is argued that by implementing the decomposed specification with the local supervisors, the control actions of the central coordinator are significantly reduced. As a motivation for their work, the authors notice the significant decrease in complexity (number of states) of the decentralized supervisor as compared with the centralized supervisor. The decomposed specification is given in the form of languages A_i and E_i, where a supervisor $\mathcal{S}_i$ is to satisfy $A_i \subseteq \mathcal{L}(\mathcal{S}_i/G_i) \subseteq E_i$ and G_i is G restricted to the events accessible to $\mathcal{S}_i$. The problem of finding each of $\mathcal{S}_i$ can then be reduced to the centralized control problem [128]. The same approach can also be applied to the coordinator. However, note that this solution is not guaranteed to be nonblocking.[1]

---

[1]Note that the nonblocking property is different from deadlock-freedom or liveness. A closed-loop system is nonblocking if from any reachable state a marked state can be reached. This implies neither liveness nor deadlock-freedom. Furthermore, the

In [35] a sufficient condition is given for the existence of decentralized supervisors $\mathcal{S}_i$ such that $\mathcal{S}_1 \wedge \mathcal{S}_2 \wedge \ldots \mathcal{S}_n$ implements a given language exactly. Note that the framework of [35] is more general, as a supervisor $\mathcal{S}_i$ does not observe the events directly, but through an observation function M_i. The observation function maps plant events to supervisor events or to the empty string. This allows the possibility that two or more observable events create the same supervisor event; that is, the supervisor cannot distinguish between them. (The observable events are the events that are not mapped to the empty string.)

In [207] a particular model of the system is assumed, for which the paper presents conditions guaranteeing that decentralized supervision can achieve the same performance as the optimal centralized supervision. The assumptions are as follows. The system consists of subsystems operating in parallel. Each subsystem i has a local set of events Σ_i. The set Σ_{iu} of uncontrollable local events of subsystem i is assumed to satisfy $\Sigma_{iu} \cap \Sigma_j = \emptyset$ for all $j \neq i$. All events available for observation to a local supervisor are local. The case in which there are local unobservable events is not studied.

In [167] it is shown that a necessary and sufficient condition for the existence of decentralized supervisors exactly implementing a given language $E \subseteq \mathcal{L}_m(G)$ is that E is *controllable* and *co-observable*. Controllability is defined as in the centralized case [159] and is taken with respect to the set Σ_{uc} that contains the events that are uncontrollable to all supervisors $\mathcal{S}_i$. Essentially,[2] co-observability corresponds to the property of the system that for all $s \in \overline{E}$ and all events σ such that $s\sigma \in \mathcal{L}(G) \setminus \overline{E}$, there is an i such that in the subsystem i the observation of s is distinct from the observations of all $s' \in \overline{E}$ with $s'\sigma \in \overline{E}$. The authors also study the problem of the existence of decentralized supervisors implementing a language $\mathcal{L}(\bigwedge_i \mathcal{S}_i/G)$ bounded by lower and upper bounds A and E: $A \subseteq \mathcal{L}(\bigwedge_i \mathcal{S}_i/G) \subseteq E$, for $A \subseteq E \subseteq \mathcal{L}(G)$. Naturally, a solution exists if and only if there is a K controllable and co-observable such that $A \subseteq K \subseteq E$.

The complexity involved in the decentralized control problems of [167] is studied in [166]. The article [166] considers the case of two local supervisors and shows that co-observability can be decided in polynomial time. However, it is shown that decentralized supervisors enforcing a co-observable and controllable language may not be computable in polynomial time. Furthermore, the authors notice that the existence of a co-observable language K bounded by $A \subseteq K \subseteq E$ cannot be decided in polynomial time. Note that complexity results for *centralized* control appear in [197].

Another version of a problem of [167] has been shown in [195] to be undecidable. The problem is to check whether there are two supervisors $\mathcal{S}_1$ and $\mathcal{S}_2$

---

nonblocking property is implied by liveness, but is not implied by deadlock-freedom. We can conclude that supervision with liveness enforcement in our Petri net setting is a stronger requirement than nonblocking supervision in the automata setting.

[2]An additional requirement is necessary when $E \neq \overline{E} \cap \mathcal{L}_m(G)$.

such that $\mathcal{S}_1 \wedge \mathcal{S}_2$ is nonblocking and $\mathcal{L}_m(\mathcal{S}_1 \wedge \mathcal{S}_2/G) \subseteq E$, for some regular language E. Note that in view of [167, 166], the problems of [167] are decidable, and so we can conclude that this problem is also decidable when it can be reduced to one of the problems of [167]. As an example, this is the case when $\overline{E} \cap \mathcal{L}_m(G) = E$.

The article [119] is written in the framework of supervisory control for specifications represented as infinite traces of events [191, 192] (ω-languages). The following decentralized control problem is shown to be undecidable. Let $S(G)$ denote the set of all infinite strings that can be executed by G. Given $A \subseteq E \subseteq S(G)$, the problem is to find the supervisors $\mathcal{S}_1$ and $\mathcal{S}_2$ such that $\mathcal{S}_1 \wedge \mathcal{S}_2$ avoids deadlock, $A \subseteq S(\mathcal{S}_1 \wedge \mathcal{S}_2/G) \subseteq E$, and other technical conditions are satisfied.

Given a co-observable and controllable language, the decentralized supervisor enforcing it can be constructed as in [167]. An alternative method appears in [117]. The main result of [117] is that the infimal prefix-closed controllable and co-observable[3] superlanguage of a language K is the intersection of the infimal prefix-closed controllable and observable superlanguages of K with respect to each of the subsystems i. Then, given $K \subseteq \mathcal{L}(G)$ controllable, co-observable, and prefix-closed, K is exactly implemented when the local supervisors implement the infimal prefix-closed controllable and observable superlanguages of K. The authors also show how to separate the computation of an infimal prefixed-closed controllable and observable superlanguage into the computations of an infimal prefix-closed controllable superlanguage and of an infimal prefix-closed observable superlanguage. As in [35], the results are given in a more general setting in which the local supervisors observe the events through masks (observation functions).

The main result of [117] is applied in [101] to various decentralized control problems. To reduce the computational burden, the authors propose to work with local models G_i of the plant G. A plant G_i contains only the dynamics of G that is related to the subset of events relevant in subsystem i. Then, the supervisors $\mathcal{S}_i$ can be computed as the infimal prefix-closed controllable and observable superlanguages with respect to the plants G_i, rather than with respect to the larger plant G. Further developments appear in [100].

Most papers on decentralized control consider an architecture in which a controllable event is enabled if and only if it is enabled by all local supervisors. Different enabling rules have been considered in [154, 215, 216]. Thus [215, 216] consider an architecture in which some controllable events are enabled by the conjunction of the decisions of the supervisors, and the other controllable events by the disjunction of the supervisor decisions. This partition of the controllable events is part of the design process and can be optimally done in polynomial time [215]. Co-observability in the context of this architecture has been defined and is also verifiable in polynomial time [216]. This architecture

---

[3]The co-observability definition of [117] is slightly less general than that of [167], as it is not effective to specifications $K \subseteq \mathcal{L}_m(G)$ with $K \neq \overline{K} \cap \mathcal{L}_m(G)$.

generalizes the conventional architecture in which conjunction is the enabling rule. The benefit of the more general architecture is that the set of languages that can be implemented by decentralized supervisors is increased [216].

In [146], the decentralized control problem is considered for two local supervisors in a cooperative game-theoretic framework. The supervisors are to enforce liveness and a safety specification. The design approach is semidecidable.

In [113], the fully decentralized control problem is studied. In this problem, there is no communication between the local supervisors and between the agents designing the supervisors. This means that a local supervisor does not rely on other supervisors to disable illegal sequences, as the way the other supervisors are designed is unknown. The requirement on the local supervisors is that they enforce together a language K satisfying $K = \overline{K \cap \mathcal{L}_m(G)}$ and $A \subseteq K \cap \mathcal{L}_m(G) \subseteq E$, where $A \subseteq E \subseteq \mathcal{L}_m(G)$ are given. Note that a fully decentralized solution is less likely to exist than a decentralized solution. Furthermore, fully decentralized supervision tends to be overrestrictive. The main advantage of the approach is reduced computational complexity.

Decentralized supervision with communication between supervisors has also been considered. Communication between supervisors allows a larger class of languages to be implemented. In [208] an asymmetric supervision problem involving two supervisors is studied. Given a specification and a unidirectional communication channel from one supervisor to the other, the problem is to decide whether the supervisors can be designed such that the specification is implemented. A necessary and sufficient condition is found. The approach of [208] is extended in [200]. In [200], information structures are associated with each supervisor in order to help them achieve the same performance as a given centralized supervisor. Note that the centralized supervisor is designed under the assumption that all locally controllable/observable events are globally controllable/observable. The information structure specifies for each local supervisor the supervisors that send information to it, and the kind of information they send. Thus a local supervisor is able to receive observation strings from selected other local supervisors. The design problem is to obtain a set of minimal information structures for which the decentralized solution is equivalent to the centralized solution.

Decentralized supervision with communication has also been considered in [16]. In this approach, the supervisors broadcast estimates of their state. An optimal communication problem is set up, in which the cost reflects how often the supervisors send their state estimates. The following is the design problem addressed in the paper. Given a specification language, find control policies and communication policies for the supervisors such that the specification is exactly implemented.

The article [165] considers also decentralized supervision with communication. The work can be applied when the control policies of the supervisors are given, and the communication between them is to be minimized. In the setting of the paper, event occurrences are communicated, rather than state estimates or observation strings.

Decentralized control with communication and delays is studied in [196]. That paper considers the case of two supervisors with a fixed communication policy and a simplified type of specifications. The communication policy is that each supervisor broadcasts the events it observes. When the communication delay is bounded by k, the set of problems for which a solution exists is denoted by $\mathcal{DCC}_k$. When the delays are allowed to be unbounded, the set of problems for which a solution exists is denoted by $\mathcal{DCUC}$. The paper shows that $\mathcal{DC} \subset \mathcal{DCUC} \subset \cdots \mathcal{DCC}_2 \subset \mathcal{DCC}_1 \subset \mathcal{DCC}_0 = \mathcal{CC}$, where $\mathcal{DC}$ is the corresponding set of problems for the case of control with no communication, and $\mathcal{CC}$ for the case of centralized control. Undecidability is proved for the existence of controllers in the cases $\mathcal{DCUC}$ and $\mathcal{DC}$. Note that the undecidability result relies on the type of specification that has been chosen. For instance, the more common versions of the decentralized control problem with no communication are decidable [167, 166].

Decentralized supervision with specification given in the form of state predicates has been studied in [189]. Note that the supervision of DES based on state predicates is described in [127, 124]. A sufficient condition is given for the existence of a solution. According to [189], when the plant is modeled by a finite automaton, it takes polynomial time to check the condition and to construct the solution.

In [190] a problem of reliable decentralized supervisory control is considered. A decentralized supervisor consisting of n local supervisors is said to be k-reliable if it exactly achieves a specification under possible failures of $n - k$ supervisors. Necessary and sufficient conditions are given for a supervisor to be k-reliable. Then algorithms to synthesize a k-reliable supervisor for a sublanguage of the specification are given, when no k-reliable supervisor exactly implements the specification.

Examples found in the decentralized control literature are as follows. In the context of [129], a manufacturing system model is used to illustrate the supervisor design approach. In [35], a model of the alternating bit protocol is considered together with a specification to be enforced via supervision. Results of the paper are applied to verify that the specification can be implemented. Other versions of the alternating bit protocol are considered in [156]. A simple manufacturing system is used in [100] to illustrate existence results developed in that paper.

5.3 Preliminaries

In this chapter, a Petri net structure is denoted by $\mathcal{N} = (P, T, F, W)$, where P is the set of places, T the set of transitions, F the set of transition arcs, and W the weight function. The incidence matrix of $\mathcal{N}$ is denoted by D. A place (transition) denoted by p_j (t_i) is the place (transition) corresponding to the jth $(i$th$)$ row (column) of the incidence matrix.

Recall that the specification of the SBPI [57, 142, 214] consists of the state constraints

$$L\mu \leq b, \tag{5.1}$$

where $L \in \mathbb{Z}^{n_c \times |P|}$, $b \in \mathbb{Z}^{n_c}$, and μ is the marking of $\mathcal{N}$. To distinguish between the case $n_c = 1$ and $n_c > 1$, we say that (5.1) represents *a constraint* when $n_c = 1$ and that (5.1) represents *a set of constraints* when $n_c > 1$. Note that $\mathcal{N}$ represents the **plant**. Recall also that the SBPI provides a supervisor in the form of a Petri net $\mathcal{N}_s = (P_s, T, F_s, W_s)$ with

$$D_s = -LD, \tag{5.2}$$

$$\mu_{0,s} = b - L\mu_0, \tag{5.3}$$

where D_s is the incidence matrix of the supervisor, $\mu_{0,s}$ the initial marking of the supervisor, and μ_0 the initial marking of $\mathcal{N}$.

Let μ_c be the marking of the closed loop, and let $\mu_c|_{\mathcal{N}}$ denote μ_c restricted to the plant $\mathcal{N}$. Let $t \in T$ be a transition. t is **closed-loop enabled** if μ_c enables t. t is **plant enabled**, if $\mu_c|_{\mathcal{N}}$ enables t in $\mathcal{N}$. The supervisor **detects** t if t is closed-loop enabled at some reachable marking μ_c and firing t changes the marking of some control place. The supervisor **controls** t if there is a reachable marking μ_c such that t is plant enabled but not closed-loop enabled. Given μ_c, the supervisor **disables** t if there is a control place C such that $(C, t) \in F_s$ and $\mu_c(C) < W_s(C, t)$.

Recall also that in Petri nets with uncontrollable and unobservable transitions, a supervisor generated as above may include control places preventing plant-enabled uncontrollable transitions to fire and may contain control places with marking varied by firings of closed-loop enabled unobservable transitions. Such a supervisor is clearly not implementable. A supervisor is feasible if it only controls controllable transitions and it only detects observable transitions. The constraints $L\mu \leq b$ are **admissible** if the supervisor defined by (5.2)–(5.3) is feasible. When inadmissible, the constraints $L\mu \leq b$ are transformed (if possible) to an admissible form $L_a\mu \leq b_a$ such that $L_a\mu \leq b_a \Rightarrow L\mu \leq b$ (section 3.5). Then, the supervisor enforcing $L_a\mu \leq b_a$ is feasible and enforces $L\mu \leq b$ as well. In this (centralized) setting, a Petri net $\mathcal{N}$ with sets of uncontrollable and unobservable transitions T_{uc} and T_{uo} will be denoted by $(\mathcal{N}, T_{uc}, T_{uo})$. The decentralized setting is defined next.

5.4 The Model

The system is given as a Petri net model $\mathcal{N} = (P, T, F, W)$. A decentralized supervisor consists of a set of supervisors $\mathcal{S}_1, \mathcal{S}_2, \ldots, \mathcal{S}_n$, each able to control and observe given sets of transitions of the plant. To distinguish between a decentralized supervisor and its components $\mathcal{S}_1 \ldots \mathcal{S}_n$, we will say that

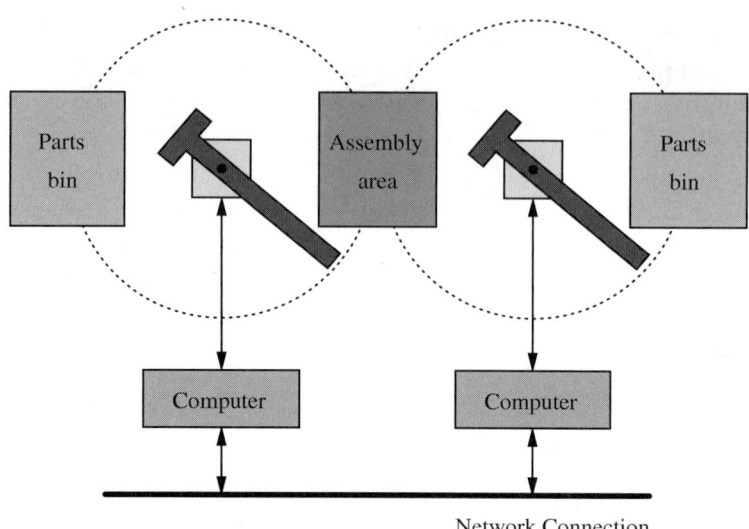

Fig. 5.1. Robotic manufacturing system.

each $\mathcal{S}_i$ is a *local* supervisor. Further, we denote by $T_{o,i}$ $(T_{c,i})$ the subset of transitions of $\mathcal{N}$ that $\mathcal{S}_i$ can observe (control). The tuple $(\mathcal{N}, T_{c,i}, T_{o,i})$ will be called the *subsystem i*, as it represents the object controlled and observed by the local supervisor $\mathcal{S}_i$. Note that we do not assume that the subsystems correspond to distinct physical entities of a system. We will also call $T_{o,i}$ $(T_{c,i})$ the set of locally observable (controllable) transitions of subsystem i. The set of unobservable (uncontrollable) transitions of $\mathcal{S}_i$ is denoted by $T_{uo,i} = T \setminus T_{o,i}$ $(T_{uc,i} = T \setminus T_{c,i})$. A system $\mathcal{N}$ with subsystems of uncontrollable and unobservable transitions $T_{uc,i}$ and $T_{uo,i}$ will be denoted by $(\mathcal{N}, T_{uc,1}, \ldots, T_{uc,n}, T_{uo,1}, \ldots, T_{uo,n})$.

The decentralized supervision problem can be formulated as follows: *Given a global specification and the sets of uncontrollable and unobservable transitions $T_{uc,1}$, $T_{uc,2}$, ..., $T_{uc,n}$ and $T_{uo,1}$, $T_{uo,2}$, ..., $T_{uo,n}$, find a set of local supervisors $\mathcal{S}_1$, $\mathcal{S}_2, \ldots$, $\mathcal{S}_n$ whose simultaneous operation guarantees that the global specification is satisfied, where each $\mathcal{S}_i$ can control $T \setminus T_{uc,i}$ and observe $T \setminus T_{uo,i}$.* As an illustration, consider a manufacturing example in which two robots transport parts to a common assembly area [106]. The system is shown in Fig. 5.1. The Petri net model of the system is shown in Fig. 5.2(a). Note that $\mu_2 = 1$ $(\mu_4 = 1)$ when the left (right) robot is in the assembly area, and $\mu_1 = 1$ $(\mu_3 = 1)$ when the left (right) robot is in the parts bin. The set of controllable transitions of the left (right) subsystem may be taken as $T_{c,1} = \{t_1, t_2\}$ $(T_{c,2} = \{t_3, t_4\})$. Assume that the subsystem of each robot knows when the other robot enters or leaves the parts bin. Then each subsystem contains the controllable transitions of the other subsystem as observable

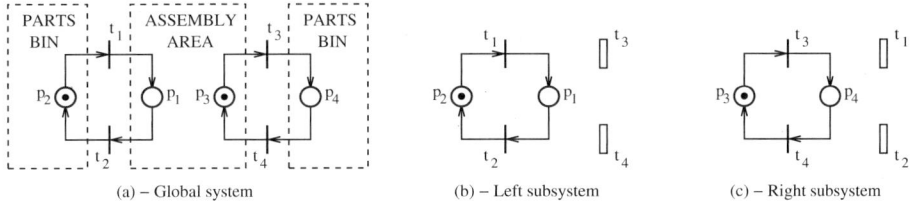

(a) – Global system   (b) – Left subsystem   (c) – Right subsystem

Fig. 5.2. A Petri net model of the robotic manufacturing system.

transitions; a possible graphical representation of the subsystems is shown in Fig. 5.2(b) and (c). Further, a possible global specification could be that the two robots should not be at the same time in the working area.

5.5 Decentralized Admissibility

5.5.1 Definition and Application

This section introduces an admissibility concept for the decentralized setting, called *d-admissibility*. Admissibility in the centralized case will be denoted here as **c-admissibility**. Thus, *c-admissibility* is taken with respect to a Petri net $(\mathcal{N}, \mu_0)$ of controllable transitions T_c and observable transitions T_o. Recall that this is the significance of c-admissibility: a c-admissible set of constraints (5.1) can be implemented with the simple construction of (5.2)–(5.3), as in the fully controllable and observable case.

In the decentralized case, we are still given a global (i.e., not distributed) specification, and we are interested in defining admissibility with respect to a Petri net $(\mathcal{N}, \mu_0)$ and the sets of controllable and observable transitions of the subsystems: $T_{c,1} \ldots T_{c,n}$ and $T_{o,1} \ldots T_{o,n}$. Admissibility in the decentralized case is called **d-admissibility** . As in the case of c-admissibility, we would like d-admissibility to guarantee that the (decentralized) supervisor can be easily constructed. This is achieved by the following definition. Recall that a *constraint* is an inequality $l\mu \leq c$, $l \in \mathbb{Z}^{1 \times |P|}$ and $c \in \mathbb{Z}$, while a *set of constraints* denotes $L\mu \leq b$, with $L \in \mathbb{Z}^{n_c \times |P|}$, $b \in \mathbb{Z}^{n_c}$, and $n_c \in \mathbb{Z}$.

Definition 5.1. *Given* $(\mathcal{N}, \mu_0, T_{c,1} \ldots, T_{c,n}, T_{o,1} \ldots, T_{o,n})$, *a constraint is* **d-admissible** *if there is a collection of subsystems* $\mathcal{C} \subseteq \{1, 2, \ldots, n\}$, $\mathcal{C} \neq \emptyset$, *such that the constraint is c-admissible with respect to* $(\mathcal{N}, \mu_0, T_c, T_o)$, *where* $T_c = \bigcup_{i \in \mathcal{C}} T_{c,i}$ *and* $T_o = \bigcap_{i \in \mathcal{C}} T_{o,i}$. *A set of constraints is* **d-admissible** *if each of its constraints is d-admissible.*

Note that the d-admissibility definition for a single constraint differs from that of a set of constraints, as the constraints $L(i, \cdot)\mu \leq b(i)$ of a set $L\mu \leq b$ are not required to be d-admissible with the same set $\mathcal{C}$. To illustrate the

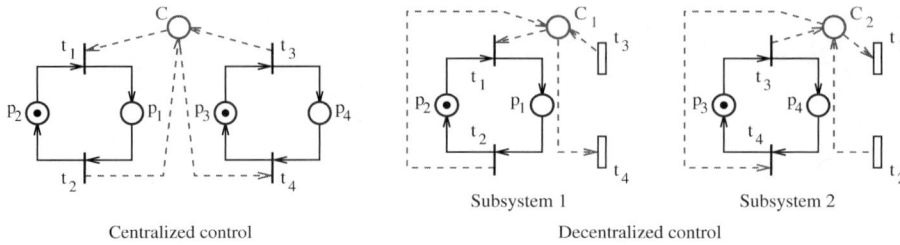

Fig. 5.3. Centralized control versus decentralized control.

definition, assume that we have a constraint that is c-admissible only with respect to subsystem 1. Then, it is d-admissible, as we can select $\mathcal{C} = \{1\}$. Further, when each constraint of $L\mu \leq b$ is c-admissible with respect to some subsystem, $L\mu \leq b$ is d-admissible. In particular, when each subsystem has full observability of the net and every transition is controllable with respect to some subsystem, any constraint is d-admissible.

The construction of a decentralized supervisor, given a d-admissible set of constraints, is illustrated on the Petri net of Fig. 5.2. The mutual exclusion constraint

$$\mu_1 + \mu_3 \leq 1 \tag{5.4}$$

is to be enforced. The centralized control solution is shown in Fig. 5.3. In the case of decentralized supervision, there are two subsystems: the first one has $T_{uo,1} = \emptyset$ and $T_{uc,1} = \{t_3, t_4\}$, and the other has $T_{uo,2} = \emptyset$ and $T_{uc,2} = \{t_1, t_2\}$. Note that (5.4) is not c-admissible with respect to any of $(\mathcal{N}, T_{c,1}, T_{o,1})$ or $(\mathcal{N}, T_{c,2}, T_{o,2})$. However, it is d-admissible for $\mathcal{C} = \{1, 2\}$. Given two variables $x_1, x_2 \in \mathbb{N}$, a decentralized supervisor $\mathcal{S}_1 \wedge \mathcal{S}_2$ enforcing (5.4) can be defined by the following rules:

The supervisor $\mathcal{S}_1$:

- Initialize x_1 to 0.
- Disable t_1 if $x_1 = 0$.
- Increment x_1 if t_2 or t_3 fires.
- Decrement x_1 if t_1 or t_4 fires.

The supervisor $\mathcal{S}_2$:

- Initialize x_2 to 0.
- Disable t_4 if $x_2 = 0$.
- Increment x_2 if t_2 or t_3 fires.
- Decrement x_2 if t_1 or t_4 fires.

Note that $\mathcal{S}_1$ and $\mathcal{S}_2$ differ only in the second rule: one disables t_1, while the other t_4. A graphical representation of $\mathcal{S}_1$ and $\mathcal{S}_2$ is possible, as shown in Fig. 5.3. Thus, $\mathcal{S}_1$ is represented by C_1 and $\mathcal{S}_2$ by C_2; x_1 is the marking of C_1 and x_2 the marking of C_2. Graphically, C_1 and C_2 are copies of the control place C of the centralized supervisor. Note that (C_1, t_4) and (C_2, t_1) model observation, not control. This is due to the fact that $\mathcal{S}_1$ never disables t_4 and $\mathcal{S}_2$ never disables t_1. As C_1 and C_2 have the same initial marking as C, their markings stay equal at all times. So, whenever t_1 should be disabled, the disablement action is implemented by C_1, and whenever t_4 is to be disabled, the disablement action is implemented by C_2.

In the general case, the construction of a supervisor enforcing a d-admissible constraint $l\mu \leq c$ ($l \in \mathbb{N}^{1 \times |P|}$ and $c \in \mathbb{N}$) is as follows. (Note that the notation of Definition 5.1 is used.)

ALGORITHM 5.1   Supervisor Design for a D-admissible Constraint

1. Let μ_0 be the initial marking of $\mathcal{N}$, C the control place of the centralized SBPI supervisor $\mathcal{N}_s = (P_s, T, F_s, W_s)$ enforcing $l\mu \leq c$, and $\mathcal{C}$ the set of Definition 5.1.

2. For all $i \in \mathcal{C}$, let $x_i \in \mathbb{N}$ be a state variable of $\mathcal{S}_i$.

3. Define $\mathcal{S}_i$, for $i \in \mathcal{C}$, by the following rules:

   • Initialize x_i to $c - l\mu_0$.

   • If $t \in T_{c,i}$, $t \in C\bullet$ and $x_i < W_s(C, t)$, then $\mathcal{S}_i$ disables t.

   • If t fires, $t \in T_{o,i}$ and $t \in \bullet C$, then $x_i = x_i + W_s(t, C)$.

   • If t fires, $t \in T_{o,i}$ and $t \in C\bullet$, then $x_i = x_i - W_s(C, t)$.

To enforce a d-admissible set of constraints $L\mu \leq b$, the construction above is repeated for each constraint $l\mu \leq c$. Note that in the graphical representation the supervisors $\mathcal{S}_i$ correspond to $|\mathcal{C}|$ copies of the control place C of the centralized supervisor, where each copy has the same initial marking as C. Next we prove that the resulting decentralized supervisor of Algorithm 5.1 is feasible (physically implementable) and as permissive as possible.

Remark 5.1. The SBPI construction (5.2)–(5.3) presents the best supervision can achieve: a transition t is disabled if and only if firing t leads to a marking μ for which $L\mu \not\leq b$. Therefore, we are interested in identifying conditions under which this supervision performance can be achieved under decentralization (with partial controllability and partial observability). As it turns out, d-admissibility is a sufficient condition for this optimal performance, as proven in the next result.

Remark 5.2. As we will prove next, every d-admissible specification is feasible. However, it is possible to have a specification that is feasible and yet not d-admissible. For instance, consider the Petri net shown in Fig. 5.4 and the constraint

$$\mu_1 + \mu_3 + \mu_7 \leq 1. \tag{5.5}$$

Assume that we have two subsystems, the first with $T_{c,1} = \{t_1\}$ and $T_{o,1} = \{t_5, t_6\}$, and the second with $T_{c,2} = \{t_4, t_6\}$ and $T_{o,2} = \{t_1, t_2, t_3, t_4, t_5, t_6\}$. Note that (5.5) is not d-admissible. However, the constraint (5.5) can be implemented with the same permissiveness as in the centralized case by using the supervisors Ξ_1 and Ξ_2 defined as follows. Ξ_1 disables t_1 after t_6 fires, and

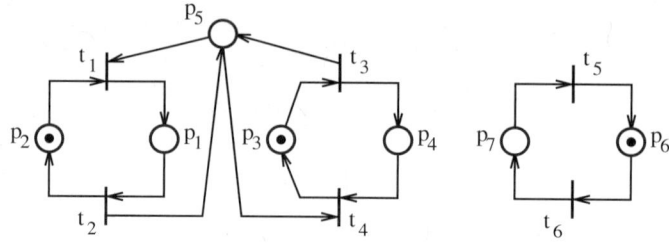

Fig. 5.4. Example for illustrating the limitations of d-admissibility.

enables t_1 after t_5 fires. Ξ_2 allows t_4 and t_6 to fire only when $\mu_1 + \mu_3 + \mu_7 = 0$, where Ξ_2 can keep track of the value of $\mu_1 + \mu_3 + \mu_7$ by observing the firings of $t_1, t_2, \ldots, t_6$.

In the following, we denote by $\mathcal{S}$ the supervisor of (5.2)–(5.3). For simplicity, the same notation $\mathcal{S}$ is used for both the enforcement of sets of constraints $L\mu \leq b$ and the enforcement of single constraints $l\mu \leq c$. Further, let $\mathcal{S}_d$ denote the decentralized supervisor. For instance, in Algorithm 5.1 $\mathcal{S}_d$ is the conjunction $\bigwedge_{i \in \mathcal{C}} \mathcal{S}_i$ of the supervisors $\mathcal{S}_i$ for $i \in \mathcal{C}$. The feasibility of a decentralized supervisor $\mathcal{S}_d$ is defined as follows. $\mathcal{S}_d$ is **feasible** if all $\mathcal{S}_i$ are feasible. In our context, the feasibility of $\mathcal{S}_i$ means that $\mathcal{S}_i$ only observes transitions $t \in T_{o,i}$ and disables transitions t only if not plant enabled or if $t \in T_{c,i}$. Let $P_i : T^* \to T_{o,i}^*$ denote the projection of a firing sequence σ on $T_{o,i}$. The fact that $\mathcal{S}_i$ observes only $t \in T_{o,i}$ means that for all firing sequences $\sigma_1, \sigma_2 \in T^*$, if $P_i(\sigma_1) = P_i(\sigma_2)$ then $\mathcal{S}_i(\sigma_1) = \mathcal{S}_i(\sigma_2)$. Note that $\mathcal{S}_i(\sigma)$ denotes the set of transitions enabled by $\mathcal{S}_i$ once σ has been fired from the initial state of the system.

Theorem 5.2. *The decentralized supervisor $\mathcal{S}_d$ constructed in Algorithm 5.1 is feasible, enforces the desired constraint, and is as permissive as the centralized supervisor $\mathcal{S}$.*

Proof. Feasibility is an immediate consequence of the construction of Algorithm 5.1. To prove the remaining part of the theorem, we consider firing sequences σ that are plant enabled at an initial marking μ_0, and we show that σ is enabled by $\mathcal{S}$ at the initial marking μ_0 if and only if σ is enabled by $\mathcal{S}_d$ at the initial marking μ_0. The proof uses the notation of Algorithm 5.1 and of Definition 5.1. Given a firing sequence $\sigma = t_{i_1} t_{i_2} \ldots t_{i_k}$ enabled from μ_0, let's denote by μ_j the markings reached in $\mathcal{N}$ while firing σ: $\mu_0 \xrightarrow{t_{i_1}} \mu_1 \xrightarrow{t_{i_2}} \mu_2 \xrightarrow{t_{i_3}} \ldots \mu_k$.

First, note that for all firing sequences $\sigma = t_{i_1} t_{i_2} \ldots t_{i_k}$ enabled by both $\mathcal{S}$ and $\mathcal{S}_d$ from μ_0, we have that at all markings μ_j reached while firing σ,

$$x_i = c - l\mu_j \quad \forall i \in \mathcal{C}. \tag{5.6}$$

This is proven by induction. For $j = 0$, (5.6) is satisfied, due to the way the variables x_i are initialized. Assume (5.6) is satisfied for $j < k$. According to the SBPI, when the plant has the marking μ_j the marking of C is $c - l\mu_j$, the same as x_i $\forall i \in C$. There are two cases: (a) $l\mu_j = l\mu_{j+1}$ and (b) $l\mu_j \neq l\mu_{j+1}$. In case (a), in view of (5.2), $W_s(C, t_{i_j}) = W_s(t_{i_j}, C) = 0$. Therefore, neither the marking of C, nor any of the x_i's is changed by firing t_{i_j}. Hence, (5.6) is satisfied at μ_{j+1}. In case (b), note that by Definition 5.1, the d-admissibility of $l\mu \leq c$ implies that S is c-admissible with respect to $(\mathcal{N}, \mu_0, T_c, T_o)$. Then, since t_{i_j} is not dead and $l\mu_j \neq l\mu_{j+1}$: $t_{i_j} \in T_o$. However, $t_{i_j} \in T_o \Rightarrow (\forall i \in C)$ $t_{i_j} \in T_{o,i}$. Hence, t_{i_j} is observable to all S_i, and so all x_i are changed in the same way. Moreover, according to the SBPI, firing t_{i_j} changes the marking of C the same way as x_i are changed. From the SBPI we know that the new marking of C is $c - l\mu_{j+1}$. It follows that when μ_{j+1} is reached, $x_i = c - l\mu_{j+1}$ $\forall i \in C$.

Finally, we prove by contradiction that the firing sequences σ enabled by S from μ_0 are the firing sequences enabled by S_d from μ_0. Assume the contrary, that there is a σ that is enabled by one supervisor and not enabled by the other. We decompose σ into $\sigma = \sigma_x t_x \sigma_y$, $t_x \in T$, where σ_x is enabled by both supervisors and $\sigma_x t_x$ is not. If $\mu_0 \xrightarrow{\sigma_x} \mu_x$, then (5.6) is satisfied at $\mu_j = \mu_x$; the marking of C is also $c - l\mu_x$. There are two cases: (a) t_x enabled by C; (b) t_x not enabled by C. Case (a) implies $t_x \notin C\bullet$ or $W_s(C, t_x) \leq c - l\mu_x$. Then, by construction, S_d must also enable t_x, which contradicts the assumption that not both S and S_d enable t_x. In case (b), according to the SBPI, we have that $W_s(C, t_x) > c - l\mu_x$ and $t_x \in T_c$, by the d-admissibility of $l\mu \leq c$. It follows that there is an $i \in C$ such that S_i disables t_x, and hence that S_d does not enable t_x. This contradicts the assumption that one of S and S_d enables the transition t_x.   □

Let T_o^M be the set of transitions detected by S and T_c^M the set of transitions controlled by S. For instance, in Fig. 5.3 $T_c^M = \{t_1, t_4\}$ and $T_o^M = \{t_1, t_2, t_3, t_4\}$. The d-admissibility of a constraint can be tested as follows.

ALGORITHM 5.2   Checking whether a Constraint is D-Admissible

1. Find T_o^M and T_c^M.

2. Let C be the set of subsystems i satisfying $T_{o,i} \supseteq T_o^M$.

3. If $C = \emptyset$, declare that the constraint is not d-admissible and exit.

4. Define $T_c = \bigcup_{i \in C} T_{c,i}$.

5. Does T_c satisfy $T_c \supseteq T_c^M$? If yes, declare the constraint d-admissible. Otherwise, declare that the constraint is not d-admissible.

Note that a d-admissible constraint can be implemented for a minimal set $C_{min} \subseteq C$ containing the minimal number of subsystems such that

$T_c^M \subseteq \bigcup_{i \in \mathcal{C}_{min}} T_{c,i}$. Note also that checking whether a set of constraints is d-admissible involves checking each constraint individually.

Proposition 5.3. *The algorithm checking d-admissibility is correct.*

Proof. A constraint is declared d-admissible if $\mathcal{C} \neq \emptyset$ and $T_c \supseteq T_c^M$. The definition of T_o^M and T_c^M implies that the constraint is c-admissible with respect to $(\mathcal{N}, T_c, T_o)$ (where $T_o = \bigcap_{i \in \mathcal{C}} T_{o,i}$). Then, in view of Definition 5.1, the algorithm is right to declare the constraint d-admissible.

Next, assume a d-admissible constraint. Then, there is a set of subsystems $\mathcal{C}' \neq \emptyset$ such that the constraint is c-admissible with respect to $(\mathcal{N}, T_c', T_o')$ (where $T_c' = \bigcup_{i \in \mathcal{C}'} T_{c,i}$ and $T_o' = \bigcap_{i \in \mathcal{C}'} T_{o,i}$). Then $T_o' \supseteq T_o^M$ and $T_c' \supseteq T_c^M$; $T_o' \supseteq T_o^M \Rightarrow \mathcal{C}' \subseteq \mathcal{C} \Rightarrow T_c \supseteq T_c' \Rightarrow T_c \supseteq T_c^M$. Consequently, the algorithm declares the constraint to be d-admissible. $\qquad\square$

In general, it may be difficult to compute the sets T_c^M and T_o^M, as this may involve some reachability analysis. Alternatively, estimates $T_c^e \supseteq T_c^M$ and $T_o^e \supseteq T_o^M$ can be used instead of T_c^M and T_o^M. However, in this case the algorithm only checks a sufficient condition for d-admissibility, and so it can no longer detect constraints that are not d-admissible. In the case of the SBPI, a constraint $l\mu \leq c$ is implemented by a control place C, as described by (5.2)–(5.3). Obviously, some estimates T_c^e and T_o^e are $T_c^e = C\bullet$ and $T_o^e = \bullet C \cup C\bullet$. Here, T_c^e differs from T_c^M if there is a $t \in C\bullet$ that is never both plant enabled and closed-loop disabled in $(\mathcal{N}, \mu_0, \mathcal{S})$. Also, T_o^e differs from T_o^M if there is some $t \in \bullet C \cup C\bullet$ that is dead in $(\mathcal{N}, \mu_0, \mathcal{S})$.

Remark 5.3. When the estimates $T_c^e = C\bullet$ and $T_o^e = \bullet C \cup C\bullet$ are used instead of T_c^M and T_o^M, the Algorithm 5.2 is the decentralized equivalent of the structural admissibility conditions $l D_{uc} \leq 0$ and $l D_{uo} = 0$ of [142] for centralized supervision. In this book, these conditions are introduced in section 3.2.2. D_{uc} and D_{uo} are the restrictions of the incidence matrix of the plant to the sets of uncontrollable and unobservable transitions, respectively.

Note that when it is possible and convenient to communicate in a reliable fashion with each subsystem of a decentralized system, a centralized solution with $T_c = \bigcup_{i=1}^n T_{c,i}$ and $T_o = \bigcup_{i=1}^n T_{o,i}$ is possible. Note also that in the implementation of d-admissible constraints, each supervisor $\mathcal{S}_i$ with $i \in \mathcal{C}$ relies on the proper operation of the other supervisors $\mathcal{S}_j$ with $j \in \mathcal{C}$. By itself, a local supervisor may not be able to implement a d-admissible constraint or its implementation may be overrestrictive. For instance, in the example of Fig. 5.2, the supervisor of the first subsystem can only enforce $\mu_1 + \mu_3 \leq 1$ by itself by imposing $\mu_1 = 0$. However, this solution is overrestrictive. D-admissibility illustrates the fact that more can be achieved when supervisors cooperate to achieve a given task, rather than when a supervisor tries on its own to achieve it (cf. "two heads better than one" in [167]).

5.5.2 Significance of D-Admissibility

D-admissibility is clearly a sufficient condition for a specification to be exactly implementable by some supervisor. D-admissibility appears to also be necessary under most circumstances, however the conditions under which this is true is a matter of further research. We provide here a necessity proof for a stronger d-admissibility concept, called *global* d-admissibility, and a related supervision problem. Thus, instead of looking at the relation between d-admissibility and the feasibility of the decentralized supervision for a given initial marking μ_0, we look at global d-admissibility and the decentralized supervision problem for all markings μ_0 and all free terms b in (5.1). That is, $\mathcal{N}$, $T_{c,i}$, $T_{o,i}$, $i = 1 \ldots n$, and L are given and fixed, while μ_0 and b are seen as variables. This approach allows us not only to shed some light on the necessity of d-admissibility, but also to establish the sense in which optimality is achieved when constraints are identified as d-admissible based on the simplified test of Remark 5.3.

Definition 5.4. *The family of single constraints $l\mu \leq c$ with fixed l is* **globally d-admissible** *if there is $\mathcal{C} \subseteq \{1, 2, \ldots, n\}$ such that for all μ_0 and $c \geq l\mu_0$, $l\mu \leq c$ is d-admissible with respect to $(\mathcal{N}, \mu_0, T_{c,1}, \ldots, T_{c,n}, T_{o,1}, \ldots, T_{o,n})$ and $\mathcal{C}$. The family of constraints $L\mu \leq b$ with fixed L is* **globally d-admissible** *if every single constraint $l\mu \leq c$ of $L\mu \leq b$ is globally d-admissible.*

Given $L\mu \leq b$, let $\mathcal{S}$ denote the supervisor corresponding to the construction of (5.2)–(5.3). When $L\mu \leq b$ consists of a single constraint $l\mu \leq c$, the supervisor $\mathcal{S}$ is implemented by a single control place C. To simplify our notation, the marking of C at the plant marking μ is denoted by $\mu(C)$ instead of $\mu_s(C)$. If σ is the transition sequence fired since the initialization of the system and $\mathcal{S}$ enables t, we write $\mathcal{S}(\sigma)[t$; otherwise, we write $\mathcal{S}(\sigma)\!\!\not[t$. For simplicity, let's write $\mathcal{S}[t$ (respectively, $\mathcal{S}\!\!\not[t$) if no transition has been fired (i.e., if σ is empty).

Let $\mathcal{S}_d = \mathcal{S}_1 \wedge \mathcal{S}_2 \wedge \ldots \wedge \mathcal{S}_n$ denote the decentralized supervisor consisting of local supervisors $\mathcal{S}_i$.

Lemma 5.5. *If $C\bullet \neq \emptyset$ and for all μ_0 and $c \geq l\mu_0$ there is a feasible $\mathcal{S}_d$ as permissive as $\mathcal{S}$, then $C\bullet \subseteq \bigcup_{i=1}^{n} T_{c,i}$ and $\bullet C \cup C\bullet \subseteq \bigcup_{i=1}^{n} T_{o,i}$.*

Proof. The proof relies on the following two observations, identifying general controllability and observability requirements for an exact implementation of a specification. First, if there are $t \in T$ and $\sigma \in T^*$ such that σt is plant enabled and σ is legal (according to the specification $l\mu \leq c$) and σt is not legal, then t must be controllable. Second, if there are $t \in T$ and $\sigma, \sigma' \in T^*$ such that $\sigma t \sigma'$ is plant enabled, $\sigma \sigma'$ is plant enabled, and one and only one of $\sigma t \sigma'$ and $\sigma \sigma'$ is legal, then t must be observable.

Let $T_c = \bigcup_{i=1}^{n} T_{c,i}$ and $T_o = \bigcup_{i=1}^{n} T_{c,i}$. We first prove that $t \in C\bullet \Rightarrow t \in T_c$ and $t \in T_o$. Given $t \in C\bullet$, by selecting an appropriate μ_0 the plant enables $\sigma = tt$. Further, we can select $\mu_0(C)$ such that $\mathcal{S}[t$ and $\mathcal{S}\not[tt$. (Note that c can be found as $c = \mu_0(C) + l\mu_0$.) Since $\mathcal{S}_d$ is as permissive as $\mathcal{S}$: $\mathcal{S}_d[t$ and $\mathcal{S}_d\not[tt$. Then, based on our two observations we find that t must be controllable and observable. Since $T_{c,i}$ and $T_{o,i}$, $i = 1 \ldots n$, are the transitions $\mathcal{S}_d$ may control and observe, it follows that $t \in T_c$ and $t \in T_o$. It remains to show that $\bullet C \subseteq T_o$. We prove $t \in \bullet C \Rightarrow t \in T_o$. Let $t \in \bullet C$ and $t' \in C\bullet$. For appropriate μ_0 and $\mu_0(C)$, tt' and t' are plant enabled, but $\mathcal{S}\not[t'$ and $\mathcal{S}[tt'$. Since $\mathcal{S}_d$ is as permissive as $\mathcal{S}$, it follows that t is observable. This concludes the proof.   □

The family of constraints $l\mu \leq c$ (with l fixed) is said to be **proper** if $C\bullet \neq \emptyset$. Further, $L\mu \leq b$ is **proper** if each of its constraints $l\mu \leq c$ is proper. By (5.2), $l\mu \leq c$ is proper if and only if $lD \not\leq 0$, where D is the incidence matrix. Thus, the constraints that are not proper do not need supervision: they stay enforced at all times, if the initial marking satisfies them.

Proposition 5.6. *Let $l\mu \leq c$ proper be given. $l\mu \leq c$ is globally d-admissible if and only if there is $\mathcal{C} \subseteq \{1, 2, \ldots, n\}$ such that $C\bullet \subseteq \bigcup_{i \in \mathcal{C}} T_{c,i}$ and $\bullet C \cup C\bullet \subseteq \bigcap_{i \in \mathcal{C}} T_{o,i}$.*

Proof. "⇐" The d-admissibility of $l\mu \leq c$ at any μ_0 and $c \geq l\mu_0$ follows trivially by the definition of d-admissibility (Definition 5.1) and the definition of c-admissibility.

"⇒" Let $\mathcal{C}$ be the set of Definition 5.4, $T_c = \bigcup_{i \in \mathcal{C}} T_{c,i}$ and $T_o = \bigcap_{i \in \mathcal{C}} T_{o,i}$. By Definition 5.1, $l\mu \leq c$ is c-admissible with respect to $(\mathcal{N}, \mu_0, T_c, T_o)$ for all μ_0 and $c \geq l\mu_0$. Then, $\mathcal{S}$ is admissible with respect to $(\mathcal{N}, \mu_0, T_c, T_o)$ for all μ_0 and $c \geq l\mu_0$. The rest of the proof is similar to the proof of Lemma 5.5.   □

Theorem 5.7. *Let $L\mu \leq b$ proper be given. $L\mu \leq b$ is globally d-admissible if and only if a feasible $\mathcal{S}_d$ as permissive as $\mathcal{S}$ exists for all μ_0 and $b \geq L\mu_0$.*

Proof. Note that it is enough to prove this result on single constraints $l\mu \leq c$. Then, the necessity is an immediate consequence of Proposition 5.6, since $\mathcal{S}_d$ can be constructed as in Algorithm 5.1.

The proof of the sufficiency is by contradiction. Assume that $l\mu \leq c$ is not globally d-admissible. Let $\bullet C = \{t_1^o, t_2^o, \ldots, t_f^o\}$ and $C\bullet = \{t_1, t_2, \ldots, t_g\}$. In view of the SBPI, $\bullet C \cap C\bullet = \emptyset$. Note that $g \geq 1$, as $l\mu \leq c$ is proper. Let k_1, k_2, …, k_g and r_1, r_2, …, r_f be positive integers such that

$$k_1 W(C, t_1) = k_2 W(C, t_2) = \cdots = k_g W(C, t_g),$$
$$r_1 W(t_1^o, C) = r_2 W(t_2^o, C) = \cdots = r_f W(t_f^o, C),$$

where, to simplify our notation, we have dropped the index s from the weights $W_s(C, t)/W_s(t, C)$ of the arcs connecting C to the plant. We also constrain the coefficients k_i to $k_i \geq 2$, $i = 1 \ldots g$. Further, if $f \geq 1$, k_i and r_i are also constrained to

$$k_2 W(C, t_2) + k_3 W(C, t_3) + \cdots + k_g W(C, t_g)$$
$$= r_1 W(t_1^o, C) + r_2 W(t_2^o, C) + \cdots + r_f W(t_f^o, C).$$

Note that $k_1 W(C, t_1)$ is missing in the previous expression. t_1 will have a special role in the proof, as described in the following. In view of Lemma 5.5, $C\bullet \subseteq \bigcup_{i=1}^{n} T_{c,i}$ and $\bullet C \cup C\bullet \subseteq \bigcup_{i=1}^{n} T_{o,i}$. Since $l\mu \leq c$ is not globally d-admissible, by Proposition 5.6, there is $t \in C\bullet$ such that for all $i = 1\ldots n$, if $t \in T_{c,i}$ then $\bullet C \cup C\bullet \not\subseteq T_{o,i}$. Without loss of generality, let t_1 be that transition t. Let the initial marking of C be

$$\mu_{s0}(C) = \begin{cases} k_2 W(C, t_2) + \cdots + k_g W(C, t_g) + W(C, t_1) & \text{if } f = 0, \\ W(C, t_1) & \text{if } f \geq 1. \end{cases}$$

Let $\sigma_i^o = t_i^o t_i^o \ldots t_i^o$ (t_i^o is repeated r_i times) and $\sigma_j = t_j t_j \ldots t_j$ (t_j is repeated k_j times), for $i = 1\ldots f$ and $j = 1\ldots g$. Assume that the initial marking μ_0 of the plant is large enough to enable all firing sequences that will be considered in the following. (Given μ_0 and $\mu_{s0}(C)$, the parameter c can be found as $c = l\mu_0 + \mu_{s0}(C)$.) Note that $\mathcal{S}[\sigma_1^o \sigma_2^o \ldots \sigma_f^o \sigma_2 \sigma_3 \ldots \sigma_g t_1$, but $\mathcal{S}[\sigma_1^o \sigma_2^o \ldots \sigma_f^o \sigma_2 \sigma_3 \ldots \sigma_g \sigma_1$. (In the case $f = 0$, we assume that $\sigma_1^o \sigma_2^o \ldots \sigma_f^o = \varepsilon$, where ε is the empty sequence.) Next, we reach a contradiction by showing that $\mathcal{S}_d[\sigma_1^o \sigma_2^o \ldots \sigma_f^o \sigma_2 \sigma_3 \ldots \sigma_g \sigma_1$. First, note that only $\mathcal{S}_i$ with $T_{c,i} \ni t_1$ may disable σ_1 after $\sigma_1^o \sigma_2^o \ldots \sigma_f^o \sigma_2 \sigma_3 \ldots \sigma_g$ is fired. For all such $\mathcal{S}_i$, we have the following cases.

Case 1: $t_1 \in T_{o,i}$. Then, there is $t_j \notin T_{o,i}$ and $t_j \in \bullet C \cup C\bullet$. There are two possible situations, as follows:

Case 1a: $t_j \in C\bullet$. Then, note that $\mathcal{S}_i$ cannot distinguish between $d = \sigma_1^o \sigma_2^o \ldots \sigma_f^o \sigma_2 \ldots \sigma_g$ and $a = \sigma_1^o \sigma_2^o \ldots \sigma_f^o \sigma_2 \ldots \sigma_{j-1} \sigma_{j+1} \ldots \sigma_g$. Hence, $\mathcal{S}_i(a) = \mathcal{S}_i(d)$. Therefore, since $\mathcal{S}(a)[\sigma_1$ and $\mathcal{S}_d$ is as permissive as $\mathcal{S}$, we have that $\mathcal{S}_i(d)[\sigma_1$.

Case 1b: $t_j \in \bullet C$. Then, $\mathcal{S}_i$ cannot distinguish between d (defined above) and $h = \sigma_1^o \sigma_2^o \ldots \sigma_{j-1}^o e \sigma_{j+1}^o \ldots \sigma_f^o \sigma_2 \ldots \sigma_g \sigma_1$, where $e = \sigma_j^o \sigma_j^o \ldots \sigma_j^o$ and σ_j^o appears enough times in e to ensure $\mathcal{S}[h\sigma_1$. Now, $\mathcal{S}_i(d) = \mathcal{S}_i(h)$, and so $\mathcal{S}_i(d)[\sigma_1$.

Case 2: $t_1 \notin T_{o,i}$. Then, $\mathcal{S}_i$ cannot distinguish between $d\sigma_1$ and dt_1. Since $\mathcal{S}[dt_1$, it follows that $\mathcal{S}_i(d)[\sigma_1$.

The cases above show that all supervisors $\mathcal{S}_i$ that are able to disable t_1 enable σ_1 after $\sigma_1^o \sigma_2^o \ldots \sigma_f^o \sigma_2 \sigma_3 \ldots \sigma_g$. However, since $\mathcal{S}$ does not enable it, it follows that $\mathcal{S}_d$ does not implement the same specification (it is not as permissive as $\mathcal{S}$). This is a contradiction. Therefore, $l\mu \leq c$ must be globally d-admissible. $\qquad \square$

Global d-admissibility for $l\mu \leq c$ means that there is a set $\mathcal{C}$ such that for all μ_0 and $c \geq l\mu_0$, $l\mu \leq c$ is d-admissible with respect to $\mathcal{C}$. A natural question is whether global d-admissibility is equivalent to d-admissibility at all μ_0 and $c \geq l\mu_0$ with sets $\mathcal{C}$ that may depend on μ_0 and c. The answer is positive, as stated by the following consequence of Theorem 5.7.

Corollary 5.8. *Let $L\mu \leq b$ proper be given. $L\mu \leq b$ is globally d-admissible if and only if for all μ_0 and $b \geq L\mu_0$, $L\mu \leq b$ is d-admissible.*

Proof. The necessity is obvious. For the sufficiency proof, the supervisor $\mathcal{S}_d$ of Theorem 5.7 can be constructed using Algorithm 5.1. Then, Theorem 5.2 guarantees that $\mathcal{S}_d$ is feasible and as permissive as $\mathcal{S}$. Finally, the conclusion follows by Theorem 5.7. □

Remark 5.4. Theorem 5.7 reveals the significance of global d-admissibility. It shows that supervisors achieving the optimal performance of $\mathcal{S}$ exist for all μ_0 and b (with $b \geq L\mu_0$) if and only if $L\mu \leq b$ is globally d-admissible. In fact, given a specification $L\mu \leq b$ that is not globally d-admissible, the proof of Theorem 5.7 can also be used to identify an infinite set of pairs (μ_0, b) for which $L\mu \leq b$ cannot be exactly implemented by any supervisors of $(\mathcal{N}, \mu_0, T_{c,1}, \ldots, T_{c,n}, T_{o,1}, \ldots, T_{o,n})$.

Remark 5.5. Recall that Algorithm 5.2 has a negligible complexity when the estimates $T_c^e = C\bullet$ and $T_o^e = \bullet C \cup C\bullet$ are used instead of T_c^M and T_o^M. Note that under this circumstance the class of constraints identified by Algorithm 5.2 as d-admissible is precisely the class of globally d-admissible constraints. In this light, the results of this section reveal the significance of a class of constraints that is important from a computational viewpoint.

5.6 Distributing a Centralized Supervisory Policy

The previous section has shown that the design of supervisors enforcing d-admissible constraints can be done easily, as in Algorithm 5.1. It remains to consider the enforcement of constraints that are not d-admissible. Two main approaches are possible here. One is to solve the problem first in a centralized setting, by assuming all locally observable and controllable transitions as observable and controllable to a central supervisor. Then, a communication policy could be used for a decentralized implementation of the centralized solution. Alternatively, another approach is to solve the problem directly in the decentralized setting. While the first approach is expected to result in more permissive supervisors, the second approach could be used to obtain solutions with less or no communication. The first approach is considered in this section, while the second will be treated in section 5.7.

In our decentralized setting, communication can be used to increase the sets $T_{c,i}$ and $T_{o,i}$ of controllable and observable transitions in a subsystem i. This is achieved by transmitting control decisions/transition firings to/from a subsystem in which the transition of interest is controllable/observable. Any transition added by communication to $T_{c,i}$ ($T_{o,i}$) and used by $\mathcal{S}_i$ for control (observation) is said to be remotely controlled (observed). Note that given a set $\mathcal{C}$, communication cannot increase $T_o = \bigcap_{i \in \mathcal{C}} T_{o,i}$ above the attainable upper bound $\overline{T_o} \supseteq T_o$, where $\overline{T_o} = \bigcup_{i=1}^{n} T_{o,i}$. In the same way, $T_c = \bigcup_{i \in \mathcal{C}} T_{c,i}$ cannot be increased above $\overline{T_c} = \bigcup_{i=1}^{n} T_{c,i}$. Indeed, $T \setminus \overline{T_c}$ ($T \setminus \overline{T_o}$) is the set of transitions uncontrollable (unobservable) in all subsystems.

As an illustration, consider the system of Fig. 5.2 with $T_{c,1} = T_{o,1} = \{t_1, t_2\}$ and $T_{c,2} = T_{o,2} = \{t_3, t_4\}$. Then, the constraint $\mu_1 + \mu_3 \leq 1$ is clearly not d-admissible. However, by communicating the firings of the transitions t_1 and t_2 to the right subsystem and of t_3 and t_4 to the left subsystem, the sets of locally observable transitions become $T'_{o,1} = T'_{o,2} = \{t_1, t_2, t_3, t_4\}$, the constraint becomes d-admissible, and the supervisory solution of Fig. 5.3 can be used again.

We begin with an algorithm that uses only the communication of transition firings, without resorting to the transmission of control decisions. In the algorithm, a set $\mathcal{C}$ is found such that T_c contains all transitions that need to be controlled. Then, T_o is increased, as needed, by communication.

ALGORITHM 5.3   Decentralized Supervisor Design with Local Control

1. *Is the specification c-admissible with respect to $(\mathcal{N}, \overline{T_c}, \overline{T_o})$? If not, transform it to be c-admissible (see section 3.5).*

2. *Let $\mathcal{S}$ be the centralized SBPI supervisor enforcing the specification. Let T_{cs} be the set of transitions controlled by $\mathcal{S}$ and T_{os} the set of transitions detected by $\mathcal{S}$.*

3. *Find a set $\mathcal{C}$ such that $T_c = \bigcup_{i \in \mathcal{C}} T_{c,i} \supseteq T_{cs}$.[4]*

4. *In view of the d-admissibility requirement that $\bigcap_{i \in \mathcal{C}} T_{o,i} \supseteq T_{os}$, the communication is designed as follows: $\forall t \in T_{os} \cap (\bigcup_{i \in \mathcal{C}} T_{uo,i})$, a subsystem j such that $t \in T_{o,j}$ transmits the firings of t to all supervisors $\mathcal{S}_k$ with $t \in T_{uo,k}$ and $k \in \mathcal{C}$.*

5. *Design the decentralized supervisor by applying Algorithm 5.1 to $\mathcal{N}$, $\mathcal{C}$, and $T_{o,i} = T_{o,i} \cup T_{os}$ $\forall i \in \mathcal{C}$.*

Remark 5.6. No communication arises when $T_{os} \cap (\bigcup_{i \in \mathcal{C}} T_{uo,i}) = \emptyset$.

Remark 5.7. The algorithm does not take into account communication limitations, such as bandwidth limitations of the communication channel. Bandwidth limitations can be considered in the approach described in the next section.

Remark 5.8. In this solution, communication is used only to make some locally unobservable transitions observable; there is no remote control of locally uncontrollable transitions.

---

[4]At least one solution exists, $\mathcal{C} = \{1 \ldots n\}$. This can be seen from the fact that $\mathcal{S}$ admissible w.r.t. $(\mathcal{N}, \overline{T_c}, \overline{T_o})$ implies $T_{cs} \subseteq \overline{T_c}$, and from $\overline{T_c} = \bigcup_{i=1 \ldots n} T_{c,i}$.

Remark 5.9. In the case of broadcast communication, this solution requires less communication than a centralized solution. This can be seen from the following. A centralized supervisor needs communication links to each subsystem in order to observe and control the transitions of the subsystems. Thus, in the centralized case, both observations and control decisions are communicated. On the other hand, in the decentralized design of Algorithm 5.3, only observations are communicated.

Remark 5.10. The only way the algorithm can fail is at step 1, if the specification is inadmissible and the transformations to an admissible form fail.

Proposition 5.9. *The decentralized supervisor is feasible and equally permissive to the centralized supervisor $\mathcal{S}$ enforcing the specification on $(\mathcal{N}, \overline{T}_c, \overline{T}_o)$.*

Proof. $\mathcal{S}$ is admissible, so $T_{cs} \subseteq \overline{T}_c$ and $T_{os} \subseteq \overline{T}_o$. Communication ensures that the sets of locally unobservable transitions become $T'_{o,i} = T_{o,i} \cup T_{os}$. Thus, the specification is d-admissible with respect to $(\mathcal{N}, T_{c,1}, \ldots, T_{c,n}, T'_{o,1}, \ldots, T'_{o,n})$, and so the conclusion follows by Theorem 5.2.   □

Note that equivalent solutions that use a different communication strategy are possible. For instance, in the example of Fig. 5.2, assume $T_{c,1} = T_{o,1} = \{t_1, t_2\}$ and $T_{c,2} = T_{o,2} = \{t_3, t_4\}$. To enforce (5.4), the Algorithm 5.3 produces $\mathcal{C} = \{1, 2\}$ and requires subsystem 1 to communicate t_1 and t_2 to subsystem 2, while subsystem 2 is required to communicate t_3 and t_4 to subsystem 1. This solution is illustrated in Fig. 5.5(a). However, the solution of Fig. 5.5(b) is also possible. In Fig. 5.5(b), subsystem 1 remotely controls t_4 and subsystem 2 communicates the transitions t_3 and t_4 to subsystem 1. Either of the two solutions could be better depending on the relative cost of communicating transitions versus the remote control of locally inaccessible transitions. Next, we show how the best solution can be found by minimizing a cost function.

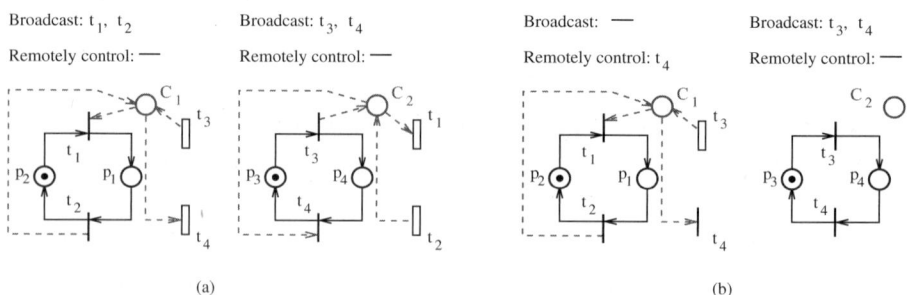

Fig. 5.5. Decentralized control with communication.

To characterize communication, let α_{ij} and ε_{ij} be binary variables defined as follows:

$\alpha_{ij} = 1$ if and only if the transition t_j is communicated to $\mathcal{S}_i$.

$\varepsilon_{ij} = 1$ if and only if the transition t_j is remotely controlled by $\mathcal{S}_i$.

Recall that $\mathcal{S}_i$ denotes the supervisor of the subsystem i, $i = 1 \ldots n$. Note that in the broadcast case $\alpha_{ij} = \alpha_j$ for all i, and $\varepsilon_{ij} = \varepsilon_j$ for all i, where the latter means that either all or none of the supervisors $\mathcal{S}_i$ are allowed to remotely control t_j. Note that in practice remote control could be implemented by allowing the supervisors to announce when their control decision (t_j enabled or t_j disabled) changes.

Assume that the specification $L\mu \leq b$ is proper and has n_c constraints. Let T_c^k (T_o^k) be the set of transitions that the centralized SBPI supervisor enforcing the kth constraint controls (observes). Note that $T_c^k \subseteq T_{cs}$ and $T_o^k \subseteq T_{os}$. Let δ_{ik} denote binary variables indicating whether $\mathcal{S}_i$ implements a control action for the constraint k. For instance, in Fig. 5.5 there is only one constraint (namely, (5.4)), so $k = 1$. In Fig. 5.5(a) $\delta_{11} = \delta_{21} = 1$, while in Fig. 5.5(b), $\delta_{11} = 1$ and $\delta_{21} = 0$, because $\mathcal{S}_2$ controls nothing.

If $\mathcal{S}_i$ implements a control action for the constraint k (i.e., if $\delta_{ik} = 1$), then d-admissibility requires it to observe all transitions in T_o^k. This is written as

$$\alpha_{ij} \geq \delta_{ik} \quad \forall j \in \{j : t_j \in T_o^k \setminus T_{o,i}\}, \forall i = 1 \ldots n, \forall k = 1 \ldots n_c. \quad (5.7)$$

Further, every transition $t \in T_c^k$ needs to be controlled by some $\mathcal{S}_i$. If $\mathcal{S}_i$ controls $t_j \in T_c^k$ and $t_j \notin T_{c,i}$, then we need $\varepsilon_{ij} = 1$. Formally, $\forall t_j \in T_c^k$ $\exists i = 1 \ldots n$: $\delta_{ik} = 1 \wedge [t_j \in T_{c,i} \vee (t_j \notin T_{c,i} \wedge \varepsilon_{ij} = 1)]$. This can be written as:

$$\sum_{i=1}^{n} \delta_{ik} \geq 1 \quad \forall k = 1 \ldots n_c, \quad (5.8)$$

$$\delta_{xk} \leq \varepsilon_{xj} + \sum_{i \in I_j} \delta_{ik} \quad \forall x = 1 \ldots n, \forall j \in \{j : t_j \in T_c^k\}, \forall k = 1 \ldots n_c, \quad (5.9)$$

where $I_j = \{i : t_j \in T_{c,i}\}$. Then we can use integer linear programming to minimize a cost of the form

$$\min \sum_{i,j} \alpha_{ij} c_{ij} + \sum_{i,j} \varepsilon_{ij} f_{ij} + \sum_{i,k} \delta_{ik} h_{ik}, \quad (5.10)$$

which penalizes communication and the number of supervisors implementing control. Now we can enhance Algorithm 5.3 with the following optimal communication policy.

ALGORITHM 5.4   Design with Optimal Communication Strategy

1. *Is the specification c-admissible with respect to $(\mathcal{N}, \overline{T_c}, \overline{T_o})$? If not, transform it to be c-admissible (see section 3.5).*

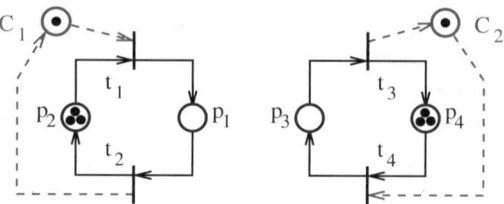

Fig. 5.6. Decentralized control example.

2. *Solve (5.10) subject to (5.7)–(5.9).*

3. *For each $k = 1 \ldots n_c$, apply Algorithm 5.1 on $\mathcal{N}$ with $\mathcal{C} = \{i : \delta_{ik} = 1\}$, $T_{c,i} = T_{c,i} \cup \{t_j : \varepsilon_{ij} = 1\}$, and $T_{o,i} = T_{o,i} \cup \{t_j : \alpha_{ij} = 1\}$.*

Remark 5.11. The decentralization approach of this section can be used for more general specifications. For example, for a modular language specification $\mathcal{L}_1 \wedge \mathcal{L}_2 \wedge \ldots \wedge \mathcal{L}_p$, Algorithm 5.4 changes as follows. The first step would insure the specification is controllable and observable with respect to $(\mathcal{N}, \overline{T_c}, \overline{T_o})$. The second step would be identical, once we let T_c^k (T_o^k) denote the transitions that need to be controlled (observed) for the centralized enforcement of $\mathcal{L}_k$. Then, at step 3, an algorithm similar to Algorithm 5.1 would be used to create "copies" of the centralized supervisor enforcing $\mathcal{L}_k$ at each subsystem i with $i \in \mathcal{C}$.

5.7 Design with Constraint Transformations

5.7.1 Supervision without Communication

In this section we propose a method for the transformation of constraints that are not d-admissible to constraints that are d-admissible. To ensure the specification will be satisfied, the new d-admissible constraints are to be at least as restrictive as the original constraints. As an illustration, consider the Petri net of Fig. 5.2(a), this time with the initial marking $\mu_0 = [0, 3, 0, 3]^T$, $T_{c,1} = T_{o,1} = \{t_1, t_2\}$, $T_{c,2} = T_{o,2} = \{t_3, t_4\}$, and the desired constraint $\mu_1 + \mu_3 \leq 2$. There is no way to transform $\mu_1 + \mu_3 \leq 2$ to a single d-admissible constraint. However, we can transform it to two d-admissible constraints: $\mu_1 \leq 1$ and $\mu_3 \leq 1$, where the first is d-admissible for $\mathcal{C} = \{1\}$ and the second for $\mathcal{C} = \{2\}$. This solution is shown in Fig. 5.6.

In the general case, the problem can be stated as follows: *Given a set of constraints $L\mu \leq b$ that is not d-admissible and the subsystem clusters $\mathcal{C}_1, \mathcal{C}_2,$ $\ldots, \mathcal{C}_m$, find sets of constraints $L_1\mu \leq b_1 \ldots L_m\mu \leq b_m$ d-admissible with respect to $\mathcal{C}_1, \mathcal{C}_2, \ldots, \mathcal{C}_m$, respectively, such that*

$$(L_1\mu \leq b_1 \wedge L_2\mu \leq b_2 \wedge \ldots L_m\mu \leq b_m) \Rightarrow L\mu \leq b. \tag{5.11}$$

Remark 5.12. This framework includes the case when not all constraints $L_i\mu \leq b_i$ are necessary to implement $L\mu \leq b$, by allowing $L_i = 0$ and $b_i = 0$.

Remark 5.13. Note that $\mathcal{C}_1 \ldots \mathcal{C}_m$ are given, rather than calculated. This is not really a limitation. Indeed, there is a finite number of possible groups $\mathcal{C}_i$, namely $2^p - 1$ for p subsystems. So, including all groups would guarantee that no possible solution of the form (5.11) is excluded. However, note that we may not need to include all $\mathcal{C}_i$'s. Indeed, it is to be expected that in practice most $\mathcal{C}_i$'s would have $T_o^{(i)} = \bigcap_{j \in \mathcal{C}_i} T_{o,j} = \emptyset$. (A supervisor of such a $\mathcal{C}_i$ observes no transitions, and so its control may only consist of disabling at all times some transitions; such a supervisor may often be undesirable.) For instance, if $T_{o,i} \cap T_{o,j} = \emptyset$ for all $i \neq j$, at most n groups $\mathcal{C}_i$ have $T_o^{(i)} \neq \emptyset$, namely $\mathcal{C}_i = \{i\}$ for $i = 1 \ldots n$.

Remark 5.14. More restrictive solutions than in the centralized case are expected here. Note that any solution of (5.11) can be implemented in a centralized fashion. Indeed, if $L_i\mu \leq b_i$ is d-admissible with respect to $\mathcal{C}_i$, then $L_i\mu \leq b_i$ is c-admissible in $(\mathcal{N}, \bigcup_{i=1}^n T_{c,i}, \bigcup_{i=1}^n T_{o,i})$.

Our problem is more tractable if we replace (5.11) with the stronger condition below:[5]

$$(\exists \alpha_1, \alpha_2, \ldots, \alpha_m \geq 0) \left[\sum_{i=1}^m \alpha_i L_i\mu \leq \sum_{i=1}^m \alpha_i b_i \right] \Rightarrow L\mu \leq b. \tag{5.12}$$

Without loss of generality, (5.12) assumes that $L_1 \ldots L_m$ have the same number of rows. Again, without loss of generality, (5.12) can be replaced by

$$\left[\sum_{i=1}^m L_i\mu \leq \sum_{i=1}^m b_i \right] \Rightarrow L\mu \leq b. \tag{5.13}$$

We further simplify our problem to

$$L_1 + L_2 + \ldots L_m = R_1 + R_2 L, \tag{5.14}$$
$$b_1 + b_2 + \ldots b_m = R_2(b+1) - 1 \tag{5.15}$$

for R_1 with nonnegative integer elements and R_2 diagonal with positive integers on the diagonal. Note that $[(R_1 + R_2 L)\mu \leq R_2(b+1) - 1] \Rightarrow L\mu \leq b$ has been proved in [142].

---

[5]In the literature, a relaxation of a hard problem that is similar to the relaxation from (5.11) to (5.12) is the S-procedure mentioned in [202] on page 62.

It is known that a sufficient condition for the c-admissibility of a set of constraints $L\mu \leq b$ is that $LD_{uc} \leq 0$ and $LD_{uo} = 0$, where D_{uc} and D_{uo} are the restrictions of the incidence matrix D to the sets of uncontrollable and unobservable transitions [142]. The admissibility requirements in our setting can then be written as

$$L_i D_{uc}^{(i)} \leq 0, \tag{5.16}$$
$$L_i D_{uo}^{(i)} = 0, \tag{5.17}$$

where $D_{uc}^{(i)}$ and $D_{uo}^{(i)}$ are the restrictions of D to the sets $T_{uc}^{(i)} = \bigcap_{i \in \mathcal{C}_i} T_{uc,i}$ and $T_{uo}^{(i)} = \bigcup_{i \in \mathcal{C}_i} T_{uo,i}$.

Integer programming can be used to find a feasible solution to (5.14)–(5.17), where the unknowns are R_1, R_2, L_i, and b_i. In general it is difficult to find constraints or a cost function that guarantee that the least restrictive solution is found, when a least restrictive solution exists. However, given a finite set $\mathcal{M}_I$ of markings of interest, it is possible to insure that the feasible space of the solution will include the markings of $\mathcal{M}_I$ by using the constraints

$$L_i M \leq b_i \mathbf{1}^T, \qquad i = 1 \ldots m, \tag{5.18}$$

where $\leq$ means that each element of $L_i M$ is less than or equal to the element of the same indices in $b_i \mathbf{1}^T$, M is a matrix whose columns are the markings of $\mathcal{M}_I$, and $\mathbf{1}^T$ is a row vector of appropriate dimension in which all elements are 1. The next result is an immediate consequence of our previous considerations.

Proposition 5.10. *Any sets of constraints $L_i \mu \leq b_i$ satisfying (5.14)–(5.18) are d-admissible and* $\bigwedge_{i=1 \ldots n} [L_i \mu \leq b_i] \Rightarrow L\mu \leq b$.

5.7.2 Supervision with Communication

Here we extend the procedure of section 5.7.1 to the case in which communication is possible. Communication is used to relax the admissibility constraints (5.16) and (5.17) by reducing the number of locally uncontrollable or unobservable transitions. However, this reduction may be limited by various communication constraints, such as bandwidth limitations. The framework of this section allows communication constraints to be incorporated in the design process and can be used to minimize communication by defining a cost function.

As in section 5.6, we use the binary variables α_{ij} and ε_{ij} to describe the communication. Recall that $\alpha_{ij} = 1$ if and only if the firings of t_j are communicated to $\mathcal{S}_i$, and $\varepsilon_{ij} = 1$ if and only if $\mathcal{S}_i$ can remotely control the firings of t_j. Note that we have the following constraints:

$$\forall t_j \in T \setminus \overline{T_o} : \alpha_{ij} = 0 \tag{5.19}$$

for $\overline{T_o} = \bigcup_{i=1}^n T_{o,i}$, where $T \setminus \overline{T_o}$ is the set of transitions that cannot be observed anywhere in the system. Similarly,

$$\forall t_j \in T \setminus \overline{T_c} : \varepsilon_{ij} = 0 \qquad (5.20)$$

for $\overline{T_c} = \bigcup_{i=1}^n T_{c,i}$. For any practical purpose, we are not interested in unbounded solutions. So, we assume some upper and lower bounds of $L_i D$ are imposed. Let B_L^i and B_U^i be the lower and upper bounds of $L_i D$. Note that B_L^i and B_U^i bound the weights of the arcs by which control places can be connected to the Petri net. In particular, if we desire the supervisor Petri net to be ordinary, we can set all elements of B_L^i to -1 and all elements of (B_U^i) to 1. However, in general, B_L^i and B_U^i can be set to some arbitrarily small/large numbers. Given the bounds, (5.17) can be relaxed to

$$\begin{aligned} L_i D(\cdot, t_j) &\leq B_U^i(\cdot, t_j)\alpha_{xj} \\ L_i D(\cdot, t_j) &\geq B_L^i(\cdot, t_j)\alpha_{xj} \end{aligned} \quad \forall t_j \in T_{uo}^{(i)}, \, \forall x \in \{x \in \mathcal{C}_i : t_j \notin T_{o,x}\}. \qquad (5.21)$$

This relaxes $L_i D_{uo}^{(i)} = 0$ by eliminating the constraints corresponding to the transitions of $T_{uo}^{(i)}$ that have their firings communicated to the supervisors of $\mathcal{C}_i$.

Similarly, (5.16) can also be relaxed by allowing the supervisors to remotely control transitions. Thus, if $t_j \in T_{uc}^{(i)}$, the admissibility requirement with respect to t_j can be relaxed when the remote control of t_j is allowed. Then, instead of (5.16) we have

$$L_i D(\cdot, t_j) \leq B_U^i(\cdot, t_j) \sum_{x \in \mathcal{C}_i} \varepsilon_{xj} \quad \text{if } \mathcal{C}_i \cap \{x : t_j \in T_{c,x}\} = \emptyset. \qquad (5.22)$$

Communication constraints stating that certain transitions cannot be remotely observed or controlled can be incorporated by setting coefficients α_{ij} and ε_{ij} to zero. Constraints limiting the average network traffic can be incorporated as constraints of the form

$$\sum_{i,j} \alpha_{ij} g_{ij} + \sum_{i,j} \varepsilon_{ij} h_{ij} \leq p, \qquad (5.23)$$

where g_{ij}, h_{ij}, and p are scalars. As an example, the coefficients g_{ij} could reflect average firing counts of the transitions over the operation of the system.

We may also choose to minimize the amount of communication involved in the system. Then we can formulate our problem as

$$\min \sum_{i,j} \alpha_{ij} c_{ij} + \sum_{i,j} \varepsilon_{ij} f_{ij}, \qquad (5.24)$$

where the variables are L_i, b_i, α_{ij}, ε_{ij}, R_1, and R_2. The coefficients c_{ij} and f_{ij} are given. The minimization is subject to the constraints (5.14)–(5.15), (5.18), (5.19)–(5.22), and $\alpha_{ij}, \varepsilon_{ij} \in \{0, 1\}^{|T|}$. This problem can be solved using integer linear programming.

5.7.3 Liveness Constraints

A difficulty of this approach is that the permissiveness of the generated constraints can be hard to control. In the worst case, the generated constraints may cause parts of the system to unavoidably deadlock. Such a situation can be prevented by using additional constraints, which we call liveness constraints.

A liveness constraint consists of a vector x such that for all i: $L_i x \leq 0$. A possible way to obtain such constraints is described next. Given a finite firing sequence σ, let x_σ be a vector such that $x_\sigma(i)$ is the number of occurrences of the transition t_i in σ. Given the Petri net of incidence matrix D and the constraints $L\mu \leq b$, let y be a nonnegative integer vector such that $Dy \geq 0$ and $-LDy \geq 0$. A vector y satisfying these inequalities has the following property. If σ is a firing sequence such that (i) σ can be fired without violating $L\mu \leq b$ and (ii) $x_\sigma = y$, then σ can be fired infinitely often without violating $L\mu \leq b$. However, if the decentralized control algorithm generates a constraint $L_i\mu \leq b_i$ such that $L_i Dy \nleq 0$, then any firing sequence σ having $x_\sigma = y$ cannot be infinitely often fired in the closed loop. If such a situation is undesirable, the matrices L_i can be required to satisfy $L_i x \leq 0$ for $x = Dy$. An illustration will be given in section 5.8.

5.8 Example

This section illustrates the approach of section 5.7 on a manufacturing example adapted[6] from [129]. The system is shown in Fig. 5.7. It consists of two machines (M_1 and M_2), four robots ($H_1 \ldots H_4$), and four buffers of finite capacity ($B_1 \ldots B_4$). The events associated with the movement of the parts within the system are marked with Greek letters. There are two types of parts. The manufacturing process of the first type of part is represented by the following sequence of events: $\gamma_1\tau_1\pi_1\alpha_3\tau_3\pi_3\alpha_1\eta_1$. The manufacturing process of the second type of part is represented by $\gamma_2\tau_4\pi_4\alpha_2\tau_2\pi_2\alpha_4\eta_2$. These processes can be represented by the Petri net of Fig. 5.8. In the Petri net, the transitions are labeled by the events they represent, and the places by the names of the manufacturing components. For instance, a token in p_{16} indicates that M_2 is idle, and a token in p_8 indicates that M_2 is working on a part of type 2 that has just entered the system. Furthermore, the number of parts in a buffer is

---

[6]Compared to the original example of [129], some changes have been made here to better illustrate our approach. We have closely followed the original example of [129] in [86], for (a) the overflow specification and (b) the fairness specification (with η_1, α_1 uncontrollable). By solving the integer programs of section 5.7, a solution as permissive as that of [129] was found for (a), but a more permissive solution was found for (b). (This is explained by the fact that the solution proposed in [129] for (b) was based on intuition and not on the computation of the supremal controllable sublanguage.)

the marking of the place modeling the buffer; for instance, μ_{13} represents the number of parts in B_2 at the marking μ. The number of parts the machines M_1 and M_2 can process at the same time is $\mu_1 + \mu_7 + \mu_{11} + \mu_{15} = n_1$ and $\mu_4 + \mu_8 + \mu_{14} + \mu_{16} = n_2$, respectively. In [129], $n_1 = n_2 = 1$.

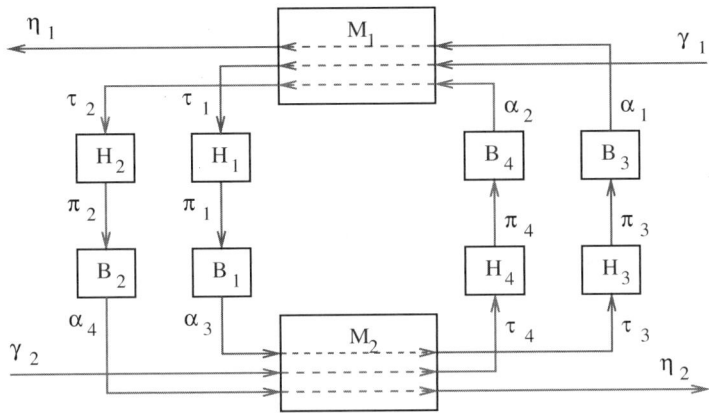

Fig. 5.7. A manufacturing system.

The first supervisory requirements are that the buffers do not overflow. Assuming that the buffers B_1 and B_2 share common space, the requirement can be written as

$$\mu_3 + \mu_{13} \le 2k, \tag{5.25}$$

where $2k$ is the maximum number of parts that can be in B_1 and B_2 at the same time. Similarly, if the buffers B_3 and B_4 share a common space of the same capacity, the constraint is

$$\mu_6 + \mu_{10} \le 2k. \tag{5.26}$$

Another requirement is that the number of completed parts of type 1 is about the same as the number of completed parts of type 2:

$$v_8 - v_{16} \le u, \tag{5.27}$$
$$v_{16} - v_8 \le u, \tag{5.28}$$

where v_8 and v_{16} denote the number of firings of t_8 and t_{16}, respectively. In [129], $u = 2$. Note that constraints involving the vector v can easily be represented as marking constraints in a transformed Petri net, as shown in section 4.2.5.

The constraints (5.25)–(5.26) are to be enforced assuming the following subsystems: $T_{c,1} = \{t_1\}$ and $T_{o,1} = \{t_1, t_2, t_3, t_4\}$, $T_{c,2} = \{t_4\}$ and $T_{o,2} =$

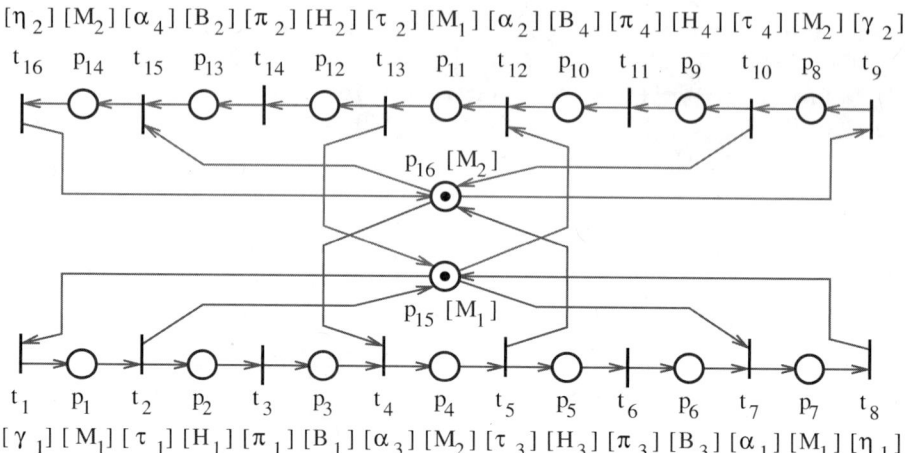

Fig. 5.8. Petri net model of the system.

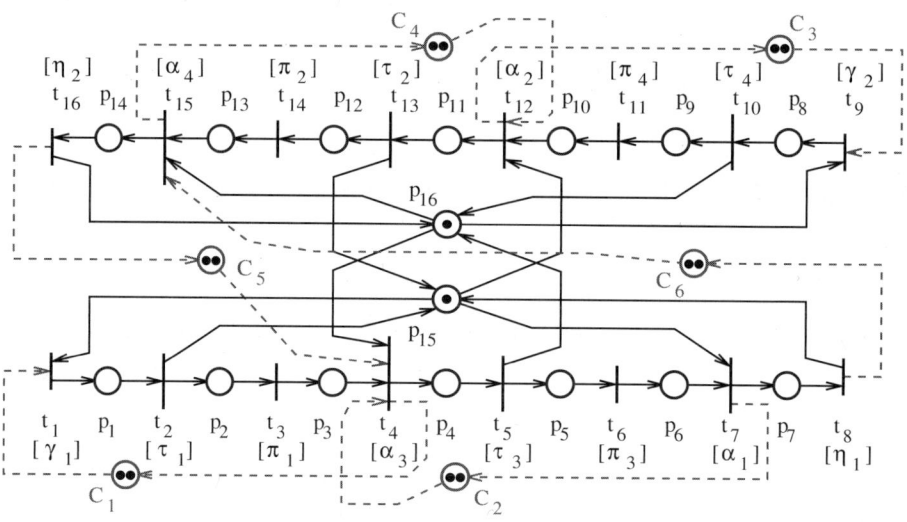

Fig. 5.9. Decentralized supervision.

$\{t_4, t_5, t_6, t_7, t_8\}$, $T_{c,3} = \{t_9\}$ and $T_{o,3} = \{t_9, t_{10}, t_{11}, t_{12}\}$, $T_{c,4} = \{t_{12}, t_{15}\}$ and $T_{o,4} = \{t_{12}, t_{13}, t_{14}, t_{15}, t_{16}\}$. We take $\mathcal{C}_i = \{i\}$ for $i = 1 \ldots 4$. Enforcing (5.25)–(5.26) for $k = 2$ results in the control places C_1, C_2, C_3, and C_4 shown in Fig. 5.9. They correspond to the subsystems 1, 2, 3, and 4, respectively, and enforce $\mu_1 + \mu_2 + \mu_3 \leq 2$, $\mu_4 + \mu_5 + \mu_6 \leq 2$, $\mu_8 + \mu_9 + \mu_{10} \leq 2$, and $\mu_{11} + \mu_{12} + \mu_{13} \leq 2$.

In order to enforce (5.27)–(5.28), we need communication of events. Indeed, without communication there is no acceptable solution. For instance, a solution is to enforce $\mu_4 + \mu_5 + \mu_6 + \mu_7 + v_8 \leq u$ in subsystem 2 and $\mu_{14} + v_{16} \leq u$ in subsystem 4. However, this implies that the manufacturing system is constrained to produce no more than $2u$ parts! To exclude such solutions to the integer program (IP), we can introduce liveness constraints. In this example, we can add the liveness constraints $L_i x \leq 0$ for $x = Dy$ and $y = [1, 1, \ldots, 1]^T$. This is to prevent the constraints generated by the algorithm from blocking the firing sequence $t_1 t_2 \ldots t_{16}$ to occur infinitely often. However, with this liveness constraint and no communication, the problem becomes infeasible. Therefore, since communication is necessary, we are interested in minimizing it. Assuming broadcast ($\alpha_{ij} = \alpha_j$, $\varepsilon_{ij} = \varepsilon_j$, for all i) and that the cost of remote control and remote observation is nonzero and equal (i.e., in (5.24) $c_{ij} = f_{ij}$), the following is an optimal solution:

$$\mu_4 + \mu_5 + \mu_6 + \mu_7 + v_8 - v_{16} \leq 2, \qquad (5.29)$$

$$\mu_{14} + v_{16} - v_8 \leq 2, \qquad (5.30)$$

which involves communicating the occurrences of t_8 and t_{16}. Note that the constraint (5.29) is implemented in the subsystem 2, and the constraint (5.30) in the subsystem 4. In Fig. 5.9, the two constraints are enforced by the control places C_5 and C_6.

Finally, note that in general the IP may have several solutions of the same cost. Further, some may be more restrictive than others. For instance, in our example we could have $\mu_{11} + \mu_{12} + \mu_{13} + \mu_{14} + v_{16} - v_8 \leq 2$ instead of (5.30). Then, a second IP could be used to select a better solution, by minimizing the sum of the positive coefficients of the constraints, while requiring the other coefficients to stay less than or equal to zero (the second IP is also subject to the constraints of the first IP and to a constraint that fixes the communication cost to the minimal value previously computed).

6

Deadlock and Liveness Properties of Petri Nets

6.1 Introduction

This chapter presents new results characterizing deadlock, liveness, and $\mathcal{T}$-liveness in Petri nets. These results can be useful when dealing with the corresponding supervision problems: deadlock prevention, liveness enforcement, and $\mathcal{T}$-liveness enforcement. $\mathcal{T}$-liveness enforcement means ensuring that all transitions in a transition subset $\mathcal{T}$ of a Petri net are live. Deadlock prevention corresponds to preventing the system from reaching a state of total deadlock. Liveness corresponds to the stronger requirement that no local deadlock occurs, or in other words, all transitions are live. $\mathcal{T}$-liveness means that all transitions in the set $\mathcal{T}$ are live. The concept of $\mathcal{T}$-liveness is useful in problems in which some transitions correspond to undesirable system events (such as faults) or when the system model contains transitions modeling an initialization process. Unless otherwise stated, supervision in this chapter assumes all transitions are controllable and observable. Note that this does not affect the generality of the main results of the chapter, as they deal with structural net properties rather than supervision.

A possible way to study the liveness properties of a Petri net is to use the reachability graph. However, such an approach can only handle bounded Petri nets, requires the initial marking to be known, and can be applied to reasonably small Petri nets, due to the state explosion problem. Unfolding has been proposed to reduce the computational burden [66], however the other two limitations remain. In this chapter we consider a structural approach to the liveness problem, based on the algebraic properties of the incidence matrix. This allows obtaining results that apply to both bounded and unbounded Petri nets, while the initial marking can be regarded as a parameter, not as a given input. This work has been inspired by the incidence matrix properties of repetitive Petri nets (e.g., [145]). Related work includes [15], presenting among others an extension of the relation between deadlocked Petri nets and siphons for generalized Petri nets, and a generalization of the extension to asymmetric choice Petri nets of the Commoner's theorem. However, our supervisory perspective, our concern on $\mathcal{T}$-liveness, and our consideration of arbitrary Petri nets, including nonrepetitive Petri nets, differentiate this work from previous results.

The contribution of this chapter is described in sections 6.3, 6.4, and 6.5. We begin in section 6.3.1 by characterizing the relation between deadlock prevention, $\mathcal{T}$-liveness enforcement, and liveness enforcement. Thus we answer the following questions: (a) Which are the Petri nets in which deadlock prevention, or $\mathcal{T}$-liveness enforcement, or liveness enforcement is possible? And (b) when is deadlock prevention equivalent to $\mathcal{T}$-liveness enforcement or liveness enforcement? We answer question (a) in Proposition 6.6, and question (b) in Theorems 6.8 and 6.9. Theorem 6.8 considers the case of the deadlock prevention supervisors which are not more restrictive than liveness or $\mathcal{T}$-liveness supervisors; Theorem 6.9 considers the general case. We conclude the first part of the chapter with Theorem 6.10, which states that the transitions of a Petri net can be divided into two classes: transitions that can be made live under an appropriate supervisor for some initial markings, and transitions which cannot be made live under any circumstances. Theorem 6.10 is very important for the further developments of the chapter.

The most important part of the chapter is section 6.3.2. In this section we show how to characterize Petri nets for deadlock prevention and liveness enforcement based on a special type of subnets. Thus we begin by defining what we call the *active subnets* of a Petri net. Then we define a special class of siphons, which we call *active siphons*. Proposition 6.15 is a necessary condition for deadlock which generalizes the known result that a deadlocked ordinary Petri net contains an empty siphon. Proposition 6.16 is a further extension, as it gives a sufficient condition in terms of empty active siphons for deadlock to be unavoidable. Commoner's theorem on free-choice Petri nets has been extended to asymmetric-choice Petri nets [42]; see also [15]. We further extend the result in Theorem 6.17: we show that each dead transition is in the postset of an uncontrolled siphon. Then in Theorem 6.19 we give a necessary condition and a sufficient condition for $\mathcal{T}$-liveness in an asymmetric choice Petri net. These results apply also to EAC Petri nets, a new extension of asymmetric choice Petri nets defined in this chapter. Polynomial complexity algorithms for the computation of the active subnets are included in section 6.5.

In section 6.4 we discuss the significance of our results for deadlock prevention and $\mathcal{T}$-liveness enforcement. Examples are included. In sections 6.4.1 and 6.4.3 we consider deadlock prevention and $\mathcal{T}$-liveness enforcement. In section 6.4.2 we include Theorem 6.23, which shows how to do least restrictive deadlock prevention.

We conclude this chapter with section 6.5, that presents a number of algorithms. Thus, section 6.5.1 presents algorithms for the computation of the various types of active subnets. Then, sections 6.5.2 and 6.5.3 present algorithms for the transformation of general Petri nets into the form required by some of the results introduced in section 6.3.2. All these algorithms are used in chapter 7, which presents procedures for deadlock prevention and $\mathcal{T}$-liveness enforcement. Finally, note that the material of this chapter represents the theoretical background of chapter 7.

6.2 Preliminaries

Throughout the chapter, Petri nets are denoted by $\mathcal{N} = (P, T, F, W)$, where P is the set of places, T the set of transitions, F the set of transition arcs, and W the transition arc weight function. The incidence matrix is denoted by D, where the rows of D correspond to places and the columns to transitions. Also, by denoting a place by p_i or a transition by t_j, we assume that p_i corresponds to the ith row of D and t_j to the jth column of D. Recall that a Petri net $\mathcal{N} = (P, T, F, W)$ is **ordinary** if $\forall f \in F : W(f) = 1$. We will refer to slightly more general Petri nets in which only the arcs from places to transitions have weights equal to 1. We are going to call such Petri nets *PT-ordinary*, because all arcs (p, t) from a place p to a transition t satisfy the requirement of an ordinary Petri net that $W(p, t) = 1$.

Definition 6.1. *A Petri net* $\mathcal{N} = (P, T, F, W)$ *is* **PT-ordinary** *if* $\forall p \in P$, $\forall t \in T$, $(p, t) \in F \Rightarrow W(p, t) = 1$.

An **asymmetric choice** Petri net is defined by the property that $\forall p_1, p_2 \in P$, if $p_1 \bullet \cap p_2 \bullet \neq \emptyset$ then $p_1 \bullet \subseteq p_2 \bullet$ or $p_2 \bullet \subseteq p_1 \bullet$. A **siphon** is a set of places $S \subseteq P$, $S \neq \emptyset$, such that $\bullet S \subseteq S \bullet$. A siphon S is **minimal** if there is no siphon $S' \subset S$. A siphon is **empty** at a marking μ if it contains no tokens. Given a Petri net $(\mathcal{N}, \mu_0)$, a **controlled siphon** is a siphon which is not empty at any reachable marking. A well-known necessary condition for deadlock in an ordinary Petri net is that there is at least one empty siphon [160]. It can easily be seen that the proof of this result also is valid for PT-ordinary Petri nets.

Proposition 6.2. *A deadlocked PT-ordinary Petri net contains at least one empty siphon.*

In general we may not want all transitions to be live. For instance, some transitions of a Petri net may model faults or initialization processes. Thus, we define next the liveness property of a Petri net in which only transitions in a set $\mathcal{T}$ are live.

Definition 6.3. *Let* $(\mathcal{N}, \mu_0)$ *be a Petri net and* $\mathcal{T}$ *a subset of the set of transitions. We say that the Petri net is* $\mathcal{T}$**-live** *if all transitions* $t \in \mathcal{T}$ *are live.*

A live transition is not the opposite of a dead transition. That is, a transition may be neither live nor dead. Indeed, a transition is live if there is no reachable marking for which it is dead. Note also that $\mathcal{T}$-liveness corresponds to liveness when the set $\mathcal{T}$ equals the set of all Petri net transitions.

Supervisors have been formally defined in section 3.3. In this chapter we restrict our attention to deterministic supervisors in the no concurrency setting. Further, issues on partial controllability and observability are not considered. Therefore, we can simplify our definition of **supervisors** to maps $\Xi : \mathcal{M} \times T^* \to 2^T$, where[1] $\mathcal{M}$ is a set of initial markings, and $\Xi(\mu_0, \sigma)$ identifies the supervisor-enabled transitions after the sequence σ is fired from the initial state μ_0. Marking-based supervisors are a particular type of supervisors for which the enabled transitions are determined only by the marking of the Petri net. Given a Petri net $\mathcal{N}$, Ξ is a **marking-based supervisor** if there is a map $\overline{\Xi} : \mathbb{N}^{|P|} \to 2^T$ such that for all initial markings $\mu_0 \in \mathcal{M}$, if σ is enabled from μ_0 and $\mu \xleftarrow{\sigma} \mu_0$, then $\Xi(\mu_0, \sigma) = \overline{\Xi}(\mu)$ (the set of supervisor-enabled transitions depends only on the marking of the Petri net).

A Petri net $(\mathcal{N}, \mu_0)$ supervised by Ξ operates as follows: at every marking μ reached by firing some σ from μ_0 ($\mu_0 \xrightarrow{\sigma} \mu$), only transitions in $\Xi(\mu_0, \sigma)$ may fire. We denote by $(\mathcal{N}, \mu_0, \Xi)$ the supervised Petri net and by $\mathcal{R}(\mathcal{N}, \mu_0, \Xi)$ its set of reachable markings. Recall that Ξ_1 is **less restrictive** (or **more permissive**) than Ξ_2 with respect to $(\mathcal{N}, \mu_0)$ if the set of firing sequences fireable from μ_0 in $(\mathcal{N}, \mu_0, \Xi_2)$ is a proper subset of the set of firing sequences fireable from μ_0 in $(\mathcal{N}, \mu_0, \Xi_1)$. We say that **deadlock can be prevented** in a Petri net $\mathcal{N}$ if there is an initial marking μ_0 and a supervisor Ξ such that $(\mathcal{N}, \mu_0, \Xi)$ is deadlock-free. We say that **liveness ($\mathcal{T}$-liveness) can be enforced** in $\mathcal{N}$ if there is an initial marking μ_0 and a supervisor Ξ such that $(\mathcal{N}, \mu_0, \Xi)$ is live ($\mathcal{T}$-live). It is known that if $(\mathcal{N}, \mu_0)$ is live, then $(\mathcal{N}, \mu)$ with $\mu \geq \mu_0$ may not be live. The same is true for deadlock-freedom, as shown in Fig. 6.1. However, it can easily be seen that if liveness ($\mathcal{T}$-liveness) is enforcible at marking μ or if deadlock can be prevented at μ, then the same is true for all markings $\mu' \geq \mu$.

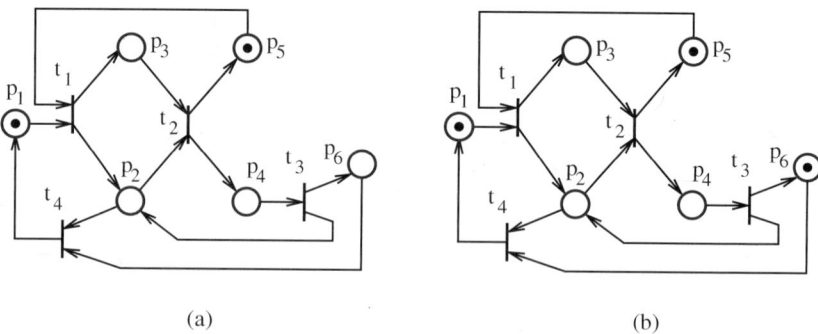

(a)   (b)

Fig. 6.1. A Petri net which is live for the initial marking μ_0 shown in (a) and not even deadlock-free for the initial marking $\mu \geq \mu_0$ shown in (b).

---

[1] T^* is the set of all firing sequences with transitions in T.

As we prove in the next section, the Petri net structures in which liveness can be enforced (for some initial markings) are the *repetitive* Petri nets, and the Petri net structures in which deadlock can be prevented are the *partially repetitive* Petri nets. We include below their formal definition from [145].

Definition 6.4. *A Petri net is said to be* **(partially) repetitive** *if there is a marking μ_0 and a firing sequence σ from μ_0 such that every (some) transition occurs infinitely often in σ.*

The following theorem can be found in [145]. It provides a convenient means to check whether a Petri net is (partially) repetitive, based on the incidence matrix D. Linear programming techniques can be used to implement the test.

Theorem 6.5. *A Petri net is (partially) repetitive if and only if a vector x of positive (nonnegative) integers exists, such that $Dx \geq 0$ and $x \neq 0$.*

6.3 Results

6.3.1 Deadlock Prevention and Liveness Enforcement

The analysis of this section ignores controllability and observability issues. Therefore, some of the results introduced here are restricted to fully controllable and observable Petri nets.

In general it may not be possible to enforce liveness or to prevent deadlock in an arbitrary given Petri net. This may happen because the initial marking is inappropriate or because the structure of the Petri net is incompatible with such a supervision purpose. The next proposition characterizes the structure of Petri nets which allow supervision for deadlock prevention and liveness enforcement, respectively. It shows that Petri nets in which liveness is enforcible are repetitive, and Petri nets in which deadlock is avoidable are partially repetitive. Part (b) of the proposition also appears in [174].

Proposition 6.6. *Let $\mathcal{N} = (P, T, F, W)$ be a Petri net.*

(a) *Initial markings μ_0 exist such that deadlock can be prevented in $(\mathcal{N}, \mu_0)$ if and only if $\mathcal{N}$ is partially repetitive.*

(b) *Initial markings μ_0 exist such that liveness can be enforced in $(\mathcal{N}, \mu_0)$ if and only if $\mathcal{N}$ is repetitive.*

(c) *Initial markings μ_0 exist such that $\mathcal{T}$-liveness can be enforced in $(\mathcal{N}, \mu_0)$ if and only if there is an initial marking μ_0 enabling an infinite firing sequence in which all transitions of $\mathcal{T}$ appear infinitely often.*

Proof. (a) If deadlock can be avoided in $(\mathcal{N}, \mu_0)$ then μ_0 enables some infinite firing sequence σ, and by definition $\mathcal{N}$ is partially repetitive. If $\mathcal{N}$ is partially repetitive, then let μ_0 and σ be as in Definition 6.4; we define Ξ such that it only allows σ to fire from μ_0. Then Ξ prevents deadlock.

(b) and (c) The proof is similar to (a). □

If $\mathcal{N}$ is partially repetitive, a constructive way to obtain an initial marking for which deadlock can be prevented or ($\mathcal{T}$-)liveness can be enforced is implied by Theorem 6.5. Let x be as in Theorem 6.5 and $\sigma_x = t_{x,1} \ldots t_{x,k}$ a firing sequence associated to a Parikh vector $v = x$. Let v_1 denote the Parikh vector after the first transition of σ_x fired, v_2 after the first two fired, and so on to $v_k = v$. If the rows of D are $d_1^T, d_2^T, \ldots, d_{|P|}^T$, $\delta_{i,j} = W(p_i, t_{x,j})$ if $t_{x,j} \in p_i \bullet$ and $\delta_{i,j} = 0$ otherwise, then a marking which enables σ_x is

$$\mu_0(p_i) = \max\{0, \delta_{i,1}, \max_{j=1\ldots k-1}(\delta_{i,j+1} - d_i^T v_j)\}, \quad i = 1 \ldots |P|. \quad (6.1)$$

At least one deadlock prevention strategy exists for μ_0: to allow only the firing sequence $\sigma_x, \sigma_x, \sigma_x, \ldots$ to fire. This infinite firing sequence is enabled by μ_0 because $\mu_0 + Dx \geq \mu_0$ and μ_0 enables σ_x.

Note that if a deadlock prevention supervisor Ξ exists for $(\mathcal{N}, \mu_0)$, then a *marking-based* deadlock prevention supervisor Ξ_m exists for $(\mathcal{N}, \mu_0)$ such that Ξ is at least as restrictive as Ξ_m. The same is true for liveness and $\mathcal{T}$-liveness-enforcing supervisors. Indeed, let $\sigma^{(j)} = t_1^{(j)} t_2^{(j)} t_3^{(j)} \ldots$, for $j = 1, 2, \ldots$, be the infinite firing sequences which can fire from μ_0 in $(\mathcal{N}, \mu_0, \Xi)$; for all $i, j = 1, 2, \ldots$ let $\mu_i^{(j)}$ be the marking reached after firing $t_1^{(j)} \ldots t_i^{(j)}$ from μ_0 and $\sigma_{i,\infty}^{(j)} = t_i^{(j)} t_{i+1}^{(j)} \ldots$. We take $\Xi_m(\mu) = \{t : \exists i, j \geq 1 \text{ such that } \mu = \mu_{i-1}^{(j)} \text{ and } t = t_i^{(j)}\}$. Hence $\forall \mu \in \mathcal{R}(\mathcal{N}, \mu_0, \Xi_m)$: $\exists i, j$ such that $\sigma_{i,\infty}^{(j)}$ is fireable from μ in $(\mathcal{N}, \mu_0, \Xi_m)$.

For marking-based supervisors it is known that if a liveness-enforcing supervisor exists, the least restrictive liveness-enforcing supervisor also exists [174]. The same is true for deadlock prevention and $\mathcal{T}$-liveness-enforcing supervisors. Further, this is also true for the more general supervisors. This follows easily from the fact that given Ξ_1 and Ξ_2, a supervisor at least as permissive as each of Ξ_1 and Ξ_2 is $\Xi = \Xi_1 \vee \Xi_2$ which allows a transition to fire if either of Ξ_1 or Ξ_2 allows it.

The next result is necessary for the proofs of some of the main results of the chapter.

Lemma 6.7. *Let $\mathcal{N} = (P, T, F, W)$ be a Petri net of incidence matrix D. Assume that there is an initial marking μ_I which enables an infinite firing sequence σ. Let $U \subseteq T$ be the set of transitions which appear infinitely often in σ. There is a nonnegative integer vector x satisfying (a) and (b) below.*

(a) *$Dx \geq 0$, $\forall t_i \in U$: $x(i) \neq 0$ and $\forall t_i \in T \setminus U$: $x(i) = 0$.*

(b) *There is a firing sequence σ_x containing only the transitions with $x(i) \neq 0$, such that $\exists \mu_1^*, \mu_2^* \in \mathcal{R}(\mathcal{N}, \mu_I)$: $\mu_1^* \xrightarrow{\sigma_x} \mu_2^*$, each transition t_i appears $x(i)$ times in σ_x, σ can be written as $\sigma = \sigma_a \sigma_x \sigma_b$, and $\mu_I \xrightarrow{\sigma_a} \mu_1^*$.*

Proof. Note that σ can be written as $\sigma_0\sigma'$, where σ_0 is finite and σ' contains only transitions in U. Let μ_0 be the marking such that $\mu_I \xrightarrow{\sigma_0} \mu_0$. We further decompose σ' in $\sigma_1\sigma_2\ldots\sigma_k\ldots$ such that each σ_k is finite and in each σ_k all transitions of U appear at least once. Let $\mu_1, \mu_2, \ldots, \mu_k, \ldots$ be such that $\mu_{k-1} \xrightarrow{\sigma_k} \mu_k$ for $k = 1, 2, \ldots$. By Dickson's lemma (see Lemma 17 in [43]) $\exists j, k, j < k$, such that $\mu_j \leq \mu_k$. Let v_j and v_k be the firing count vectors: $\mu_j = \mu_0 + Dv_j$ and $\mu_k = \mu_0 + Dv_k$; let $x = v_k - v_j$. Then $\mu_k - \mu_j \geq 0 \Rightarrow Dx \geq 0$, and by construction $x \geq 0$, $x(i) > 0 \ \forall t_i \in U$ and $x(i) = 0 \ \forall t_i \in T \setminus U$. Also we take $\sigma_a = \sigma_0\sigma_1\ldots\sigma_j$, $\sigma_x = \sigma_{j+1}\ldots\sigma_k$, $\sigma_b = \sigma_{k+1}\sigma_{k+2}\ldots$, $\mu_1^* = \mu_j$, and $\mu_2^* = \mu_k$. $\qquad\square$

In order to characterize the supervisors which prevent deadlock, or enforce liveness or $\mathcal{T}$-liveness, we define the properties P_1, P_2, and P_3 below, in which $\mathcal{N} = (P, T, F, W)$ is a Petri net, $\mathcal{T} \subseteq T$, and σ denotes a nonempty firing sequence.

(P_1) ($\exists\sigma \ \exists\mu_1', \mu_1 \in \mathcal{R}(\mathcal{N}, \mu)$: $\mu_1 \xrightarrow{\sigma} \mu_1'$ and $\mu_1' \geq \mu_1$)

(P_2) ($\exists\sigma \ \exists\mu_1', \mu_1 \in \mathcal{R}(\mathcal{N}, \mu)$: $\mu_1 \xrightarrow{\sigma} \mu_1'$, $\mu_1' \geq \mu_1$ and all transitions of T appear in σ)

(P_3) ($\exists\sigma \ \exists\mu_1', \mu_1 \in \mathcal{R}(\mathcal{N}, \mu)$: $\mu_1 \xrightarrow{\sigma} \mu_1'$, $\mu_1' \geq \mu_1$ and all transitions of $\mathcal{T}$ appear in σ)

In general, supervisors guaranteed to prevent deadlock are easier to obtain than supervisors guaranteed to enforce liveness or $\mathcal{T}$-liveness. For some problems, obtaining certain deadlock prevention supervisors is enough to guarantee they are also liveness or $\mathcal{T}$-liveness-enforcing supervisors. The following theorem addresses this situation, by characterizing the relations existing between supervisors preventing deadlock and supervisors enforcing ($\mathcal{T}$-)liveness. In general, we may expect deadlock prevention supervisors to be at least as permissive as the supervisors enforcing the stronger requirement of liveness or $\mathcal{T}$-liveness. These are the kinds of deadlock prevention supervisors considered in parts (d) and (e) of the following theorem.

Theorem 6.8. *Let $\mathcal{N} = (P, T, F, W)$ be a Petri net and $\mathcal{T} \subseteq T$.*

(a) *Deadlock can be prevented in $(\mathcal{N}, \mu)$ if and only if (P_1) is true.*

(b) *Liveness can be enforced in $(\mathcal{N}, \mu)$ if and only if (P_2) is true.*

(c) *$\mathcal{T}$-liveness can be enforced in $(\mathcal{N}, \mu)$ if and only if (P_3) is true.*

(d) *Let μ_0 be an arbitrary marking for which liveness can be enforced, Ξ_L the least restrictive liveness-enforcing supervisor of $(\mathcal{N}, \mu_0)$, and $\mathcal{S}$ the set of all deadlock prevention supervisors of $(\mathcal{N}, \mu_0)$ at least as permissive as Ξ_L. Then all $\Xi \in \mathcal{S}$ enforce liveness in $(\mathcal{N}, \mu_0)$ if and only if $\forall\mu \in \mathcal{R}(\mathcal{N}, \mu_0)$: ($P_1$) $\Rightarrow$ (P_2).*

(e) *Let μ_0 be an arbitrary marking for which $\mathcal{T}$-liveness can be enforced, Ξ_L the least restrictive $\mathcal{T}$-liveness-enforcing supervisor of $(\mathcal{N}, \mu_0)$, and $\mathcal{S}$ the set of all deadlock prevention supervisors of $(\mathcal{N}, \mu_0)$ at least as permissive as Ξ_L.*

Then all $\Xi \in \mathcal{S}$ enforce $\mathcal{T}$-liveness in $(\mathcal{N}, \mu_0)$ if and only if $\forall \mu \in \mathcal{R}(\mathcal{N}, \mu_0)$:
$(P_1) \Rightarrow (P_3)$.

Proof. (a) If (P_1) is true, then a deadlock prevention strategy is to first allow only a firing sequence that leads from μ to μ_1, and then only the infinite firing sequence $\sigma, \sigma, \sigma, \ldots$. On the other hand, if deadlock can be prevented, there is an infinite firing sequence enabled by the initial marking. Then, by Lemma 6.7, it follows that (P_1) is true.

(b) This is a particular case of (c) for $\mathcal{T} = T$.

(c) The first part of the proof is similar to (a). If $\mathcal{T}$-liveness can be enforced, there is an infinite firing sequence σ enabled by the initial marking, and the transitions in $\mathcal{T}$ appear infinitely often in σ. Then, (P_3) follows by Lemma 6.7.

(d) This is a particular case of (e) for $\mathcal{T} = T$.

(e) ($\Rightarrow$) Assume the contrary: $\exists \mu \in \mathcal{R}(\mathcal{N}, \mu_0)$ such that (P_1) is true and (P_3) is not. Note that the least restrictive deadlock prevention supervisor of $(\mathcal{N}, \mu_0)$, Ξ_D, is in $\mathcal{S}$. By part (a), deadlock can be prevented at the marking μ, so $\mu \in \mathcal{R}(\mathcal{N}, \mu_0, \Xi_D)$. However, by part (c), $(\mathcal{N}, \mu)$ cannot be made $\mathcal{T}$-live, so Ξ_D does not enforce $\mathcal{T}$-liveness, which is a contradiction.

($\Leftarrow$) Since $\mathcal{T}$-liveness can be enforced at μ_0, deadlock can be prevented at μ_0, so $\mathcal{S}$ is nonempty. Let $\Xi \in \mathcal{S}$. The proof checks that for all $\mu \in \mathcal{R}(\mathcal{N}, \mu_0, \Xi)$ there is a firing sequence enabled by μ, accepted by Ξ, and which includes all transitions in $\mathcal{T}$. Let $\mu \in \mathcal{R}(\mathcal{N}, \mu_0, \Xi)$. Since deadlock is prevented, (P_3) is true as (P_1) is true. Let Ξ_x be the supervisor that enforces $\mathcal{T}$-liveness in $(\mathcal{N}, \mu_0)$ by firing $\sigma_1 \sigma_2 \sigma \sigma \sigma \ldots$, where $\mu_0 \xrightarrow{\sigma_1} \mu \xrightarrow{\sigma_2} \mu_1$, and σ, μ, and μ_1 are the variables from (P_3). Since Ξ is at least as permissive as Ξ_L, Ξ is at least as permissive as Ξ_x. Thus Ξ allows $\sigma_2 \sigma$ to fire from μ. Therefore all transitions of $\mathcal{T}$ appear in some firing sequence enabled by μ and allowed by Ξ. $\square$

In practice it may be difficult to check $(P_1) \Rightarrow (P_2)$ or $(P_1) \Rightarrow (P_3)$ in order to see whether a deadlock prevention supervisor will also enforce liveness or $\mathcal{T}$-liveness. In contrast, the conditions of the next theorem can easily be verified using linear programming.

Theorem 6.9. *Let $\mathcal{N} = (P, T, F, W)$ be a Petri net, D its incidence matrix, $\mathcal{T} \subseteq T$, $n = |T|$ the number of transitions, $M = \{x \in \mathbb{N}^n : x \neq 0, Dx \geq 0\}$, $N = \{x \in M : \forall i = 1 \ldots n : x(i) \neq 0\}$, and $P = \{x \in M : \forall t_i \in \mathcal{T} : x(i) \neq 0\}$.*

(a) *The following statements are equivalent:*
  (i) *$M \neq \emptyset$ and $M = N$.*
  (ii) *Supervisors that prevent deadlock exist for some initial marking, and for all such initial markings μ_0 all supervisors preventing deadlock in $(\mathcal{N}, \mu_0)$ also enforce liveness in $(\mathcal{N}, \mu_0)$.*
(b) *The following statements are equivalent:*
  (i) *$M \neq \emptyset$ and $M = P$.*

(ii) *Supervisors that prevent deadlock exist for some initial marking, and for all such initial markings μ_0 all supervisors preventing deadlock in $(\mathcal{N}, \mu_0)$ also enforce $\mathcal{T}$-liveness in $(\mathcal{N}, \mu_0)$.*

(c) *The following statements are equivalent:*
   (i) $N \neq \emptyset$ *and* $N = P$.
   (ii) *Supervisors which enforce $\mathcal{T}$-liveness exist for some initial marking, and for all such initial markings μ_0 all supervisors enforcing $\mathcal{T}$-liveness in $(\mathcal{N}, \mu_0)$ also enforce liveness in $(\mathcal{N}, \mu_0)$.*

Proof. (a) This is a particular case of (b) for $\mathcal{T} = T$.

(b) [(i)$\Rightarrow$(ii)] Since $M \neq \emptyset$, a marking μ_0 for which a deadlock prevention supervisor exists can be found as in equation (6.1). Let μ_0 be an initial marking for which deadlock prevention supervisors exist and Ξ a deadlock prevention supervisor of $(\mathcal{N}, \mu_0)$. We show that there is no reachable marking such that a transition in $\mathcal{T}$ is dead. Let $\mu \in \mathcal{R}(\mathcal{N}, \mu_0, \Xi)$. Since Ξ prevents deadlock, there is an infinite firing sequence σ enabled by μ and allowed by Ξ. Using Lemma 6.7 for $\mu_I = \mu$, we see that while firing σ a marking μ_1^* is reached such that μ_1^* enables σ_x corresponding to $x \in M$. But $M = P$, so all transitions in $\mathcal{T}$ appear in σ_x. Therefore no transition in $\mathcal{T}$ is dead at μ, so Ξ also enforces $\mathcal{T}$-liveness.

[(ii)$\Leftarrow$(i)] Assume the contrary. Then there is a nonnegative integer vector x, $x \neq 0$, such that $Dx \geq 0$ and $x(i) = 0$ for some $t_i \in \mathcal{T}$. Let Ξ be a deadlock prevention supervisor for $(\mathcal{N}, \mu_0)$, where μ_0 is such that it enables a firing sequence σ_x defined as follows: t_i appears in σ_x if and only if $x(i) \neq 0$, in which case it appears $x(i)$ times. If Ξ is defined to only allow firing $\sigma_x \sigma_x \sigma_x \ldots$, then deadlock is prevented but $\mathcal{T}$-liveness is not enforced, as σ_x does not include all transitions of $\mathcal{T}$. Contradiction.

(c) The proof is identical to (b) if we substitute in (b) $\mathcal{T}$ with T, deadlock prevention with $\mathcal{T}$-liveness enforcement, M with P, and P with N. $\square$

Fig. 6.2(a) shows an example for Theorem 6.9(a): all nonnegative vectors x such that $Dx \geq 0$ are a linear combination with nonnegative coefficients of $[1, 2, 1, 1]^T$ and $[2, 3, 3, 3]^T$. Fig. 6.2(b) shows an example for Theorem 6.8(d). Indeed, all markings μ that enable any of t_1, t_2, or t_4 satisfy (P_2). Also, a marking that enables only t_3 either leads to deadlock or enables the sequence t_3, t_4 and hence satisfies (P_2). Note that the deadlock prevention supervisor that repeatedly fires t_2, t_1 does not enforce liveness because it does not satisfy the requirement of Theorem 6.8(d) to be at least as permissive as any liveness-enforcing supervisor.

With regard to Theorem 6.8(d)–(e), note that designing deadlock prevention supervisors at least as permissive as liveness-enforcing supervisors has been demonstrated, for instance, in [90, 91, 93, 120]. A deadlock prevention technique satisfying this property will also be defined in chapter 7.

The following theorem is a technical result necessary in the developments of the next subsection. The theorem shows that given a Petri net structure that is not repetitive, the transitions of the net can be divided into two categories:

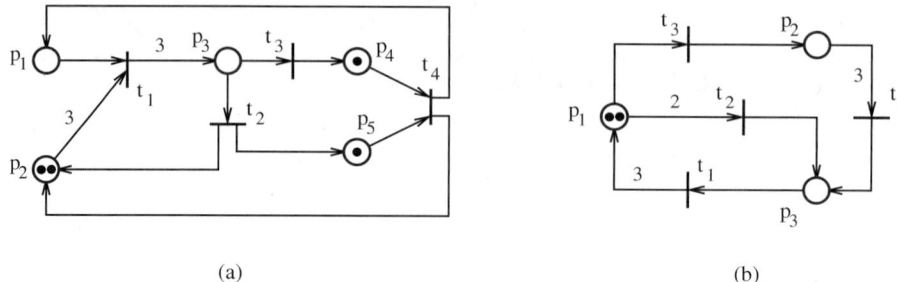

(a)
(b)

Fig. 6.2. Examples for Theorems 6.8 and 6.9.

1. transitions that cannot be made live regardless of the initial marking;
2. transitions that can together be made live for appropriate initial markings.

This means that it is impossible to have that (a) two transitions t_1 and t_2 exist such that t_1 can be made live for some initial marking, t_2 can be made live for some initial marking, but t_1 and t_2 cannot together be made live for any initial marking; or (b) for all transitions t there is an initial marking allowing t to be made live (while not all transitions can be made live at the same time, as the net is nonrepetitive).

Theorem 6.10. *Consider a Petri net $\mathcal{N} = (P, T, F, W)$ which is not repetitive. At least one transition exists such that for any initial marking it cannot fire infinitely often. Let T_D be the set of all such transitions. There are initial markings μ_0 and a supervisor $\varXi$ such that $\forall \mu \in \mathcal{R}(\mathcal{N}, \mu_0, \varXi)$ no transition in $T \setminus T_D$ is dead.*

Proof. Let $\|x\|$ be the *support* of the vector x, that is, $\|x\| = \{i : x(i) \neq 0\}$. There is an integer vector $x \geq 0$ with *maximum support* such that $Dx \geq 0$; that is, for all integer vectors $w \geq 0$ such that $Dw \geq 0$: $\|w\| \subseteq \|x\|$. Indeed if $y, z \geq 0$ are integer vectors and $Dy \geq 0$ and $Dz \geq 0$, then $D(z + y) \geq 0$, $y + z \geq 0$, and $\|y\|, \|z\| \subseteq \|y + z\|$.

If $t_j \in T$ can be made live, there is a marking that enables an infinite firing sequence σ such that t_j appears infinitely often in σ. Therefore, by Lemma 6.7 $\exists y \geq 0$ such that $Dy \geq 0$ and $y(j) > 0$. Since x has maximum support, $\|y\| \subseteq \|x\|$ and so $t_j \in \|x\|$. This proves that all transitions that can be made live are in $\|x\|$. Moreover, only the transitions that can be made live are in $\|x\|$. Indeed, let σ_x be a firing sequence such that (a) t_i appears in σ_x if and only if $x(i) \neq 0$; and (b) t_i appears $x(i)$ times in σ_x if $x(i) \neq 0$. Then, there is a marking μ_0 given by equation (6.1) that enables the infinite firing sequence $\sigma_l = \sigma_x \sigma_x \sigma_x \ldots$. We may choose $\varXi$ to only allow σ_l to fire from μ_0, and we note that all transitions in $\|x\|$ are live. However, $T \not\subseteq \|x\|$, or else σ_l contains all transitions of T and so $\mathcal{N}$ is repetitive. Therefore, $T \setminus \|x\| \neq \emptyset$. Since $\|x\|$ contains the transitions that can be made live, $T \setminus \|x\| \neq \emptyset$ contains the transitions that cannot be made live under any circumstances. So we have $T_D = T \setminus \|x\|$ and $T_D \neq \emptyset$. $\square$

A special case in Theorem 6.10 is $T \setminus T_D = \emptyset$, when the Petri net is not even partially repetitive, and so deadlock cannot be avoided for any marking. It was already shown that only repetitive Petri nets can be made live (Proposition 6.6). Theorem 6.10 shows that the set of transitions of a partially repetitive Petri net can be uniquely divided in transitions that can be made live and transitions that cannot be made live. So the liveness property of partially repetitive Petri nets is that all transitions that can be made live are live ($\{T \setminus T_D\}$-liveness). For an example, consider the Petri nets of Fig. 6.6(a) and (b). For the first one $T_D = \{t_4, t_5\}$, and for the second one $T_D = \{t_1, t_2, t_3\}$.

6.3.2 Deadlock and ($\mathcal{T}$-)Liveness Characterization Based on Active Subnets

We denote by the *active subnet* a part of a Petri net that can be made live for appropriate markings by supervision. In the following definition we use the notation from Theorem 6.10.

Definition 6.11. *Let $\mathcal{N} = (P, T, F, W)$ be a Petri net, D the incidence matrix, and $T_D \subseteq T$ the set of all transitions that cannot be made live for any initial marking. $\mathcal{N}^A = (P^A, T^A, F^A, W^A)$ is an **active subnet** of $\mathcal{N}$ if $P^A = T^A\bullet$, $F^A = F \cap \{(T^A \times P^A) \cup (P^A \times T^A)\}$, W^A is the restriction of W to F^A, and T^A is the set of transitions with nonzero entry in some non-negative vector $x \neq 0$ satisfying $Dx \geq 0$. $\mathcal{N}^A$ is the **maximal active subnet** of $\mathcal{N}$ if $T^A = T \setminus T_D$ and $T \setminus T_D \neq \emptyset$. $\mathcal{N}^A$ is a **minimal active subnet** if there is no other active subnet $\mathcal{N}_1^A = (P_1^A, T_1^A, F_1^A, W_1^A)$ such that $T_1^A \subseteq T^A$.*

Definition 6.12. *Given an active subnet $\mathcal{N}^A$ of a Petri net $\mathcal{N}$, a siphon of $\mathcal{N}$ is said to be an **active siphon** (with respect to $\mathcal{N}^A$) if it is or includes a siphon of $\mathcal{N}^A$. An active siphon is **minimal** if it does not include another active siphon (with respect to the same active subnet).*

In Fig. 6.3(a) and (c) two Petri nets are given. Fig. 6.3(b) shows the minimal active subnets of the Petri net in Fig. 6.3(a). The union of the two subnets is the maximal active subnet. Fig. 6.3(d) shows the only active subnet of the Petri net of Fig. 6.3(c). The minimal active siphons of the Petri net in Fig. 6.3(a) with respect to the active subnet having $T^A = \{t_6, t_7, t_9\}$ are $\{p_1, p_5, p_6, p_7\}$ and $\{p_6, p_7, p_8\}$. The minimal active siphons of the Petri net of Fig. 6.3(c) are $\{p_1, p_4, p_7\}$, $\{p_2, p_5, p_7\}$, $\{p_3, p_5, p_7\}$, and $\{p_6, p_7\}$.

Proposition 6.13. *A siphon which contains places from an active subnet is an active siphon with respect to that subnet.*

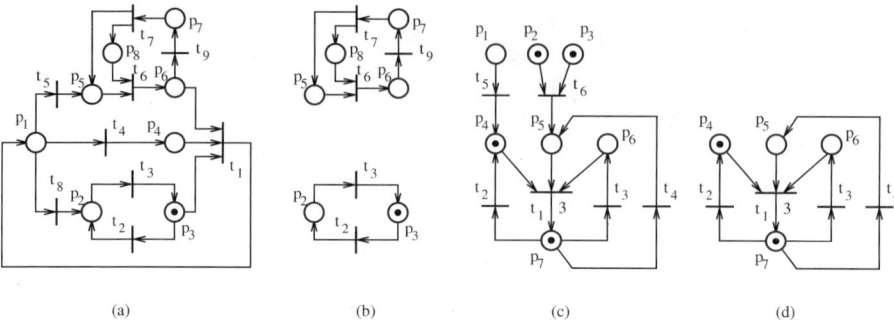

(a)   (b)   (c)   (d)

Fig. 6.3. Two Petri nets, (a) and (c), and their active subnets, (b) and (d), respectively.

Proof. We use the notation from Definition 6.11. Let σ_x be a firing sequence such that a transition t_i appears in σ_x if and only if $x(i) \neq 0$, in which case it appears $x(i)$ times. Let S be a siphon such that $S \cap P^A \neq \emptyset$. We are to prove that there is a siphon s of $\mathcal{N}^A$ such that $s \subseteq S$. $\bullet S \subseteq S \bullet$ implies that $\bullet S \cap T^A \subseteq S \bullet \cap T^A$. Using the construction of equation (6.1) there is a marking enabling $\sigma_x \sigma_x \sigma_x \ldots$. Since $T^A = \|x\|$ and $P^A = T^A \bullet$ (by definition), we have that $\forall t \in T^A$: $\bullet t \subseteq P^A$. Hence $S \bullet \cap T^A \subseteq (S \cap P^A) \bullet$ and so $S \bullet \cap T^A = (S \cap P^A) \bullet \cap T^A$. Note also that $\bullet (S \cap P^A) \cap T^A \subseteq \bullet S \cap T^A$. Therefore $\bullet S \cap T^A \subseteq S \bullet \cap T^A$ implies $\bullet (S \cap P^A) \cap T^A \subseteq (S \cap P^A) \bullet \cap T^A$, which proves that $s = S \cap P^A$ is a siphon of $\mathcal{N}^A$. $\qquad\square$

The significance of the active subnets for deadlock prevention can be seen in the following propositions. First we prove a technical result.

Lemma 6.14. *Let $\mathcal{N}^A = (P^A, T^A, F^A, W^A)$ be an active subnet of $\mathcal{N}$. Given a marking μ of $\mathcal{N}$ and μ^A its restriction to $\mathcal{N}^A$, if $t \in T^A$ is enabled in $\mathcal{N}^A$, then t is enabled in $\mathcal{N}$.*

Proof. By definition, there is a nonnegative integer vector $x \geq 0$ such that $Dx \geq 0$ (D is the incidence matrix) and $x(i) > 0$ for $t_i \in T^A$ and $x(i) = 0$ for $t_i \in T \setminus T^A$. This implies that there are markings such that the transitions of T^A can fire infinitely often, without firing other transitions (see equation (6.1)). The proof is by contradiction. Assume that t is not enabled in $\mathcal{N}$. As t is enabled in $\mathcal{N}^A$, $\exists p \in \bullet t$: $p \notin P^A$. (The preset/postset operators $\bullet$ are taken with respect to $\mathcal{N}$, not $\mathcal{N}^A$.) Note that $p \notin P^A$ implies $\bullet p \cap T^A = \emptyset$. If $\bullet p = \emptyset$, t cannot fire infinitely often, which contradicts the definition of T^A (Definition 6.11), since $t \in T^A$. If $\bullet p \neq \emptyset$, the transitions of T^A cannot fire infinitely often without firing one or more of the transitions $t_x \in \bullet p$ (infinitely often), which again contradicts the definition of T^A. Therefore t is also enabled in $\mathcal{N}$. $\qquad\square$

Note that in a repetitive Petri net all siphons are active with respect to the maximal active subnet. The next result is a generalization of the well-known Proposition 6.2. It is a more powerful result since it states not only that deadlock implies an empty siphon but also that for any active subnet $\mathcal{N}^A$ there is an empty active siphon with respect to $\mathcal{N}^A$.

Proposition 6.15. *Let $\mathcal{N}^A$ be an arbitrary active subnet of a PT-ordinary Petri net $\mathcal{N}$. If μ is a deadlock marking of $\mathcal{N}$, then there is at least one empty minimal active siphon with respect to $\mathcal{N}^A$.*

Proof. Since μ is a deadlock marking and $\mathcal{N} = (P, T, F, W)$ is PT-ordinary, $\forall t \in T \ \exists p \in \bullet t: \mu(p) = 0$. The active subnet is built in such a way that if the marking μ restricted to the active subnet enables a transition t, then μ enables t in the total net (Lemma 6.14). Therefore, because the total net $(\mathcal{N}, \mu)$ is in deadlock, the active subnet is too. In view of Proposition 6.2, let s be an empty minimal siphon of the active subnet. Consider s in the total net. If s is a siphon of the total net, then s is also a minimal active siphon; therefore the net has a minimal active siphon which is empty. If s is not a siphon of the total net $\bullet s \setminus T^A \neq \emptyset$. Let S be the set inductively constructed as follows: $S_0 = s$, $S_i = S_{i-1} \cup \{p \in \bullet(\bullet S_{i-1} \setminus S_{i-1}\bullet) : \mu(p) = 0\}$, where μ is the (deadlock) marking of the net. In other words S is a completion of s with places with null marking such that S is a siphon. By construction S is an active siphon and is empty for the marking μ. Hence an empty minimal active siphon exists. $\qquad\square$

The practical significance of Proposition 6.15 is that it can be used for deadlock prevention, since deadlock is not possible when all active siphons with respect to an active subnet cannot become empty.

Proposition 6.16. *Deadlock is unavoidable for the marking μ if for all minimal active subnets $\mathcal{N}^A$ there is an empty active siphon with respect to $\mathcal{N}^A$.*

Proof. All transitions in the postset of an empty siphon are dead. Therefore every minimal active subnet has some dead transitions. The proof is by contradiction. Assume that deadlock is avoidable. Then, in view of Lemma 6.7, after some transitions firings a marking can be reached which enables $\sigma\sigma\sigma\ldots\sigma\ldots$, where σ is a finite firing sequence. Let v be the firing count vector for σ. Then $Dv \geq 0$. If the active subnet for q is minimal, we let $x = q$, but if it is not, there is x such that $\|x\| \subset \|v\|$, $x \neq 0$, $x \geq 0$, $Dx \geq 0$, and the active subnet associated with x is minimal. But there must be an empty active siphon with regard to that active subnet, therefore not all of the transitions of $\|x\|$ can fire, which implies that not all of the transitions of σ can fire, which is a contradiction. $\qquad\square$

Propositions 6.15 and 6.16 generalize Proposition 6.2. Thus a Petri net will certainly enter deadlock if for all minimal active subnets $\mathcal{N}^A$ there is an empty active siphon with respect to $\mathcal{N}^A$. Conversely, a deadlock state implies that for each active subnet there is an empty active siphon with regard to that subnet. Propositions 6.15 and 6.16 suggest an approach for least restrictive deadlock prevention, and we consider it in section 6.4.2.

Next, the relation between $\mathcal{T}$-liveness and active subnets is considered. An ordinary Petri net $\mathcal{N} = (P, T, F, W)$ is a free-choice net if for all $p_i, p_j \in P$, $p_i \bullet \cap p_j \bullet \neq \emptyset \Rightarrow |p_i \bullet| = |p_j \bullet| = 1$. A result known as Commoner's theorem states that a free-choice net is live if and only if all its siphons contain some marked trap. While Commoner's theorem is a necessary and sufficient condition, its extension to asymmetric choice Petri nets is usually presented as a sufficient condition (e.g., Theorem 10.4 in [42]). The reason for this is that attention has been restricted to trap-controlled siphons, which is only a particular class of controlled siphons. In terms of the general notion of controlled siphons, the extension of Commoner's theorem is a necessary and sufficient condition (see Corollary 27 in [15]). The next result further extends Commoner's theorem as follows: the result not only states that a dead transition t implies an empty siphon for some reachable marking, but also that there is such an empty siphon S with $t \in S\bullet$. This fact is important when we try to verify or ensure that t is live, since we only have to look at the siphons S such that $t \in S\bullet$.

Theorem 6.17. *Consider a PT-ordinary asymmetric choice Petri net $\mathcal{N}$ and a marking μ such that a transition t is dead. Then there is $\mu' \in \mathcal{R}(\mathcal{N}, \mu)$ such that S is an empty siphon for the marking μ' and $t \in S\bullet$.*

Proof. It is known that if a transition t of an ordinary Petri net with asymmetric choice is dead at a marking μ, then $\exists \mu_1 \in \mathcal{R}(\mathcal{N}, \mu)$ and $\exists p_1 \in \bullet t$ $\forall \mu_x \in \mathcal{R}(\mathcal{N}, \mu_1): \mu_x(p_1) = 0$. This is proved, for instance, in Lemma 10.2 of [42], and the proof applies without change to PT-ordinary asymmetric choice Petri nets. We inductively use this property to construct S. Note that all transitions in $\bullet p_1$ are dead at μ_1. Let $S_0 = \emptyset$ and $S_1 = \{p_1\}$. We inductively construct S by generating $S_2, \ldots, S_{n+1}$ and the markings $\mu_2, \ldots, \mu_{n+1}$. S_i for $i \geq 1$ is such that all transitions in $\bullet S_i$ are dead for some marking μ_i. The construction in an iteration is as follows. Let $T_i = \bullet(S_i \setminus S_{i-1})$ and $\mu_{i+1} \in \mathcal{R}(\mathcal{N}, \mu_i)$ such that $\forall t_x \in T_i$ and $\forall \mu_x \in \mathcal{R}(\mathcal{N}, \mu_{i+1})$ $\exists p \in \bullet t_x$: $\mu_x(p) = 0$. Then we let $G_i = \bigcup_{t_x \in T_i} \{p \in \bullet t_x : \forall \mu_x \in \mathcal{R}(\mathcal{N}, \mu_{i+1}) : \mu_x(p) = 0\}$ and $S_{i+1} = S_i \cup G_i$. There is an n such that $S_{n+1} = S_n$, for the Petri net has a finite number of places. We let $S = S_n$ and $\mu' = \mu_n$. By construction S is a siphon (note that $\bullet S_i \subseteq S_{i+1}\bullet$ for $i = 0 \ldots n$), S is empty at μ', $\mu' \in \mathcal{R}(\mathcal{N}, \mu)$, and $t \in S\bullet$ (since $p_1 \in S$). $\qquad\square$

The next definition introduces a new class of active subnets that are very convenient in characterizing $\mathcal{T}$-liveness.

Definition 6.18. *Let $\mathcal{N}$ be a Petri net, $\mathcal{T}$ a nonempty subset of the set of transitions, and $\mathcal{N}^A = (P^A, T^A, F^A, W^A)$ an active subnet. We say that $\mathcal{N}^A$ is $\mathcal{T}$-minimal if $\mathcal{T} \subseteq T^A$ and $T_x^A \not\subseteq T^A$ for any other active subnet $\mathcal{N}_x^A = (P_x^A, T_x^A, F_x^A, W_x^A)$ such that $\mathcal{T} \subseteq T_x^A$.*

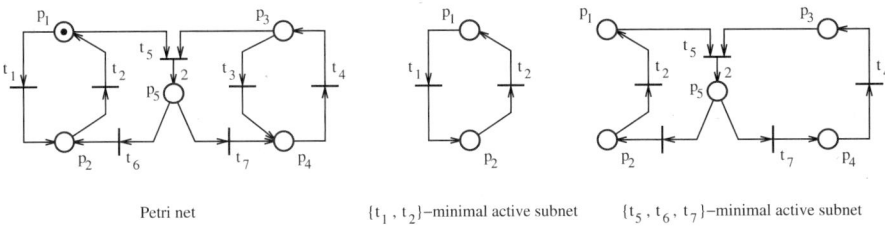

Fig. 6.4. Examples of $\mathcal{T}$-minimal active subnets.

For examples of $\mathcal{T}$-minimal active subnets, consider the Petri net shown in Fig. 6.4. $\mathcal{N}_1^A$ with $P_1^A = \{p_1, p_2\}$ and $T_1^A = \{t_1, t_2\}$ is the $\{t_1, t_2\}$-minimal active subnet. $\mathcal{N}_2^A$ with $P_2^A = \{p_1, \ldots, p_5\}$ and $T_2^A = \{t_2, t_4, t_5, t_6, t_7\}$ is the $\{t_5, t_6, t_7\}$-minimal active subnet. Note that a $\mathcal{T}$-minimal active subnet may not be unique. Indeed, both $\mathcal{N}_1^A$ and $\mathcal{N}_2^A$ are $\{t_2\}$-minimal active subnets. Nonetheless, an active subnet $\mathcal{N}^A = (P^A, T^A, F^A, W^A)$ is always the only T^A-minimal active subnet.

The next theorem gives necessary and sufficient conditions for $\mathcal{T}$-liveness in terms of the $\mathcal{T}$-minimal active subnets.

Theorem 6.19. *Given a PT-ordinary asymmetric choice Petri net $\mathcal{N}$, let $\mathcal{T}$ be a set of transitions and $\mathcal{N}^A$ a $\mathcal{T}$-minimal active subnet which contains the transitions in $\mathcal{T}$. If all the minimal siphons with respect to $\mathcal{N}^A$ are controlled, the Petri net is $\mathcal{T}$-live (and T^A-live). If the Petri net is $\mathcal{T}$-live, there is no reachable marking such that for each $\mathcal{T}$-minimal active subnet $\mathcal{N}^A$ there is an empty minimal active siphon with respect to $\mathcal{N}^A$.*

Proof. For the first part, the proof is by contradiction. Assume that there is a reachable marking such that a transition $t \in T^A$ is dead. Since $\mathcal{T} \subseteq T^A$, by Theorem 6.17 there is a reachable marking such that a siphon S is empty and $t \in S\bullet$. However, $t \in S\bullet$ implies $S \cap P^A \neq \emptyset$, and by Proposition 6.13 S is an active siphon. Finally, S empty contradicts the fact that all active siphons are controlled.

For the second part, let $\mathcal{N}_i^A$ denote a $\mathcal{T}$-minimal active subnet, $i = 1 \ldots k$, where k is the number of $\mathcal{T}$-minimal active subnets. First, note that if there is an empty active siphon S_i with respect to $\mathcal{N}_i^A$, there is a nonempty set of dead transitions in $\mathcal{N}_i^A$. Indeed, let μ be a marking such that S_i is empty. Let $T_i = S_i \bullet \cap T_i^A$, where T_i^A is the set of transitions of $\mathcal{N}_i^A$. Since S_i is active, T_i is nonempty; because S_i is empty, the transitions of T_i are dead. Next, we prove the second part of the theorem by contradiction. Assume that there is an infinite firing sequence σ_x such that all transitions of $\mathcal{T}$ appear infinitely often in σ_x, and after a part of σ_x is fired (let μ_x be the marking reached) all $\mathcal{T}$-minimal active subnets $\mathcal{N}_i^A$ have an empty active siphon S_i. Let σ be the remaining part of σ_x which is enabled by μ. All transitions of $\mathcal{T}$ appear infinitely often in σ. Therefore, by Lemma 6.7, there is an $x \geq 0$ such that $Dx \geq 0$ (D is the incidence matrix) and $\mathcal{T} \subseteq \|x\|$. However, $\|x\|$ does not contain all transitions of any of the $\mathcal{T}$-minimal subnets $\mathcal{N}_i^A$: $T_i \subseteq T_i^A \setminus \|x\|$, for $i = 1 \dots k$. This implies that $\|x\|$ defines another $\mathcal{T}$-minimal active subnet, which contradicts the fact that $\mathcal{N}_i^A$, $i = 1 \dots k$, are all the $\mathcal{T}$-minimal active subnets. $\qquad\square$

In the particular case in which there is a single $\mathcal{T}$-minimal active subnet, Theorem 6.19 shows that the net is $\mathcal{T}$-live if and only if all siphons are controlled. When $\mathcal{T}$ equals the total set of transitions of the net and the Petri net is repetitive, the $\mathcal{T}$-minimal active subnet exists, is unique, and equals the total net; in this case we obtain the extension of the Commoner's theorem to asymmetric choice nets.

A natural question is whether we can extend the results from asymmetric choice Petri nets to more general Petri nets. Theorem 6.19 relies on Theorem 6.17, while the proof of Theorem 6.17 only requires Petri nets $\mathcal{N}$ with the property that if a transition t is dead at a marking μ, then $\exists \mu_1 \in \mathcal{R}(\mathcal{N}, \mu)$ and $\exists p_1 \in \bullet t \; \forall \mu_x \in \mathcal{R}(\mathcal{N}, \mu_1)$: $\mu_x(p_1) = 0$. Verifying this property may be hard in general. A class of Petri nets more general than asymmetric choice nets satisfying this property is given in the following definition.

Definition 6.20. *Let $\mathcal{N} = (P, T, F, W)$ be a Petri net. Given an arbitrary $t \in T$, consider the following notation. Let T_D be the set of transitions which cannot be made live when t is dead (where $t \in T_D$), and $P_L = \{p \in \bullet t : \exists t' \in (\bullet t) \bullet \setminus T_D$ and $p \in \bullet t'\}$. Furthermore, let $\mathcal{N}'$ be the Petri net obtained by removing from $\mathcal{N}$ the transitions in T_D. Then $\mathcal{N}$ is said to be an* **EAC net** *if $T = \emptyset$, or for all transitions $t \in T$ one of the following conditions is satisfied:*

(a) $\forall p_i, p_j \in \bullet t$: $p_i \bullet \subseteq p_j \bullet$ *or* $p_j \bullet \subseteq p_i \bullet$.
(b) $P_L = \emptyset$, *or in $\mathcal{N}'$ it is true that* $\forall p_i, p_j \in P_L$: $p_i \bullet \subseteq p_j \bullet$ *or* $p_j \bullet \subseteq p_i \bullet$.
(c) *There is a transition t' in $\mathcal{N}'$ such that $P_L = \bullet t'$ and $\mathcal{N}'$ is an EAC net.*

In the previous definition, *EAC net* stands for *extended asymmetric choice net*, since all asymmetric choice nets are *EAC* nets. Note that a transition t' is in T_D if for all markings for which t is dead, t' is dead or will certainly die. In order to show that Theorems 6.17 and 6.19 also apply to *EAC* nets, we only need to prove the following result.

Proposition 6.21. *Let $\mathcal{N}$ be an EAC net and t a transition dead at the marking μ. Then $\exists \mu_1 \in \mathcal{R}(\mathcal{N}, \mu)$ and $\exists p_1 \in \bullet t \; \forall \mu_x \in \mathcal{R}(\mathcal{N}, \mu_1)$: $\mu_x(p_1) = 0$.*

Proof. We consider the markings reachable from $\mu' \in \mathcal{R}(\mathcal{N}, \mu)$ such that at μ' all transitions in T_D are dead. If t satisfies (a) in Definition 6.20, then the proof is the same as that in Lemma 10.2 of [42].

If t satisfies (b), first consider the situation in which $P_L = \emptyset$. Then the marking of the places in $\bullet t$ can never decrease, and since t is dead, there is a place $p \in \bullet t$ such that for all reachable markings μ_x we have that $\mu_x(p) = 0$. Now consider $P_L \neq \emptyset$. To prove the proposition in this case it is enough to show that if there is no reachable marking μ_x such that $\exists p \in \bullet t \setminus P_L, \forall \mu_y \in \mathcal{R}(\mathcal{N}, \mu_x): \mu_y(p) = 0$, then there is a reachable marking μ_1 such that $\exists p \in P_L, \forall \mu_y \in \mathcal{R}(\mathcal{N}, \mu_1): \mu_y(p) = 0$. The marking of the places $p \in \bullet t \setminus P_L$ cannot decrease; therefore, if there is no reachable marking μ_x such that $\exists p \in \bullet t \setminus P_L, \forall \mu_y \in \mathcal{R}(\mathcal{N}, \mu_x): \mu_y(p) = 0$, there is $\mu_z \in \mathcal{R}(\mathcal{N}, \mu')$ such that $\forall p \in \bullet t \setminus P_L: \mu_z(p) \geq 1$. Furthermore, this implies $\forall p \in \bullet t \setminus P_L, \forall \mu \in \mathcal{R}(\mathcal{N}, \mu_z): \mu(p) \geq 1$. Then, note that the case when there is no marking $\mu_1 \in \mathcal{R}(\mathcal{N}, \mu_z)$ such that for some $p \in P_L \; \forall \mu_x \in \mathcal{R}(\mathcal{N}, \mu_1): \mu_x(p) = 0$ is impossible. Indeed, if it would be possible, it would imply $\exists \mu_2 \in \mathcal{R}(\mathcal{N}, \mu_z)$ such that $\mu_2(p) \geq 1 \; \forall p \in P_L$ (see the proof of Lemma 10.2 in [42]). However, since $\mu_2 \in \mathcal{R}(\mathcal{N}, \mu_z)$, t is enabled, which contradicts that t is dead. This completes the proof for case (b).

Now assume that t satisfies (c). Then $\mathcal{N}'$ is also an EAC net. Using a similar reasoning as in (b), if there is no reachable marking μ_x such that $\exists p \in \bullet t \setminus P_L, \forall \mu_y \in \mathcal{R}(\mathcal{N}, \mu_x): \mu_y(p) = 0$, then there is a $\mu_z \in \mathcal{R}(\mathcal{N}, \mu')$ such that $\forall p \in \bullet t \setminus P_L: \mu_z(p) \geq 1$. Then t' is dead (or else t would not be dead), and so the problem is reduced to proving the proposition for the smaller net $(\mathcal{N}', \mu_z)$ and the dead transition t'. Thus, we eventually reach the conclusion by backtracking. $\qquad \square$

Corollary 6.22. *Theorems 6.17 and 6.19 also apply to EAC nets.*

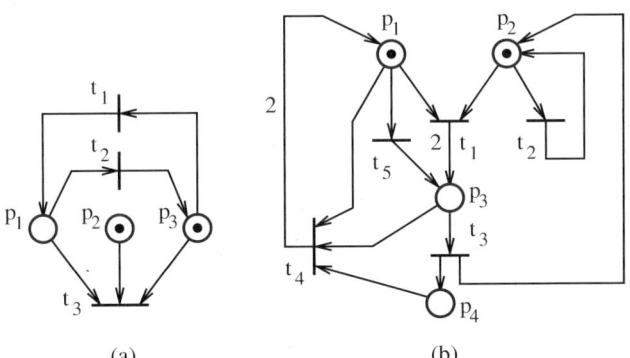

(a)   (b)

Fig. 6.5. (a) A Petri net which is not an EAC net; (b) an EAC net.

As an example, consider the Petri nets of Fig. 6.5. The Petri net in Fig. 6.5(a) is not an EAC net, as the transition t_3 does not satisfy any of the cases (a), (b), and (c) of Definition 6.20. To see this, note that for t_3 we have that $T_D = \emptyset$ and $P_L = \{p_1, p_3\}$. Case (a) is not satisfied, as $p_1, p_3 \in \bullet t_3$,

$p_1\bullet = \{t_2, t_3\}$, and $p_3\bullet = \{t_1, t_3\}$. As $P_L \neq \emptyset$, case (b) is not satisfied for the same reason as (a). Case (c) is not satisfied as there is no transition t' satisfying $\bullet t' = P_L$. The Petri net of Fig. 6.5(b) is an EAC net, as can be seen in what follows. The transitions t_2, t_3, and t_5 satisfy case (a) in Definition 6.20. The transition t_1 satisfies case (b) in Definition 6.20. Indeed, for all markings for which t_1 is dead, the transitions t_3, t_4, and t_5 are dead or eventually die. So $T_D = \{t_1, t_3, t_4, t_5\}$. The set P_L is $P_L = \{p_2\}$. Then we can see that t_1 satisfies case (b). Note that transition t_1 also satisfies case (c). In the case of t_4 we have the same T_D and $P_L = \emptyset$. So t_4 satisfies case (b). Note that the net is not an asymmetric choice net.

6.4 Implications and Discussion

In this section we discuss our results and show how they relate to the supervisory problems of deadlock prevention, liveness enforcement, and $\mathcal{T}$-liveness enforcement. Some of the theoretical results of this chapter consider only particular classes of Petri nets, specifically PT-ordinary and asymmetric choice nets. However, for our supervisory problems this is a surmountable difficulty, since it is possible to transform a Petri net to a PT-ordinary or PT-ordinary asymmetric choice Petri net; then, it is possible to derive a deadlock prevention (or a ($\mathcal{T}$-)liveness enforcement) supervisor from a supervisor for deadlock prevention (or ($\mathcal{T}$-)liveness enforcement) of the transformed net [95, 92, 172]. Briefly, a possible solution for our supervisory problems is as follows: given a target net $\mathcal{N}_0$, generate a sequence of increasingly enhanced nets $\mathcal{N}_1, \mathcal{N}_2 \ldots$, until we reach a net $\mathcal{N}_k$, such that we can use Proposition 6.15 or Theorem 6.19 on $\mathcal{N}_k$ to guarantee deadlock-freedom or ($\mathcal{T}$-)liveness; then a supervisor for $\mathcal{N}_0$ is derived based on the construction of $\mathcal{N}_k$. Such an approach has been used in [91, 92, 93, 94, 79, 95], and is presented in chapter 7.

6.4.1 Deadlock Prevention

Proposition 6.2 implies that if the marking of any of the minimal siphons of a PT-ordinary Petri net can never become zero, the Petri net is deadlock-free. This is a useful property for repetitive Petri nets, but not always for nonrepetitive Petri nets. For partially repetitive Petri nets Proposition 6.15 is much more useful. For instance, consider the Petri net of Fig. 6.6(a). The only active subnet has $T^A = \{t_1, t_2, t_3\}$. After firing t_4, $\{p_4\}$ is an empty siphon. However, there is no empty active siphon (the minimal active siphons are $\{p_1, p_3, p_4\}$, $\{p_2, p_3, p_5\}$, and $\{p_2, p_3, p_6\}$), and thus we can see from Proposition 6.15 that the Petri net is not in deadlock, while this cannot be ascertained from Proposition 6.2. The same is true for the Petri net in Fig. 6.6(b): $\{p_1, p_3\}$ is an empty siphon, but the only minimal active siphon, $\{p_4, p_5, p_6, p_7\}$, is not empty, and therefore the Petri net is not in deadlock by Proposition 6.15.

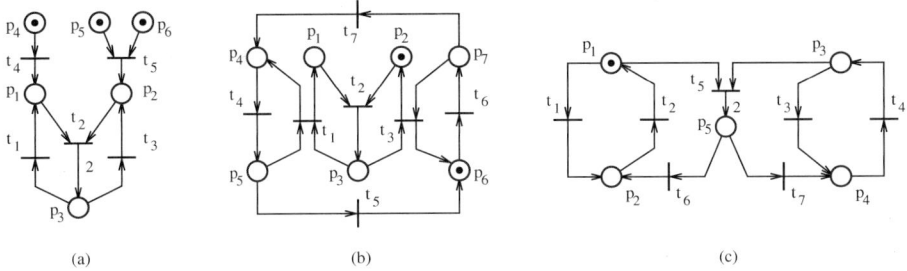

Fig. 6.6. Deadlock illustrations.

Proposition 6.15 is more useful than Proposition 6.2 even for repetitive Petri nets, as seen in Fig. 6.6(c). The Petri net of Fig. 6.6(c) has several active subnets. While with respect to some of them there are empty active siphons, if we take the active subnet $\mathcal{N}^A$ defined by $T^A = \{t_1, t_2\}$, the only minimal active siphon with respect to $\mathcal{N}^A$ is $\{p_1, p_2, p_5\}$, which is not empty. Thus we are able to detect based on Proposition 6.15 that the Petri net is not in deadlock.

In the applications in which deadlock prevention is desired to approximate liveness enforcement, Proposition 6.15 can be used for the maximal active subnet. Thus it would be desirable that no active siphon with respect to the maximal active subnet ever become empty. Indeed, if an active siphon S with respect to the maximal active subnet is empty, all transitions in $S\bullet$ are dead; this would be undesirable, as $S\bullet$ contains one or more of the transitions that could be made live.

For the applications in which least restrictive deadlock prevention is desired rather than a liveness approximation, see the next section.

Proposition 6.15 can be used for deadlock prevention by extending the target Petri net to a net in which all siphons are controlled. The usual technique for siphon control involves adding a new place to each siphon to be controlled, such that place invariants are created. Such additional places can be seen as implementing a (marking-based) supervisor for deadlock prevention. We have designed a deadlock prevention methodology based on Proposition 6.15. This methodology is presented in chapter 7 and in [91, 95]. The methodology produces two sets of constraints: $L\mu \geq b$ and $L_0\mu \geq b_0$. The constraints $L\mu \geq b$ define the supervisor (the set of additional places ensuring that all active siphons are invariant controlled). The constraints $L_0\mu \geq b_0$ are used to define the initial markings for which the supervisor operates. Thus, the supervisor is defined for all initial markings μ_0 satisfying $L\mu_0 \geq b$ and $L_0\mu_0 \geq b_0$. For an example, consider the Petri nets in Fig. 6.7(a) and (b). The additional places defining the supervisor are, in both cases, the places C_1, C_2, and C_3. It can easily be checked that all minimal active siphons are invariant controlled in both cases. In the case (a) the inequalities $L\mu \geq b$

are $\mu(p_1) + \mu(p_3) + \mu(p_4) \geq 1$ (so $\mu(C_1) = \mu(p_1) + \mu(p_3) + \mu(p_4) - 1$), $\mu(p_2) + \mu(p_3) + \mu(p_5) \geq 1$ ($\mu(C_2) = \mu(p_2) + \mu(p_3) + \mu(p_5) - 1$), and $\mu(p_2) + \mu(p_3) + \mu(p_6) \geq 1$ ($\mu(C_3) = \mu(p_2) + \mu(p_3) + \mu(p_6) - 1$); $L_0\mu_0 \geq b_0$ contains the inequalities $\mu_0(p_1) + \mu_0(p_2) + \mu_0(p_3) + \mu_0(p_4) + \mu_0(p_5) \geq 2$ and $\mu_0(p_1) + \mu_0(p_2) + \mu_0(p_3) + \mu_0(p_4) + \mu_0(p_6) \geq 2$. In the case (b), the inequalities $L\mu \geq b$ are $\mu(p_1) + \mu(p_2) \geq 1$ ($\mu(C_1) = \mu(p_1) + \mu(p_2) - 1$), $\mu(p_3) + \mu(p_4) \geq 1$ ($\mu(C_2) = \mu(p_3) + \mu(p_4) - 1$), and $\mu(p_1) + \mu(p_2) + \mu(p_3) + \mu(p_4) \geq 3$ ($\mu(C_3) = \mu(C_1) + \mu(C_2) - 1$); there are no constraints $L_0\mu_0 \geq b_0$. Moreover, by Theorem 6.9, the supervisors also enforce $\{t_1, t_2, t_3\}$-liveness in case (a), and liveness in case (b).

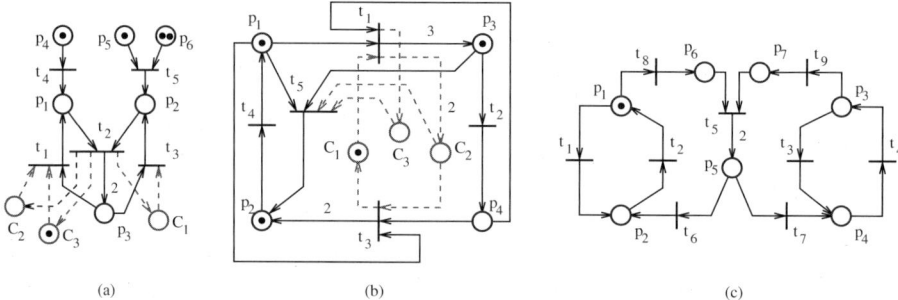

Fig. 6.7. Deadlock prevention examples.

6.4.2 Least Restrictive Deadlock Prevention

Assume that we have u supervisors for deadlock prevention in $\mathcal{N}_0$: Ξ_1, Ξ_2, ..., Ξ_u. Each supervisor can prevent deadlock if the initial marking is in the sets $\mathcal{M}_1$, $\mathcal{M}_2$, ..., $\mathcal{M}_u$, respectively. Let Ξ be the supervisor defined on $\mathcal{M} = \bigcup_{i=1}^{u} \mathcal{M}_i$, which allows a transition to fire only if at least one of the supervisors Ξ_i, defined for the current marking, allows that transition to fire. We denote the supervisor by $\Xi = \bigvee_{i=1}^{u} \Xi_i$. Obviously, Ξ is a deadlock prevention supervisor, and Ξ is at least as permissive as any of Ξ_i.

Theorem 6.23. *Let $\mathcal{N}_0$ be a Petri net and $\mathcal{N}_i^A$, for $i = 1 \ldots u$, the minimal active subnets of $\mathcal{N}_0$. Let T_i denote the set of transitions of $\mathcal{N}_i^A$ and let Ξ_i, for $i = 1 \ldots u$, be deadlock prevention supervisors. Assume that each Ξ_i is defined for all initial markings for which T_i^A-liveness can be enforced and that each Ξ_i is at least as permissive as any T_i^A-liveness-enforcing supervisor. Then $\Xi = \bigvee_{i=1}^{u} \Xi_i$ is the least restrictive deadlock prevention supervisor of $\mathcal{N}_0$.*

Proof. The only thing to be proved is that a marking unacceptable to Ξ leads to deadlock. Consider such a marking μ. Let $x_1, x_2, \ldots, x_u$ be the nonnegative integer vectors defining $\mathcal{N}_1^A, \mathcal{N}_2^A, \ldots, \mathcal{N}_u^A$ in Definition 6.11. Thus $T_i^A = \|x_i\|$ for $i = 1 \ldots u$. Since μ is unacceptable to all of Ξ_i and each Ξ_i is at least as permissive as any T_i^A-liveness-enforcing supervisors, for all $i = 1 \ldots u$ not all transitions of T_i^A can be made live given the marking μ. We prove by contradiction that μ is a marking from which deadlock cannot be avoided. Assume the contrary, that deadlock can be prevented at μ. Then, there is an infinite firing sequence σ enabled by μ. Let T_x be the set of transitions which appear infinitely often in σ. By Lemma 6.7 there is a nonnegative integer vector x such that $T_x = \|x\|$ and $Dx \geq 0$, where D is the incidence matrix. Since $\mathcal{N}_1^A, \mathcal{N}_2^A, \ldots, \mathcal{N}_u^A$ are all the minimal active subnets of $\mathcal{N}_0$, there is a $j \in \{1, 2, \ldots, u\}$ such that $\|x_j\| \subseteq \|x\|$. But this contradicts the fact that not all transitions of $\|x_j\|$ can be made live at μ. $\square$

Each of the supervisors Ξ_i satisfying the requirements of the theorem above can be found with the procedure for deadlock prevention that we present in chapter 7 and [95], by starting it with an initial active subnet $\mathcal{N}_i^A$. As an example, consider the Petri net of Fig. 6.7(c). There are three minimal active subnets $\mathcal{N}_1^A$, $\mathcal{N}_2^A$, and $\mathcal{N}_3^A$, defined by $T_1^A = \{t_1, t_2\}$, $T_2^A = \{t_3, t_4\}$, and $T_3^A = \{t_2, t_4, t_5, t_6, t_7, t_8, t_9\}$, respectively. Three deadlock prevention supervisors corresponding to $\mathcal{N}_1^A$, $\mathcal{N}_2^A$, and $\mathcal{N}_3^A$ are Ξ_1, Ξ_2, and Ξ_3, defined as follows. For simplicity of notation, we let $\mu_i = \mu(p_i)$. Ξ_1 requires $\mu_1 + \mu_2 + \mu_5 + \mu_6 \geq 1 \wedge \mu_1 + \mu_2 + \mu_3 + \mu_4 + \mu_5 + \mu_7 \geq 1$ (the inequalities correspond to the two minimal active siphons with respect to $\mathcal{N}_1^A$); Ξ_2 requires $\mu_3 + \mu_4 + \mu_5 + \mu_7 \geq 1 \wedge \mu_1 + \mu_2 + \mu_3 + \mu_4 + \mu_5 + \mu_6 \geq 1$; Ξ_3 requires $\mu_1 + \mu_2 + \mu_5 + \mu_6 \geq 1 \wedge \mu_3 + \mu_4 + \mu_5 + \mu_7 \geq 1$, and the initial marking μ_0 to satisfy in addition $\sum_{i=1}^{7} \mu_{0,i} \geq 2$. It can be easily seen that $\Xi = \Xi_1 \vee \Xi_2 \vee \Xi_3$ is the least restrictive deadlock prevention supervisor. In this particular case $\Xi_1 \vee \Xi_2 \vee \Xi_3 = \Xi_1 \vee \Xi_2$.

6.4.3 $\mathcal{T}$-liveness Enforcement

Based on Theorem 6.19, it is possible to enforce $\mathcal{T}$-liveness in a Petri net. This will be shown in detail in chapter 7. Our results can also be found in [79, 92].

Consider the Petri net of Fig. 6.8(a), in which it is desired to ensure $\mathcal{T}$-liveness for $\mathcal{T} = \{t_1, t_2, t_3\}$. For the displayed marking, all of t_1, t_2, and t_3 are dead. However, we cannot use Theorem 6.17, as the Petri net is not with asymmetric choice. Fig. 6.8(b) shows the same Petri net transformed to be with asymmetric choice. Theorem 6.17 is verified, as the minimal active siphon $S = \{p_1, p_2, p_3, p_4, p_5, p_6, p_7\}$ (with respect to the active subnet with set of transitions $\mathcal{T}$) is uncontrolled. Indeed, by firing t_4, t_5, and t_{13}, S becomes empty. The Petri net of Fig. 6.8(a) is not $\mathcal{T}$-live for most initial markings. By applying our $\mathcal{T}$-liveness enforcement approach (chapter 7 and [79, 92]), the least restrictive $\mathcal{T}$-liveness supervisor of the Petri net of Fig. 6.8(a) enforces $2\mu_1 + 2\mu_2 + 2\mu_3 + \mu_4 + \mu_5 + \mu_6 + 2\mu_7 \geq 2$.

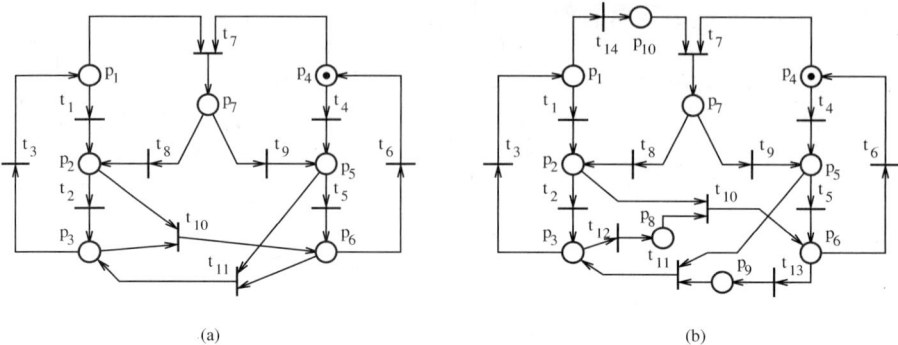

(a)   (b)

Fig. 6.8. Examples for $\mathcal{T}$-liveness enforcement.

6.5 Algorithms

6.5.1 The Computation of Active Subnets

The following algorithm computes the maximal active subnet which does not contain the transitions in a set X.

ALGORITHM 6.1   Computation of the Maximal Active Subnet

Input: *The Petri net $\mathcal{N} = (P, T, F, W)$ and its incidence matrix D; optionally, a set X of transitions to be excluded from the active subnet (default is $X = \emptyset$).*

Output: *The active subnet $\mathcal{N}^A = (P^A, T^A, F^A, W^A)$.*

Let $M = T \setminus X$ and $x_s = \mathbf{0}_{|T| \times 1}$.

While $M \neq \emptyset$ do

1. *Check feasibility of $Dx \geq 0$ subject to $x \geq 0$, $\sum_{t_i \in M} x(i) \geq 1$ and $x(i) = 0 \ \forall t_i \in X$.*

2. **If feasible then** *let x^* be a solution; $M = M \setminus \|x^*\|$ and $x_s = x^* + x_s$.*
   Else $M = \emptyset$.

End while

The active subnet is $\mathcal{N}^A = (P^A, T^A, F^A, W^A)$, $T^A = \|x_s\|$, $P^A = T^A \bullet$, $F^A = F \cap \{(T^A \times P^A) \cup (P^A \times T^A)\}$ and W^A is the restriction of W to F^A.

The next algorithm computes a $\mathcal{T}$-minimal active subnet of a Petri net. Note that this algorithm can also be used to compute the minimal active subnets of the net, since all minimal active subnets are generated by computing the t-minimal active subnets of the Petri net for all transitions t.

ALGORITHM 6.2   Computation of $\mathcal{T}$-Minimal Active Subnets

Input: *The Petri net* $\mathcal{N} = (P, T, F, W)$ *and its incidence matrix* D; *a nonempty set of transitions* $\mathcal{T} \subseteq T$; *optionally, a set* X *of transitions which must not appear in the* $\mathcal{T}$-*minimal active subnet (by default* $X = \emptyset$).

Output: *The active subnet* $\mathcal{N}^A = (P^A, T^A, F^A, W^A)$.

1. *Check the feasibility of* $Dx \geq 0$ *subject to* $x \geq 0$, $x(i) \geq 1 \ \forall t_i \in \mathcal{T}$ *and* $x(i) = 0 \ \forall t_i \in X$.

   If *feasible,* **then** *let* x_0 *be a solution;* $T^A = minactn(T, x_0, D, \mathcal{T})$.

   Else $T^A = maxactn(T, D, \mathcal{T}, X)$ *(no* $\mathcal{T}$-*minimal solution exists, and so an approximation is constructed).*

2. *The active subnet is* $\mathcal{N}^A = (P^A, T^A, F^A, W^A)$, $P^A = T^A \bullet$, $F^A = F \cap \{(T^A \times P^A) \cup (P^A \times T^A)\}$, *and* W^A *is the restriction of* W *to* F^A.

minactn$(T, \ x_0, \ D, \ \mathcal{T})$

   Let $M = \|x_0\|$ *and* $x_s = x_0$.

   For $t_i \in M \setminus \mathcal{T}$ **do**

      Check feasibility of $Dx \geq 0$ *subject to* $x \geq 0$, $x(i) = 0$, $x(j) = 0$ $\forall t_j \in T \setminus M$ *and* $x(j) \geq 1 \ \forall t_j \in \mathcal{T}$.

      If *feasible* **then** *let* x^* *be a solution;* $M = \|x^*\|$ *and* $x_s = x^*$.

   Return $\|x_s\|$.

maxactn$(T, \ D, \ \mathcal{T}, \ X)$

   Let $M = \mathcal{T}$ *and* $x_s = \mathbf{0}_{|T| \times 1}$

   While $M \neq \emptyset$ **do**

      Check feasibility of $Dx \geq 0$ *subject to* $x \geq 0$, $\sum_{t_i \in M} x(i) \geq 1$ *and* $x(i) = 0 \ \forall t_i \in X$.

      If *feasible* **then** *let* x^* *be a solution;* $M = M \setminus \|x^*\|$ *and* $x_s = x^* + x_s$.

      Else $M = \emptyset$.

   $N = minactn(T, x_s, D, \mathcal{T} \cap \|x_s\|)$.

   Return N.

The algorithms presented in this section only involve linear programming. Note that the algorithms can be performed with polynomial complexity.

6.5.2 Transformation of Petri Nets to PT-ordinary Petri Nets

We use a modified form of the similar transformation from [120], and we call it the **PT-transformation**. Let $\mathcal{N} = (P, T, F, W)$ be a Petri net. Transitions $t_j \in T$ such that $W(p, t_j) > 1$ for some $p \in \bullet t_j$ may be **split** (decomposed) in several new transitions according to the following algorithm.

A transition t_j is **split** in $m = n(t_j)$ transitions: $t_{j,0}, t_{j,1}, t_{j,2}, \ldots, t_{j,m-1}$, where $n(t_j) = \max\{W(p, t_j) : (p, t_j) \in F\}$. Also, $m - 1$ new places are added: $p_{j,1}, p_{j,2}, \ldots, p_{j,m-1}$. The connections are as follows:

(i) $\bullet p_{j,i} = t_{j,i}$, $t_{j,i}\bullet = p_{j,i}$ and $p_{j,i}\bullet = t_{j,i-1}$, for $i = 1 \ldots m - 1$.

(ii) $\bullet t_{j,i} = \{p \in \bullet t_j : W(p, t_j) > i\}$, for $i = 0 \ldots m - 1$.

(iii) $t_{j,0}\bullet = t_j\bullet$.

Note that t_j very much resembles $t_{j,0}$: $t_{j,0}$ has all the connections of t_j plus one additional transition arc. *After the split is performed, we denote $t_{j,0}$ by t_j.*

The **PT-transformation** consists in splitting all transitions t such that $W(p, t) > 1$ for some $p \in \bullet t$. In this way the transformed Petri net is PT-ordinary. Note that

$$|p_{j,i}\bullet| = |\bullet p_{j,i}| = 1, \quad i = 1 \ldots m - 1, \tag{6.2}$$

$$|t_{j,i}\bullet| = 1, \quad i = 1 \ldots m - 1. \tag{6.3}$$

We use the convention that a split transition t_j is also a transition of the PT-transformed net, since we denote $t_{j,0}$ by t_j. A transition split example is illustrated in Fig. 6.9(a)–(b).

6.5.3 Transformation of Petri Nets to EAC Nets

This section presents an algorithm that can be used to transform a Petri net to an EAC net or to an asymmetric-choice Petri net. The transformation of a Petri net to an EAC net is called the **EAC-transformation**, and the transformation to an asymmetric-choice Petri net is called the **AC-transformation**.

Let $\mathcal{N} = (P, T, F, W)$ be a Petri net and $\mathcal{N}' = (P', T', F', W')$ be the transformed Petri net, where $P \subseteq P'$, $T \subseteq T'$. The idea of the transformation is as follows. Given the transition t, the conditions (a), (b), and (c) of Definition 6.20 are checked. If none is satisfied, for all p_i and p_j such that $p_i \in \bullet t$, $p_j \in \bullet t$, $p_i\bullet \not\subseteq p_j\bullet$, and $p_j\bullet \not\subseteq p_i\bullet$, remove t from either the postset of p_i or that of p_j by adding an additional place and transition. This insures that part (a) of Definition 6.20 is satisfied for t in the transformed net. The idea is illustrated in Fig. 6.9(c)–(d) for $t = t_2$. Note that the transformation of the original net consists of performing a modified form of transition split operations (where transition splits have been defined in section 6.5.2). The same is true of the AC-transformation. In fact, the EAC-transformation and the AC-transformation only differ in the set of transitions that are split; the EAC-transformation is used to reduce the number of transitions that are split.

ALGORITHM 6.3   The AC-Transformation and the EAC-Transformation

Input: $\mathcal{N} = (P, T, F, W)$, *type* $\in \{AC, EAC\}$ *indicating the transformation type, and optionally $M \subseteq P$; the default value of M is $M = P$.*

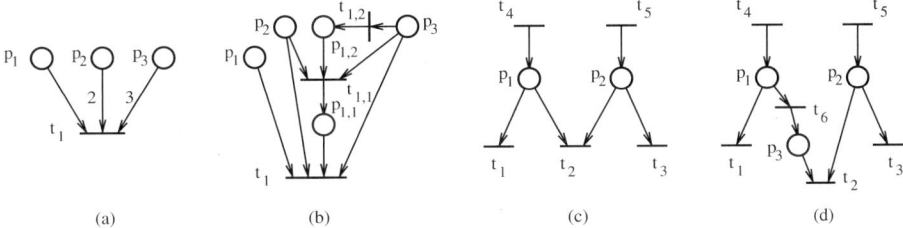

Fig. 6.9. Illustration of the transition split: (a) initial configuration; (b) the effect of the PT-transformation; (c) initial configuration; (d) the effect of the EAC-transformation.

Output: $\mathcal{N}'$

Initialize $\mathcal{N}'$ to be identical with $\mathcal{N}$.

Let x be a vector indexed by the transitions of T. For all $t \in T$ set $x(t) = 1$ if $|\bullet t| > 1$ and $x(t) = 0$ otherwise.

While $\|x\| \neq \emptyset$ **do**

1. *Select a transition $t \in \|x\|$ and set $x(t) = 0$.*

2. **If** $type = AC$, **then** let
$$U = \{(p_i, p_j) \in P \times P : p_i \in \bullet t, p_j \in \bullet t, p_i \bullet \not\subseteq p_j \bullet \text{ and } p_j \bullet \not\subseteq p_i \bullet\};$$
   Else, let $U = eactest(\mathcal{N}, t)$.

3. **If** U *is empty,* **then** *continue with the next iteration.*

4. *Let $Q := \emptyset$.*

5. **For** *every $(p_i, p_j) \in U$*

   (a) *A place $p \in \{p_i, p_j\} \cap M$ is selected. If two choices are possible:*

      (i) *$p = p_i$ (or $p = p_j$) if p_i (or p_j) has been previously selected for another element of U.*

      (ii) *Otherwise p is chosen such that p appears in another element of U. If both p_i and p_j satisfy this property, select $p \in \{p_i, p_j\}$ such that $|p \bullet| = \max\{|p_i \bullet|, |p_j \bullet|\}$.*

      (iii) *If none of p_i and p_j appears in another element of U, select $p \in \{p_i, p_j\}$ such that $|p \bullet| = \max\{|p_i \bullet|, |p_j \bullet|\}$.*

   (b) *If a place p could be selected (i.e., if $\{p_i, p_j\} \cap M \neq \emptyset$) then $Q := Q \cup \{p\}$.*

   (c) *For all $t' \in p \bullet \setminus \{t\}$ set $x(t') = 1$ if $|\bullet t'| > 1$.*

6. **For** *all $p \in Q$, delete from $\mathcal{N}'$ the transition arc (p, t) and add a new place p' and a new transition t' such that $\bullet t' = \{p\}$, $t' \bullet = \{p'\}$, $p' \bullet = \{t\}$, $W'(p, t') = W'(t', p') = 1$, $W'(p', t) = W(p, t)$, and $x(t') = 0$.*

eactest$(\mathcal{N}, t)$

Let D be the incidence matrix of $\mathcal{N}$.

Construct $U = \{(p_i, p_j) \in P \times P : p_i \in \bullet t, p_j \in \bullet t, p_i \bullet \not\subseteq p_j \bullet \text{ and } p_j \bullet \not\subseteq p_i \bullet\}$.

If U is empty, **then return** $\emptyset$.

Let $V = (\bullet t) \bullet \setminus \{t\}$ and $T_D = \{t\}$.

While $V \neq \emptyset$ **do**

Check feasibility of $Dx \geq 0$ subject to $x \geq 0$, $\sum_{t_i \in V} x(i) \geq 1$, and $x(i) = 0$ for $t_i = t$.

If feasible **then** let x^* be a solution; $V = V \setminus \|x^*\|$.

Else $T_D = T_D \cup V$, and $V = \emptyset$.

Construct $U = \{(p_i, p_j) \in P \times P : p_i \in \bullet t, p_j \in \bullet t, (p_i \bullet \setminus T_D) \not\subseteq (p_j \bullet \setminus T_D) \text{ and } (p_j \bullet \setminus T_D) \not\subseteq (p_i \bullet \setminus T_D)\}$.

If U is empty, **then return** $\emptyset$.

Let $P_L = \{p \in \bullet t : \exists t' \in (\bullet t) \bullet \setminus T_D \text{ and } p \in \bullet t'\}$.

If there is a t' such that $\bullet t' = P_L$ **then**

Let $\mathcal{N}_1$ be $\mathcal{N}$ after removing the transitions in T_D.

$U = eactest(\mathcal{N}_1, t')$.

Return U.

Note that the only difference between the AC-transformation and the EAC-transformation is the way the set U is computed at step 2. Note also that the original Petri net is only modified by transition splits. The transition splits are performed in step 6 of the algorithm. These transition splits are different from the transition splits of the PT-transformation. In fact, as seen in Fig. 6.9(d), it is more precise to say that transition arcs are split. M, the second argument of the transformation, is used to select the transition arcs to be split. Indeed, in general the choice of transitions/transition arcs to be split is not unique. In the next chapter, the parameter M will be used to prevent the algorithm from splitting certain transitions.

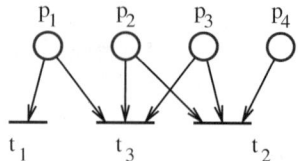

Fig. 6.10. Example for step 5(c) of the EAC-transformation.

The usefulness of step 5(c) may not be immediately apparent. To see it, consider the following scenario. Fig. 6.10 shows a part of a Petri net. The algorithm checks the transitions t_1, t_2, and t_3, in this order. The transitions t_1 and t_2 are found to be satisfactory (they yield $U = \emptyset$). However, after t_3 is checked, the arcs (p_2, t_3) and (p_3, t_3) are split. Without step 5(c), t_2 would not be rechecked, even though the split of (p_2, t_3) and (p_3, t_3) causes t_2 to no longer satisfy Definition 6.20(a)!

The fact that the algorithm terminates can be seen from the following fact: each transition split reduces the number of arcs (p, t) of the net that satisfy $|p \bullet| \geq 1$ and $|\bullet t| \geq 1$. As the initial number of such arcs is finite, the algorithm terminates.

The *eactest* subroutine operates as follows. First, it checks whether case (a) of Definition 6.20 is satisfied. If not, it checks case (b) and computes a set $U \subset P \times P$ with the following property: if for all $(p_i, p_j) \in U$ either of (p_i, t) or (p_j, t) is split, then case (b) is satisfied for t in the resulting net. Finally, if none of (a) and (b) is satisfied, *eactest* tests (c). If (c) is not satisfied, *eactest* returns the set U computed when working at case (b). Note that due to case (c), *eactest* is a recursive function. However, if we renounce checking case (c), the *eactest* becomes significantly faster, as it is no longer recursive. A computer implementation may favor the latter situation.

7

Liveness Enforcement in Petri Nets: A Structural Approach. Part I

7.1 Introduction

This chapter presents procedures for the automated design of deadlock prevention supervisors and $\mathcal{T}$-liveness-enforcing supervisors for Petri nets. **Deadlock** reflects a state of a Petri net in which no transition is enabled. A **deadlock prevention** supervisor ensures that the closed-loop system never enters a state of deadlock. A $\mathcal{T}$-**liveness-enforcing** supervisor ensures that the closed-loop system never enters a state from which a transition in $\mathcal{T}$ can never be fired. **Liveness enforcing** denotes $\mathcal{T}$-liveness enforcing in the case when the set $\mathcal{T}$ equals the total set of the Petri net transitions. Note that our use of deadlock prevention is as an approximation of $\mathcal{T}$-liveness enforcement, as it is easier to obtain a deadlock prevention supervisor. In fact, our procedure for the design of deadlock prevention supervisors may generate supervisors that enforce $\mathcal{T}$-liveness as well. However, to guarantee that $\mathcal{T}$-liveness is enforced, the $\mathcal{T}$-liveness enforcing procedure can be used instead. The procedure for the design of deadlock prevention supervisors will be called the **dp-procedure**, while the procedure for the design of $\mathcal{T}$-liveness supervisors will be called the **le-procedure**.

The procedures we propose operate on arbitrary Petri net structures. Consequently, the plant Petri net may be structurally unbounded, generalized, partially repetitive, and with uncontrollable and unobservable transitions. The procedures do not rely on initial marking knowledge; instead, they generate supervisors having a parametric dependence on the initial marking of the plant. The supervisors are described by two sets of marking constraints $L\mu \geq b$ and $L_0\mu \geq b_0$. A supervisor can be used for all initial markings μ_0 of the plant that satisfy $L\mu_0 \geq b$ and $L_0\mu_0 \geq b_0$. Furthermore, a supervisor is implemented by enforcing $L\mu \geq b$ on the plant via SBPI. Note that the procedures guarantee that $L\mu \geq b$ is admissible with respect to the set of uncontrollable and unobservable transitions of the plant.

The procedures are able to take advantage of knowledge of the reachable markings, when such information is available. This information can be communicated to the procedures via optional arguments describing *initial constraints* and/or *initial-marking constraints*. **Initial constraints** are given in the form $L_I\mu \geq b_I$, meaning that the plant will only be used for initial markings such that all reachable markings μ satisfy $L_I\mu \geq b_I$. **Initial-marking constraints** are given in the form $L_{I0}\mu \geq b_{I0}$, meaning that the plant will only be used for initial markings μ_0 satisfying $L_{I0}\mu_0 \geq b_{I0}$.

In the case of fully controllable and observable Petri nets with no initial constraints and no initial-marking constraints, the performance of the dp-procedure is as follows. If it terminates, it generates a deadlock prevention supervisor if one exists, or signals if none exists. If a supervisor is generated, the closed loop is guaranteed to be deadlock-free for all initial markings μ_0 of the plant satisfying $L\mu_0 \geq b$ and $L_0\mu_0 \geq b_0$. Finally, if deadlock prevention is used as an approximation to $\mathcal{T}$-liveness, a sufficient condition for the supervisor to be at least as permissive as the least restrictive $\mathcal{T}$-liveness-enforcing supervisor is that the plant has a unique $\mathcal{T}$-minimal active subnet.

In the case of fully controllable and observable Petri nets with no initial constraints and no initial-marking constraints, the performance of the le-procedure is as follows. Upon termination, the procedure generates the constraints $L\mu_0 \geq b$ and $L_0\mu_0 \geq b_0$, and a set $\mathcal{T}' \subseteq \mathcal{T}$. If the procedure terminates, it generates a $\mathcal{T}'$-liveness-enforcing supervisor, where (a) $\mathcal{T}' = \mathcal{T}$ if a $\mathcal{T}$-liveness-enforcing supervisor exists, (b) $\mathcal{T}' \subset \mathcal{T}$ if no $\mathcal{T}$-liveness-enforcing supervisor exists, and (c) $\mathcal{T}' = \emptyset$ if deadlock prevention is impossible. Furthermore, when a $\mathcal{T}$-liveness-enforcing supervisor exists, the supervisor is least restrictive if the plant has a single $\mathcal{T}$-minimal active subnet. In particular, this means that in the case of liveness enforcement the supervisors generated by the procedure are always least restrictive, as the whole net is the unique $\mathcal{T}$-minimal active subnet.

In the general case, the performance of the dp-procedure is as follows. If it terminates without declaring failure, it generates a deadlock prevention supervisor. The supervisor satisfies that the closed loop is guaranteed to be deadlock-free for all initial markings μ_0 of the plant satisfying $L\mu_0 \geq b$ and $L_0\mu_0 \geq b_0$. Results concerning the analysis of the cases when the procedure declares failure are provided. Finally, if deadlock prevention is used as an approximation to $\mathcal{T}$-liveness, we provide sufficient conditions ensuring that the generated supervisors are at least as permissive as the least restrictive $\mathcal{T}$-liveness-enforcing supervisor.

In the general case, the performance of the le-procedure is as follows. If it terminates without declaring failure, it generates a $\mathcal{T}'$-liveness-enforcing supervisor, for $\mathcal{T}' \subseteq \mathcal{T}$. (The procedure attempts to maximize the set $\mathcal{T}'$.) The supervisor satisfies that the closed loop is guaranteed to be $\mathcal{T}'$-live for all initial markings μ_0 of the plant satisfying $L\mu_0 \geq b$ and $L_0\mu_0 \geq b_0$. Results concerning the analysis of the cases when the procedure declares failure are provided. Sufficient conditions for $\mathcal{T}' = \mathcal{T}$ and for the supervisor to be least restrictive are provided.

The procedures may perform computationally complex operations at every iteration. However, once a supervisor has been designed, the only real-time computations are checking whether the supervisor Petri net enables a transition and updating the marking of the supervisor after a transition is fired. On the negative side, the number of places of the supervisor may be large. Moreover, the termination of the procedures cannot be guaranteed. In fact, examples are provided for which the procedures are guaranteed to iterate forever.

For the sake of simplicity, the presentation of the procedures has been divided in two parts. This chapter presents simplified procedures, developed under the assumptions that the target Petri nets are fully controllable and observable and that there are no initial constraints or initial-marking constraints. The next chapter presents the general procedures.

The simplified procedures presented in this chapter operate as follows. As they do not have the output $\mathcal{T}'$, they terminate immediately if the target net cannot be made $\mathcal{T}$-live for any initial marking. Otherwise, if the net can be made $\mathcal{T}$-live and the procedures terminate, we have the following. The dp-procedure generates the constraints $L\mu \geq b$ and $L_0\mu \geq b_0$ such that the supervisor enforcing $L\mu \geq b$ prevents deadlock for all initial markings μ_0 satisfying $L\mu_0 \geq b$ and $L_0\mu_0 \geq b_0$. Moreover, the supervisor is at least as permissive as the least restrictive $\mathcal{T}$-liveness-enforcing supervisor when the target net has a single $\mathcal{T}$-minimal active subnet. The le-procedure generates the constraints $L\mu \geq b$ and $L_0\mu \geq b_0$ such that the supervisor enforcing $L\mu \geq b$ enforces $\mathcal{T}$-liveness for all initial markings μ_0 satisfying $L\mu_0 \geq b$ and $L_0\mu_0 \geq b_0$. Further, the supervisor is the least restrictive $\mathcal{T}$-liveness-enforcing supervisor when the target net has a single $\mathcal{T}$-minimal active subnet.

The chapter is organized as follows. Section 7.2 presents related work in the literature. Section 7.3 states the deadlock prevention and $\mathcal{T}$-liveness enforcement problems of this chapter. Section 7.4 provides a motivation for the stages of the proposed procedures. Section 7.5 defines the dp-procedure and the le-procedure. Section 7.6 provides examples illustrating the operation of the procedures. Finally, section 7.7 contains theoretical results on the performance of the procedures.

7.2 Related Work

This section presents related work on deadlock prevention and liveness enforcement. In this work, deadlock prevention refers to the prevention of total deadlock and liveness enforcement to the prevention of all deadlocks. However, note that many papers in the literature do not make this distinction. In such papers deadlock prevention is the same as liveness enforcement, referring to the prevention of all deadlocks.

Historically, the study of deadlocks has arisen in the development of operating systems for computers. In this context, deadlock can arise when concurrent tasks enter a "circular wait," in which each task waits for another for the release of "resources." A survey of the early work on system deadlocks can be found in [36]. A more recent review with emphasis on database applications appeared in [98]. There has also been work on deadlock avoidance in communication networks [138, 139, 25]. A result of the early work are the conditions of Coffman [36], which are necessary and sufficient for deadlocks to occur. Based on these conditions, various methods have been proposed to ensure liveness. As models of concurrent systems, Petri nets have also been studied in the context of system deadlocks. Petri nets can conveniently represent the sequence in which tasks take and release resources. However, since Petri nets have a fixed (static) structure, they offer limited support for resource allocation in operating systems, in which tasks appear and disappear according to the programs that are executed. On the other hand, the approaches of computer operating systems, which are intended for concurrent systems with variable structure, tend to be too restrictive when applied to concurrent systems with fixed structure [13], naturally modeled by Petri nets.

The typical application for the study of system deadlocks in concurrent systems with fixed structure is the resource allocation problem in flexible manufacturing. More recently, it was shown that the supervision of railway networks can be studied in the same framework [58]. The study of deadlocks is also important when additional constraints are imposed on the behavior of concurrent systems, such as mutual exclusion; the additional constraints can be a source of deadlocks. In section 9.4.2 it will be shown that the system deadlock framework can also be applied to the supervisory enforcement of a certain real-time constraint.

In the context of manufacturing systems, the liveness enforcement problem has been approached for restricted classes of Petri net models. These models incorporate the assumptions regarding the interaction between jobs and resources and belong to subclasses of repetitive and conservative Petri nets. In [13] a liveness enforcement approach is proposed for a class of Petri nets modeling jobs that can hold one resource unit at a time. A less restrictive approach is proposed in [74] for a more general model, allowing a job to hold units from multiple resources at the same time. Another approach for liveness enforcement for a virtually identical class of Petri nets appears in [48]. This approach is further generalized in [147] to allow alternative job routes and multiple units of the same resource to be held by a single job. Notably, this results in Petri nets that are no longer ordinary (that is, they have integer weights on the transition arcs). An approach applying to systems with single unit resources and jobs holding a single resource unit at a time appears in [49]. The approach of [49] uses digraphs, but can be converted to the equivalent class of Petri nets. Another liveness enforcement procedure for Petri nets that are virtually identical to those of [13] appears in [210]. Under certain conditions, the proposed supervision is shown to be least restrictive. In a computer science context, least restrictive liveness enforcing in processes with resource allocation appears in the early works of [121, 186]. In addition to several technical conditions, it is assumed that a process does not request units of a resource again after it has started releasing units of that resource. Since multiple units of the same resources can be allocated, the Petri net models are not ordinary. Excepting the integer arc weights, the Petri net model of [121, 186] can be seen as a restriction of that of [74].

Liveness enforcement has also been studied in the general context of Petri nets, that is, without a specific application in mind. Note that for bounded Petri nets, given an initial marking, one can build the reachability graph and then design a supervisor that avoids the states from which liveness cannot be maintained. However, this approach is computationally complex, as the size of the reachability graph can be exponentially related to the size of the Petri net. Note that building the reachability graph corresponds to the construction of an automaton with the same language (DES dynamics) as the Petri net initialized at the given initial marking. To mitigate this computational problem, reachability analysis based on the unfolding of the net has been proposed in [67] for the design of liveness-enforcing supervisors for bounded Petri nets. Another approach to reduce the computational burden is found in [176]. The paper identifies a class of Petri nets that can be reduced to Petri nets involving fewer places and transitions. Liveness enforcement is then reduced to liveness enforcement in the smaller Petri net, which is naturally less complex. Other computational savings can be obtained for Petri nets consisting of subnets interconnected by single places or single transitions [177]. Alternatively, structural methods can be applied in order to avoid building the reachability graph. In [173], it is shown that for a subclass of free-choice Petri nets, called independent increasing free-choice Petri nets (II-FCPNs), liveness can be achieved by ensuring that certain siphons are properly marked. Of course, this property is typically not true for general Petri nets. The result of [173] has been extended in [175] to Petri nets with a free-choice equivalent that is a II-FCPN.

In the literature there are only a few results related to liveness enforcing in Petri nets which may have uncontrollable and unobservable transitions. The unfolding approach of [67] considers partially controllable Petri nets. In [148], liveness enforcement under partial controllability is considered for the Petri net models of [147, 194]. The existence of liveness-enforcing supervisors for Petri nets with uncontrollable transitions has been considered in [174]. In [140], it has been noticed that for some Petri nets with uncontrollable and unobservable transitions, liveness enforcement and deadlock prevention can be achieved by enforcing linear marking constraints. In [178], it is shown that in a partially controlled free-choice Petri net in which liveness is enforcible for some initial markings, there is a set of minimal markings which can be used to check whether liveness is enforcible at a given initial marking and to synthesize the least restrictive liveness-enforcing supervisor. Apparently, to date there is no work on liveness enforcement in Petri nets under partial observability.

Liveness is a special case of $\mathcal{T}$-liveness, as it means that all transitions in a Petri net are live. In the literature there is little work on $\mathcal{T}$-liveness. Note that the supervisory problem solved by our procedure cannot be solved with finite automata-based approaches. Indeed, since we consider Petri net structures rather than a Petri net with an initial marking, an automaton which would have the behavior of the Petri net for any initial marking would have an infinite number of states. Of course, this is not the case for the approaches which consider a single initial marking and a bounded Petri net. Applications which may benefit from considering the initial marking to be unknown are in the area of flexible manufacturing, as part of the initial marking corresponds to the number of available resources. The problem of characterizing the set of markings for which a Petri net can be made $\mathcal{T}$-live is decidable in the case of Petri nets with controllable and observable transitions [199]. The algorithm proposed in [199] searches the marking space to find a set of minimal markings; based on this set the least restrictive $\mathcal{T}$-liveness-enforcing supervisor can be immediately derived. However, the approach of [199] is not very practical for two reasons: (a) the coverability graph is to be evaluated for every marking considered during the search; (b) the number of minimal markings may be large (e.g., exponential in the size of the net).

The $\mathcal{T}$-liveness enforcement procedure presented here is iterative, at every iteration correcting new deadlock situations. Using iterations to correct deadlock situations has also been used in [120, 194]. In our procedure we employ the SBPI [57, 141, 214], which is also described in chapter 3. We also use a transformation to almost ordinary Petri nets and a transformation to asymmetric-choice nets. The first transformation was inspired by a similar transformation in [120]. A transformation to free-choice nets, a particular class of asymmetric-choice nets, has been used in [172, 175]. Note also that the use of control places for liveness enforcement has appeared first in [121, 186].

7.3 Problem Statement

This section presents the main problems approached in this chapter. Section 7.3.1 presents the problem statement for deadlock prevention, while section 7.3.2 presents the problem statement for $\mathcal{T}$-liveness enforcement.

7.3.1 Deadlock Prevention

The following describes the problem statement for deadlock prevention in this chapter. Given a Petri net $\mathcal{N}$ and a set of transitions $\mathcal{T}$, the problem is to find, if possible, two sets of marking constraints $L\mu \geq b$ and $L_0\mu \geq b_0$ such that:

1. For all initial markings μ_0 that satisfy $L\mu_0 \geq b$ and $L_0\mu_0 \geq b_0$, the supervisor enforcing $L\mu \geq b$ via SBPI prevents deadlock.

2. The supervisor is not overly restrictive.

3. The supervisor is a good approximation of a T-liveness enforcement supervisor.

Finally, the case when T-liveness enforcement is impossible at all initial markings should be identified.

Requirements 2 and 3 are not to be taken strictly. The third requirement states that our purpose is not to do least restrictive deadlock prevention (i.e., to allow any local deadlocks to occur as long as the system is not totally deadlocked), but rather to approximate T-liveness enforcement. Such an approximation is of interest because deadlock prevention supervisors are (computationally) easier to obtain than T-liveness enforcement supervisors. In our dp-procedure, approximating T-liveness enforcement means that if local deadlock possibilities leading to loss of T-liveness are identified, the designed supervisor will avoid them. However, the dp-procedure is not guaranteed to identify all such possibilities; the le-procedure, described next, can be used for this purpose.

7.3.2 T-liveness Enforcement

The following describes the problem statement for T-liveness enforcement in this chapter. Given a Petri net $\mathcal{N}$ and a set of transitions T, the problem is to find, if possible, two sets of marking constraints $L\mu \geq b$ and $L_0\mu \geq b_0$ such that:

1. For all initial markings μ_0 that satisfy $L\mu_0 \geq b$ and $L_0\mu_0 \geq b_0$, the supervisor enforcing $L\mu \geq b$ via SBPI enforces T-liveness.

2. The supervisor is not overly restrictive.

Finally, the case when T-liveness enforcement is impossible at all initial markings should be identified.

With regard to the second requirement, note that the le-procedure aims to generate least restrictive supervisors. The supervisors generated by the le-procedure are guaranteed for certain plant Petri nets to be least restrictive. An extension of the le-procedure will be proposed to generate the least restrictive supervisor when the supervisors generated by the le-procedure are not guaranteed to be least restrictive.

7.4 Motivation

The intention of this section is to motivate our approach to deadlock prevention and T-liveness enforcement. Our approach can be seen as an iterative procedure that controls at every iteration active siphons of the net and performs net transformations. Iterations are necessary because controlling siphons may generate new uncontrolled siphons. The procedure terminates when no more uncontrolled active siphons remain. Upon termination, it generates the two sets of linear marking constraints $L\mu \geq b$ and $L_0\mu \geq b_0$.

This section is organized as follows. Section 7.4.1 illustrates the use of linear marking constraints as a compact description of the set of markings for which a Petri net can be made live. Section 7.4.2 shows in an example why iterations are necessary. Section 7.4.3 explains the need for net transformations.

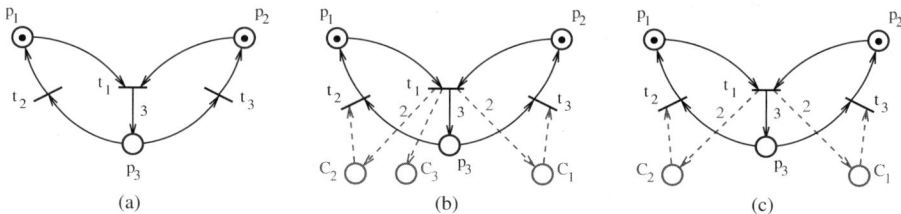

Fig. 7.1. Motivating the linear marking inequalities.

7.4.1 The Role of Linear Marking Inequalities

This section illustrates the constraints $L\mu \geq b$ and $L_0\mu \geq b_0$ on the Petri net of Fig. 7.1(a). It can be noticed that the following set of marking inequalities characterizes all initial markings for which liveness can be enforced:

$$\mu_1 + \mu_3 \geq 1, \tag{7.1}$$

$$\mu_2 + \mu_3 \geq 1, \tag{7.2}$$

$$\mu_1 + \mu_2 + \mu_3 \geq 2. \tag{7.3}$$

Furthermore, each inequality is necessary; by removing any of the inequalities we can find an initial marking satisfying the remaining inequalities for which liveness cannot be enforced. Once we have come up with the set of initial markings for which liveness can be enforced, we can create a supervisor enforcing liveness via SBPI (see Theorem 3.1). The supervised Petri net is shown in Fig. 7.1(b), where the control places C_1, C_2, and C_3 correspond to the inequalities (7.1), (7.2), and (7.3). The initial marking of the control places depends on the initial marking μ_0 of the Petri net as follows:

$$\mu_{0,C_1} = \mu_{0,1} + \mu_{0,3} - 1, \tag{7.4}$$

$$\mu_{0,C_2} = \mu_{0,2} + \mu_{0,3} - 1, \tag{7.5}$$

$$\mu_{0,C_3} = \mu_{0,1} + \mu_{0,2} + \mu_{0,3} - 2. \tag{7.6}$$

However, it can be noticed that by removing the control place C_3 liveness is still enforced (Fig. 7.1(c)) for all initial markings satisfying (7.1), (7.2), and (7.3). Then, we can write (7.1) and (7.2) as a matrix inequality $L\mu \geq b$, and (7.3) as matrix inequality $L_0\mu \geq b_0$. With these notations we can say that liveness is enforced for all initial markings μ_0 satisfying $L\mu_0 \geq b$ and $L_0\mu_0 \geq b_0$ by the supervisor enforcing $L\mu \geq b$.

Finally, note that in some problems the set of markings for which $\mathcal{T}$-liveness can be enforced cannot be represented as a conjunction of linear marking inequalities. Such a set of markings cannot be represented by the sets of inequalities $L\mu \geq b$ and $L_0\mu \geq b_0$. For such problems the le-procedure of this paper can behave in two ways: (i) it does not converge; (ii) it does not generate the least restrictive $\mathcal{T}$-liveness-enforcing supervisor. Note that we prove that behavior (ii) may happen only if the Petri net has more than one $\mathcal{T}$-minimal active subnet. As an example, consider the Petri net of Fig. 7.3(b). For both markings $\mu_0 = [2, 0, 0, 0]$ and $\mu_1 = [0, 2, 0, 0]$ liveness can be enforced, however $\mu_2 = 0.5\mu_0 + 0.5\mu_1$ is a deadlock marking; therefore no conjunction of linear marking inequalities can describe the set of initial markings for which liveness can be enforced.

7.4.2 The Role of Iterations

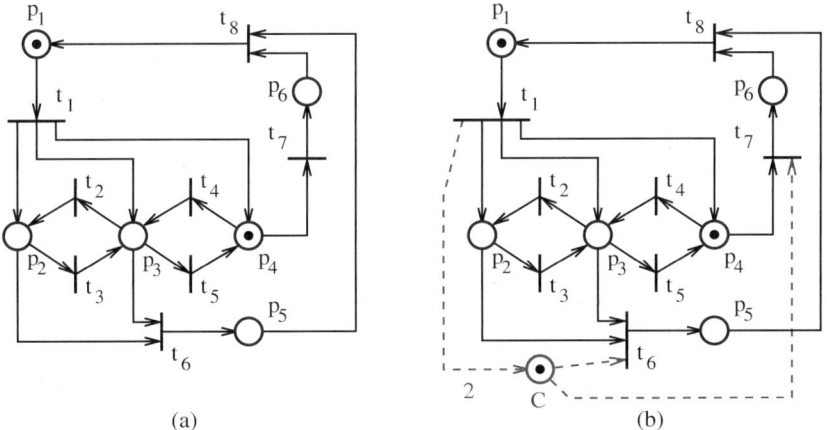

(a)   (b)

Fig. 7.2. Siphon control may cause new siphons that need control.

This section explains, based on the example of Fig. 7.2, why our approach needs to be iterative. In our approach minimal active siphons are controlled, as deadlock and loss of $\mathcal{T}$-liveness has been related to empty active siphons (see Proposition 6.15 and Theorem 6.19). However, it is known that the control of siphons can generate new siphons, and so new possibilities of deadlock [48, 120, 14]. This is shown in the example of Fig. 7.2. To control the siphon

$\{p_1, p_2, p_3, p_4, p_5\}$, the control place C is generated. By adding C to the net two new minimal siphons appear: $\{C, p_1, p_5\}$ and $\{C, p_1, p_6\}$. Note that for the marking shown in Fig. 7.2(b), the siphon $\{C, p_1, p_5\}$ is uncontrolled. So, controlling $\{p_1, p_2, p_3, p_4, p_5\}$, creates new siphons to control. Therefore, in order to obtain Petri nets with no uncontrolled active siphons, our procedure iteratively controls the minimal active siphons until no uncontrolled active siphons remain.

Finally, our interest is to find, if possible, all initial markings for which a Petri net can be made $\mathcal{T}$-live. In this context, the definition of a controlled siphon differs from that in the literature. In our case, a siphon S will be considered **controlled**, if for all markings for which the previous siphons controlled by the procedure are not empty, S is not empty.

7.4.3 The Need for Net Transformations

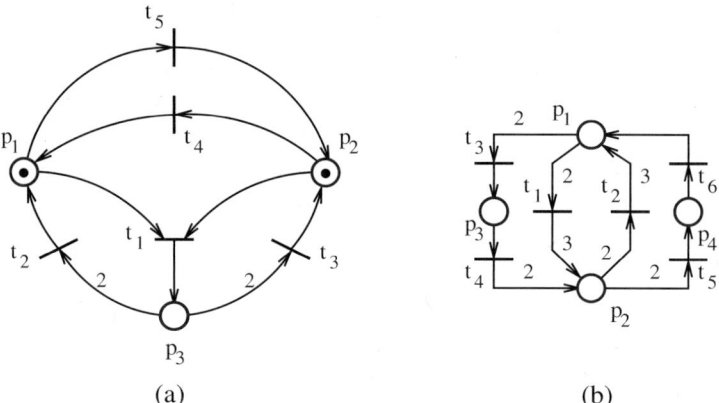

Fig. 7.3. More motivating examples.

This section explains the need for net transformations in our approach. Consider the Petri net of Fig. 7.3(a). It can be seen that only the transitions t_4 and t_5 can be made live. So there are no initial markings for which liveness is enforcible. However, there are initial markings for which $\{t_4, t_5\}$-liveness is enforcible. These initial markings can be described by the inequality

$$2\mu_1 + 2\mu_2 + \mu_3 \geq 2. \tag{7.7}$$

The only active subnet of the net is defined by the set of transitions $\{t_4, t_5\}$, and the only siphon of the net equals the total set of places of the Petri net. For all nonzero initial markings this siphon is controlled. However, a nonzero initial marking does not imply that (7.7) is always satisfied. This suggests that the empty siphon criterion for deadlock is not very useful for $\mathcal{T}$-liveness enforcement in Petri nets which are not PT-ordinary and with asymmetric choice, as is the case for our Petri net. Furthermore, this would also suggest the use of transformations to asymmetric-choice and PT-ordinary nets, in order to be able to use Theorem 6.19.

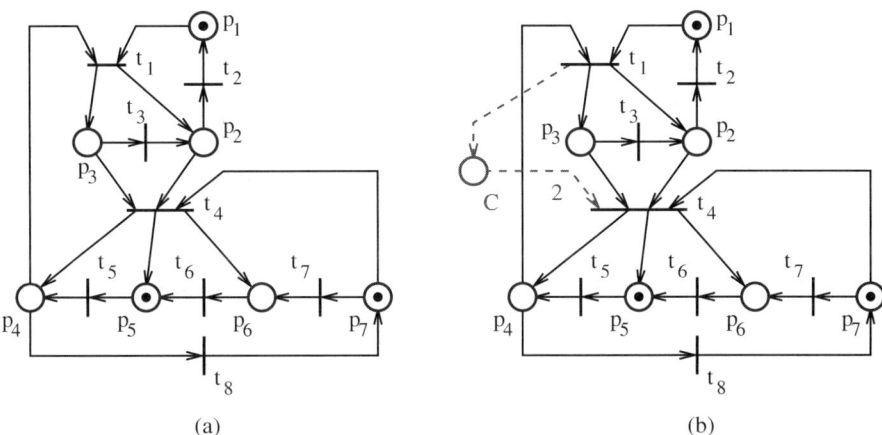

(a)   (b)

Fig. 7.4. Siphon control may change an ordinary net to a nonordinary net. Here, C controls the siphon $\{p_1, p_2, p_3\}$.

Note also that controlling siphons may cause an ordinary Petri net to become a generalized Petri net. This is an additional motivation for considering transformations to PT-ordinary nets in the procedures. An example is shown in Fig. 7.4. Controlling the siphon $\{p_1, p_2, p_3\}$ with the control place C results in the weight $W(C, t_4) = 2 > 1$, showing that the supervised net is no longer ordinary. Finally, it is also true that siphon control may cause a net to be no longer with asymmetric choice.

7.5 Procedure Definition

7.5.1 Definition

In this section the dp-procedure and the le-procedure are defined. Due to the fact that they share many common operations, they are defined as a single procedure with an input argument selecting between deadlock prevention and $\mathcal{T}$-liveness enforcement. From now on, this procedure will be denoted as *the procedure*; depending on its input argument, the procedure can be either the dp-procedure or the le-procedure.

Given a target Petri net $\mathcal{N}_0$, the procedure generates a sequence of Petri nets $\mathcal{N}_1, \mathcal{N}_2, \ldots, \mathcal{N}_k$, increasingly enhanced for $\mathcal{T}$-liveness. In the case of deadlock prevention, $\mathcal{N}_1, \mathcal{N}_2, \ldots, \mathcal{N}_k$ are PT-ordinary. In the case of $\mathcal{T}$-liveness enforcement, $\mathcal{N}_1, \mathcal{N}_2, \ldots, \mathcal{N}_k$ are with asymmetric-choice and PT-ordinary. The Petri net $\mathcal{N}_1$ is $\mathcal{N}_0$ transformed to be PT-ordinary or PT-ordinary and with asymmetric choice. The Petri nets $\mathcal{N}_i$ are defined as follows. In each iteration i, the uncontrolled minimal active siphons of $\mathcal{N}_i$ are controlled. Then, if needed, the Petri net is transformed to be PT-ordinary or PT-ordinary and with asymmetric choice. The result is the net $\mathcal{N}_{i+1}$. Controlling a siphon S involves enforcing the inequality $\sum_{p \in S} \mu(p) \geq 1$ via SBPI. A siphon S is **controlled** if for all markings for which the previously controlled siphons are not empty, S is not empty. Otherwise, S is **uncontrolled**. The active siphons are taken with respect to a $\mathcal{T}$-minimal active subnet. For each $\mathcal{N}_i$, that subnet is denoted by $\mathcal{N}_i^A$. The procedure terminates at the iteration k for which $\mathcal{N}_k$ has no uncontrolled active siphons.

The constraints $L\mu \geq b$ and $L_0\mu \geq b_0$ are obtained as follows. At every iteration i, when a siphon S is controlled, a marking constraint corresponding to $\sum_{p \in S} \mu(p) \geq 1$ is derived and added to either of $L\mu \geq b$ or $L_0\mu \geq b_0$. Let's denote by $L_i\mu \geq b_i$ and $L_{0,i}\mu \geq b_{0,i}$ the constraints $L\mu \geq b$ and $L_0\mu \geq b_0$ after the iteration i. The final constraints $L\mu \geq b$ and $L_0\mu \geq b_0$ are obtained from $L_k\mu \geq b_k$ and $L_{0,k}\mu \geq b_{0,k}$ by restricting them to the places of $\mathcal{N}_0$.

In the procedure:

- μ_p is the marking of the places which are not control places.
- μ_c is the marking of the control places.
- The Petri net of iteration i is $\mathcal{N}_i = (P_i, T_i, F_i, W_i)$.
- The active subnet of $\mathcal{N}_i$ is $\mathcal{N}_i^A = (P_i^A, T_i^A, F_i^A, W_i^A)$.

The procedure notation is such that equation

$$\mu_c = L\mu_p - b \tag{7.8}$$

describes the invariants enforced by the control places at any iteration. We denote a set of constraints $X\mu \geq x$ as (X, x). We give the detailed description of the specific steps of the procedure in the following subsections. Thus we annotate the procedure steps with the number of the subsection in which we describe in detail the specific operation.

PROCEDURE 7.1   Deadlock Prevention/$\mathcal{T}$-liveness Enforcement

Input: *The target Petri net $\mathcal{N}_0$, a nonempty set of transitions $\mathcal{T}$, and type $\in \{DP, LE\}$, identifying the design type: deadlock prevention (DP) or $\mathcal{T}$-liveness enforcement (LE).*

Output: *Two sets of constraints (L, b) and (L_0, b_0).*

A. $\mathcal{N}_0$ is transformed to be PT-ordinary if $type = DP$, and asymmetric choice PT-ordinary if $type = LE$ (the transformations appear in section 6.5.2 and in Algorithm 6.3).[1] Let the transformed net be $\mathcal{N}_1$. Let $i = 1$, $P = P_1$, and $\mathcal{C} = \emptyset$.

B. A $\mathcal{T}$-minimal active subnet $\mathcal{N}_1^A$ is computed for $\mathcal{N}_1$ (Algorithm 6.2). If none exists, the procedure terminates and declares that $\mathcal{T}$-liveness cannot be enforced for any initial marking.

C. While *true* **do**

1. Let (A, d) and (A_0, d_0) be empty sets of marking constraints.

2. If no uncontrolled minimal active siphon is found (section 7.5.2), the next step is D.[2]

3. For every uncontrolled minimal active siphon S:

   Test whether $\sum_{p \in S} \mu(p) \geq 1$ needs control place enforcement (section 7.5.2). If it does, include $\sum_{p \in S} \mu(p) \geq 1$ in (A, d). Else include $\sum_{p \in S} \mu(p) \geq 1$ in (A_0, d_0).

4. Let $\mathcal{N}_i' = (P_i', T_i', F_i', W_i')$ be the Petri net structure obtained by enforcing $A\mu \geq d$ in $\mathcal{N}_i$ via SBPI (see Theorem 3.1), and let $A^I \mu' = d$ be the corresponding place invariant equations (see equation (7.8)).

5. If $type = DP$, $\mathcal{N}_i'$ is transformed to be PT-ordinary; if $type = LE$, the Petri net is transformed to be PT-ordinary and with asymmetric choice (section 6.5.2 and Algorithm 6.3)[1]; note that the argument M of the Algorithm 6.3 is set to $M = P_i' \setminus P_i$. Let $\mathcal{N}_{i+1}$ be the transformed net.

6. Update A^I according to the net transformations performed at step 5 (section 7.5.5). Let A^u be the updated A^I (this means that $A^I \mu' = d$ in $\mathcal{N}_i'$ corresponds to $A^u \mu = d$ in $\mathcal{N}_{i+1}$, where μ' and μ are markings of $\mathcal{N}_i'$ and $\mathcal{N}_{i+1}$).

7. Let $P = P \cup (P_{i+1} \setminus P_i')$, $\mathcal{C}^o = \mathcal{C}$, and $\mathcal{C} = \mathcal{C}^o \cup (P_i' \setminus P_i)$. Let[3] $\mu_p = \mu|_P$ and $\mu_c = \mu|_{\mathcal{C}}$, for any marking μ of $\mathcal{N}_{i+1}$. For each place in $P_{i+1} \setminus P_i'$ add a null column to each of L and L_0, to match the size of μ. Similarly, add null columns to A_0 to match the size of μ. Let[4] $A_p = A^u|_P$, $A_{p0} = A_0|_P$, $A_c = A^u|_{\mathcal{C}^o}$, and $A_{c0} = A_0|_{\mathcal{C}^o}$.

8. If (L, b) is empty, include $A_p \mu_p \geq d$ in (L, b) and $A_{p0} \mu_p \geq d_0$ in (L_0, b_0).

---

[1]The transformation to PT-ordinary Petri nets has no effect if the Petri net is already PT-ordinary; the same is true of the transformation to asymmetric-choice nets.

[2]In the worst case, the number of uncontrolled minimal siphons depends exponentially on the size of the net. Checking whether a siphon is uncontrolled may involve solving a linear integer program.

[3]Given a set of places X, $\mu|_X$ is the restriction of μ to the places of X.

[4]$A^u|_P$ is the restriction of A^u to the columns corresponding to places in P; $A_0|_P$, $A^u|_{\mathcal{C}^o}$, ..., have a similar meaning.

Else, do the following

   (a) If (A_0, d_0) is not empty, include $(A_{p0} + A_{c0}L)\mu_{p0} \geq d_0 + A_{c0}b$ in (L_0, b_0).

   (b) If (A, d) is not empty, include $(A_p + A_cL)\mu_p \geq d + A_cb$ in (L, b).

9. *Compute the new active subnet $\mathcal{N}_{i+1}^A$ (section 7.5.6). Let $i = i+1$. The next step is C.1.*

D. *The constraints (L, b) and (L_0, b_0) are restricted to the columns corresponding to the places of $\mathcal{N}_0$.*

E. *Optionally, the redundant constraints of (L, b) and (L_0, b_0) are removed.*[5]

Remarks: Note that the final constraints (L, b) and (L_0, b_0) are such that:

- If *type* $= LE$, $\mathcal{T}$-liveness is enforced for all initial markings μ_0 such that $L\mu_0 \geq b$ and $L_0\mu_0 \geq b_0$ when $(\mathcal{N}_0, \mu_0)$ is supervised according to $L\mu \geq b$.
- If *type* $= DP$, deadlock is prevented for all initial markings μ_0 such that $L\mu_0 \geq b$ and $L_0\mu_0 \geq b_0$ when $(\mathcal{N}_0, \mu_0)$ is supervised according to $L\mu \geq b$.

Note also that in the case of $\mathcal{T}$-liveness enforcement, the transformations to asymmetric choice Petri nets could be replaced by transformations to EAC Petri nets (Algorithm 6.3). Experimental results indicate the procedure is more likely to converge when transformations to EAC Petri nets are used. □

We proceed by describing the specific operations involved in the procedure that have not yet been (completely) defined.

7.5.2 Siphons Not Needing Control

Here the step C.3 of the procedure is discussed. First, the uncontrolled siphons are explicitly defined. Then, the situation in which an uncontrolled siphon does not need a control place is explained. A siphon S is **uncontrolled** if

$$\sum_{p \in S} \mu(p) \geq 1 \tag{7.9}$$

is not implied by $\mu_c = L\mu_p - b$, $L_0\mu_p \geq b_0$, $A\mu \geq d$, and $A_0\mu \geq d_0$. In other words, S is uncontrolled if and only if the system of $\mu|_S = 0$, $\mu_c = L\mu_p - b$, $L_0\mu_p \geq b_0$, $A\mu \geq d$, and $A_0\mu \geq d_0$ has an integer solution $\mu \geq 0$. We design the procedure, in particular the transformation to PT-ordinary asymmetric-choice Petri nets, in such a way that an uncontrolled siphon is always a siphon which did not exist at a previous iteration. Thus at step C.3 it is enough to check only the new siphons which appeared due to the steps C.4 and C.5 of the previous iteration. It can be seen that checking whether a siphon is

---

[5] This operation may involve integer programming.

uncontrolled may involve solving an integer program. When this check is not done the procedure remains correct, however it may converge more slowly or even diverge.

There are siphons S which satisfy (7.9) at all reachable markings if (7.9) is satisfied at the initial marking. Such siphons do not need a control place to ensure that (7.9) is satisfied. We identify that an uncontrolled siphon S does not need a control place C by checking whether C would satisfy $C\bullet \subseteq \bullet S$. When this is the case, (7.9) is included in (A_0, d_0), which contains constraints on the initial marking.

EXAMPLE 7.1

Consider the Petri net of Fig. 7.1(a), and assume that we apply the dp-procedure to it and that $\mathcal{T} = \{t_1, t_2, t_3\}$. At the first iteration there are two minimal siphons: $S_1 = \{p_1, p_3\}$ and $S_2 = \{p_2, p_3\}$. Assume that the procedure considers them in this order: first S_1, then S_2. Since (L, b), (L_0, b_0), (A, d), and (A_0, d_0) are empty, S_1 is uncontrolled. The control place C_1 controlling S_1 is needed, as $C_1\bullet \not\subseteq \bullet S_1$; see Fig. 7.1(c). Therefore (A, d) becomes $\mu_1 + \mu_3 \geq 1$. Next, when S_2 is considered, S_2 is also uncontrolled, since (L, b), (L_0, b_0), and (A_0, d_0) are empty, and the system of $\mu_1 + \mu_3 \geq 1$ and $\mu_2 + \mu_3 = 0$ has a nonnegative integer solution. The control place C_2 of S_2 is also needed, as $C_2\bullet \not\subseteq \bullet S_2$. At the second iteration we have two new minimal siphons: $S_3 = \{p_1, C_2\}$ and $S_4 = \{p_2, C_1\}$. Assume that the procedure considers first S_3 and then S_4. The constraints (L, b) at the second iteration are $\mu_1 + \mu_3 \geq 1$ and $\mu_2 + \mu_3 \geq 1$, while (L_0, b_0) are empty. The siphon S_3 is uncontrolled, since the system of $\mu_1 + \mu(C_2) = 0$, $\mu(C_1) = \mu_1 + \mu_3 - 1$, $\mu(C_2) = \mu_2 + \mu_3 - 1$ has a nonnegative integer solution, and (L_0, b_0), (A, d), and (A_0, d_0) are empty. The control place C_3 that results for S_3 is not needed, as it satisfies $C_3\bullet \subseteq \bullet S_3$; see Fig. 7.1(b). Consequently, (A_0, d_0) becomes $\mu_1 + \mu(C_2) \geq 1$. Next, note that the siphon S_4 is controlled, since the system of $\mu_2 + \mu(C_1) = 0$, $\mu_1 + \mu(C_2) \geq 1$, $\mu(C_1) = \mu_1 + \mu_3 - 1$, and $\mu(C_2) = \mu_2 + \mu_3 - 1$ has no solution.

7.5.3 Generating the Sets of Inequalities (L, b) and (L_0, b_0)

This section explains the meaning of the step C.8. The marking constraints generated by the procedure correspond to the constraints (7.9) on the uncontrolled minimal active siphons of each iteration. These constraints are first stored in (A, d) and (A_0, d_0), and then the constraints of (A, d) are added to those of (L, b), and those of (A_0, d_0) to (L_0, b_0). The step C.8 also insures that the constraints are added to (L, b) and (L_0, b_0) not as they are, but after writing them in terms of the places of the net that are not control places. Indeed, the procedure is set up such that all constraints (L, b) and (L_0, b_0) are written only in terms of μ_p, the marking of the places that are not control places. In this way the Petri net of each iteration satisfies that $\mu_c = L\mu_p - b$

(and so $L\mu_p \geq b$) for all reachable markings if $\mu_c = L\mu_p - b$ is satisfied at the initial marking.

The constraints $L\mu_p \geq b$ are recursively obtained as follows. The siphons in an iteration i may contain control places added in previous iterations. Thus (7.9) may involve not only places of the target net $\mathcal{N}_0$, but also control places. However, the marking of the control places appearing in (7.9) can be eliminated by using $\mu_c = L\mu_p - b$. Thus the operations in the step C.8 correspond to adding new constraints to (L, b) and (L_0, b_0), after substituting in them the control place markings given by $\mu_c = L\mu_p - b$.

EXAMPLE 7.2

Here we continue the Example 7.1. After the siphons S_3 and S_4 have been considered, the procedure adds the constraint $\mu_1 + \mu(C_2) \geq 1$ of (A_0, d_0) to the empty set of constraints (L_0, b_0). The current constraints in (L, b) are $\mu_1 + \mu_3 \geq 1$ and $\mu_2 + \mu_3 \geq 1$, that is,

$$L = \begin{bmatrix} 1 & 0 & 1 \\ 0 & 1 & 1 \end{bmatrix} \text{ and } b = \begin{bmatrix} 1 \\ 1 \end{bmatrix}.$$

As $\mu_c = L\mu_p - b$, $\mu(C_2)$ is substituted with $\mu_2 + \mu_3 - 1$; thus $\mu_1 + \mu(C_2) \geq 1$ becomes $\mu_1 + \mu_2 + \mu_3 \geq 2$, which is the inequality added to (L_0, b_0). The constraints (L_0, b_0) are

$$L_0 = \begin{bmatrix} 1 & 1 & 1 \end{bmatrix} \text{ and } b_0 = \begin{bmatrix} 2 \end{bmatrix}.$$

7.5.4 Petri Net Transformations

This section describes the Petri net transformations performed at the step C.5. The transformation to PT-ordinary nets (the *PT-transformation*) has been described in section 6.5.2. The transformation to asymmetric-choice Petri nets (the *AC-transformation*) is described in Algorithm 6.3. Note that in the case of $\mathcal{T}$-liveness enforcement, the procedure first applies the PT-transformation and then the AC-transformation. There are many ways in which these transformations could be done. Our concern has been to design the transformations so that we can prove that the procedure generates deadlock prevention/$\mathcal{T}$-liveness-enforcing supervisors, and that the supervisors are permissive. To this end we impose three requirements **R1**, **R2**, and **R3**, which we state below. With regard to the requirements below, recall that the transformations we use employ *transition splits*, where a transition is split when decomposed into a sequence of places and transitions. The requirements we impose are written in terms of the notation of the procedure. The requirements are:

R1 No control place in $\mathcal{C}$ is in the postset of a transition created by a transition split.

R2 Any set of inequalities $X\mu \geq x$ that hold true in $\mathcal{N}_i$, hold true also in[6] $\mathcal{N}_{i+1}$, for $i \geq 1$.

R3 The constraints $A\mu \geq d$ enforced on $\mathcal{N}_i$ in step C.4 are satisfied in $\mathcal{N}_{i+1}$.[7]

The argument M of the AC-transformation is used to select the transitions to be split. Indeed, in general there are many ways to transform a net to an asymmetric-choice net by splitting transitions. The procedure sets $M = P_i' \setminus P_i$ in order that the requirement R2 may be satisfied, thus ensuring that the constraints added in the previous iterations remain enforced. Note that $P_i' \setminus P_i$ equals the set of control places that result from enforcing $A\mu \geq d$ at step C.4. The fact that the transformation of $\mathcal{N}_i'$ to an asymmetric-choice net requires only splitting arcs (p, t) with $p \in P_i' \setminus P_i$ results from the fact that $\mathcal{N}_i$ has asymmetric choice. Note that M is not used at the AC-transformation of step A of the procedure. When M is not used, Algorithm 6.3 is given no preference with regard to what transitions to split.

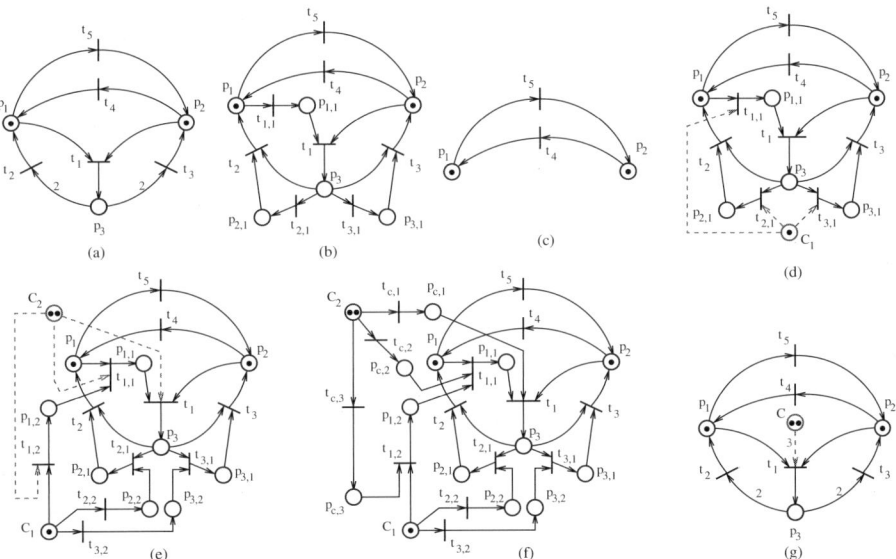

Fig. 7.5. The illustration of Examples 7.3 and 7.4. (a) $\mathcal{N}_0$; (b) $\mathcal{N}_1$; (c) $\mathcal{N}_1^A$, the same as $\mathcal{N}_2^A$ and $\mathcal{N}_3^A$; (d) $\mathcal{N}_1$ and the added control place; (e) $\mathcal{N}_2$ and the added control place; (f) $\mathcal{N}_3$; (g) $\mathcal{N}_0$ supervised for $\mathcal{T}$-liveness.

---

[6]That is, for all markings μ_0 of $\mathcal{N}_i$ satisfying $\forall \mu \in \mathcal{R}(\mathcal{N}_i, \mu_0)$: $X\mu \geq x$, we have that for all markings $\mu_{0,i+1}$ of $\mathcal{N}_{i+1}$ such that $\mu_{0,i+1}|_{\mathcal{N}_i} = \mu_0$, the following holds true: $\forall \mu_{i+1} \in \mathcal{R}(\mathcal{N}_{i+1}, \mu_{0,i+1})$, $X\mu_{i+1}|_{\mathcal{N}_i} \geq x$.

[7]If μ_{i+1} denotes a marking of $\mathcal{N}_{i+1}$, this corresponds to $\forall \mu_{i+1}$: $A^u \mu_{i+1} = d \Rightarrow A\mu_{i+1}|_{\mathcal{N}_i} \geq d$.

EXAMPLE 7.3

This example illustrates the PT- and AC-transformations on the Petri net of Fig. 7.5(a). The supervision purpose is $\{t_4, t_5\}$-liveness enforcement. The $\{t_4, t_5\}$-minimal active subnet is shown in Fig. 7.5(c). At step A, $\mathcal{N}_0$ is PT-transformed, and then AC-transformed. The resulting Petri net is $\mathcal{N}_1$, shown in Fig. 7.5(b). The places and transitions generated by the PT-transformation are $p_{2,1}$, $p_{3,1}$, $t_{2,1}$, and $t_{3,1}$. The AC-transformation generates $p_{1,1}$ and $t_{1,1}$. At the first iteration, the control place C_1 is added to control the siphon $\{p_1, p_2, p_3\}$. The net no longer has asymmetric choice, due to C_1. The following AC-transformation is applied at step C.5 with the argument $M = P_1' \setminus P_1$, namely, $M = \{C_1\}$. Therefore, the transition arcs $(C_1, t_{1,1})$, $(C_1, t_{2,1})$, and $(C_1, t_{3,1})$ are split; the places $p_{1,2}$, $p_{2,2}$, and $p_{3,2}$ and the transitions $t_{1,2}$, $t_{2,2}$, and $t_{3,2}$ result (Fig. 7.5(e)). At the second iteration, $\{p_1, p_2, p_{2,1}, p_{3,1}, p_{2,2}, p_{3,2}, C_1\}$ is the only new minimal active siphon. Controlling the siphon results in the control place C_2 violating the asymmetric-choice requirement. The following AC-transformation at step C.5 transforms the net as shown in Fig. 7.5(f).

7.5.5 The Effect of Net Transformation on Marking Constraints

This section considers the operations done at step C.6 of the procedure. It also shows that the net transformations satisfy the requirements R1, R2, and R3.

Note that the way we implement the PT- and AC-transformations ensures that for all $i \geq 1$, $\mathcal{N}_{i+1}$ can be seen as $\mathcal{N}_i$ connected to another Petri net via additional arcs to the transitions of $\mathcal{N}_i$ (not unlike the connection between a plant Petri net and a supervisor Petri net). Thus the marking constraints already enforced in $\mathcal{N}_i$ are not disturbed, and so requirement R2 is satisfied.

Let $\mathcal{N}$ be a Petri net and assume that $\mathcal{N}$ is PT-transformed and then AC-transformed; let $\mathcal{N}_t$ be the resulting Petri net. Let $l^T \mu \geq b$ be a marking constraint enforced in $\mathcal{N}$ for initial markings in some set $\mathcal{M}_I$. It can be checked that the form of $l^T \mu \geq b$ in $\mathcal{N}_t$ is $l_t^T \mu_t \geq b_t$, obtained from $l^T \mu \geq b$ with the substitution

$$\mu(p) \quad \longrightarrow \quad \mu_t(p) + \sum_{z=1}^{r} \mu_t(p_z) + \sum_{i=1}^{k} \sum_{j=1}^{m_i-1} j\mu_t(p_{i,m_i-j}) \tag{7.10}$$

for each place p of $\mathcal{N}$, where k and m_i are determined in $\mathcal{N}$: $k = |p \bullet|$, $m_i = W(p, t_i) \, \forall t_i \in p\bullet$. The places $p_{i,j}$ are the places that result from splitting the transitions $t_i \in p\bullet$, where the notation of section 6.5.2 is used. The places p_z are the places resulting from the AC-transformation which satisfy $\bullet\bullet p_z = p$.

According to SBPI, when a control place C is added to enforce $\sum_{p \in S} \mu(p) \geq 1$ in a siphon S, the following place invariant is created:

$$\mu(C) = -1 + \sum_{p \in S} \mu(p). \tag{7.11}$$

Consider an equality (7.11) enforced in step C.4. Then (7.10) can be used to derive the form of (7.11) in $\mathcal{N}_{i+1}$. Accordingly, (7.11) is transformed to

$$\mu(C) + \sum_{z=1}^{r} \mu(p_z) + \sum_{i=1}^{k} \sum_{j=1}^{m_i-1} j\mu(p_{i,m_i-j}) = -1 + \sum_{p \in S} \mu(p), \qquad (7.12)$$

where the notation is similar to (7.10): $k = |C\bullet|$, $m_i = W(C, t_i) \; \forall t_i \in C\bullet$, $p_{i,j}$ are the places that result from splitting the transitions $t_i \in C\bullet$, and p_z are the places that result from the AC-transformation such that $\bullet\bullet p_z = C$. Note that the siphon S remains controlled; that is, (7.12) implies that $\sum_{p \in S} \mu(p) \geq 1$ is still satisfied. Therefore the requirement R3 is satisfied.

The considerations above showed that the PT- and AC-transformations satisfy the requirements R2 and R3. The next result states that R1 is also satisfied.

Proposition 7.1. *At every iteration i, the requirement R1 is satisfied.*

Proof. Let $\bullet_i$ and $\bullet'_i$ denote the preset/postset operators in $\mathcal{N}_i$ and $\mathcal{N}'_i$, respectively. First, note that the transitions of $\mathcal{N}_i$ obtained through transition splits form the set $T_i \setminus T_0$. Note also that if R1 is not satisfied, there is a control place C and a transition $t \in T_i \setminus T_0$ such that $C \in t\bullet_i$. However, $C \in t\bullet_i$ implies $|t\bullet_i| \geq 2$. So we prove by induction that for all i and $\forall t \in T_i \setminus T_0$: $|t\bullet_i| = 1$. At $i = 1$ we have $\forall t \in T_1 \setminus T_0$: $|t\bullet_1| = 1$, by construction. Given an iteration number i, assume $\forall t \in T_i \setminus T_0$: $|t\bullet_i| = 1$. We prove $\forall t \in T_{i+1} \setminus T_0$: $|t\bullet_{i+1}| = 1$. Assume the contrary, that $\exists t \in T_{i+1} \setminus T_0$: $|t\bullet_{i+1}| > 1$. Then $t \in T_i \setminus T_0$ and there is a control place C added in step C.4 of iteration i such that $C \in t\bullet'_i$. Let S be the siphon controlled by C. It follows that $t \in \bullet_i S$, and firing t in $\mathcal{N}_i$ from some enabling marking increases the total marking of S. However, this contradicts $t \in S\bullet_i$ (since S is a siphon) and $|t\bullet_i| = 1$ in $\mathcal{N}_i$. The conclusion follows. $\qquad\square$

EXAMPLE 7.4

This example illustrates the constraint transformations and refers to the Petri net of Fig. 7.5(a) and to Example 7.3. At the first iteration, the siphon $\{p_1, p_2, p_3\}$ is controlled, and so $\mu_1 + \mu_2 + \mu_3 \geq 1$ is added to (A, d). The control place C_1 is added at step C.4, and the invariant $A^I \mu' = d$ is $\mu(C_1) = \mu_1 + \mu_2 + \mu_3 - 1$. The transformed net $\mathcal{N}_2$ is shown in Fig. 7.5(e). The updated invariant $A^u \mu = d$ of $\mathcal{N}_2$ is

$$\mu_1 + \mu_2 + \mu_3 - \mu(C_1) - \mu_{1,2} - \mu_{2,2} - \mu_{3,2} = 1. \qquad (7.13)$$

Note that the invariant is changed due to the AC-transformation, as the PT-transformation does not change the net. The constraint $A_p \mu_p \geq d$ added at step C.8 to (L, b) is $\mu_1 + \mu_2 + \mu_3 - \mu_{1,2} - \mu_{2,2} - \mu_{3,2} \geq 1$. At the second *iteration, the siphon* $\{p_1, p_2, p_{2,1}, p_{3,1}, p_{2,2}, p_{3,2}, C_1\}$ *is controlled, and so* $\mu_1 + \mu_2 + \mu_{2,1} + \mu_{3,1} + \mu_{2,2} + \mu_{3,2} + \mu(C_1) \geq 1$ *is added to* (A, d) *and the control place* C_2 *to the net (Fig. 7.5(e)). The invariant of* C_2 *in* $\mathcal{N}'_2$ *is* $\mu_1 + \mu_2 + \mu_{2,1} + \mu_{3,1} + \mu_{2,2} + \mu_{3,2} + \mu(C_1) - \mu(C_2) = 1$. *Then, the updated invariant* $A^u \mu = d$ *of* $\mathcal{N}_3$ *is*

$$\mu_1 + \mu_2 + \mu_{2,1} + \mu_{3,1} + \mu_{2,2} + \mu_{3,2} + \mu(C_1) - \mu(C_2) - \mu_{c,1} - \mu_{c,2} - \mu_{c,3} = 1, \quad (7.14)$$

where the changes are due to the PT-transformation (the AC-transformation has no effect on the net). Therefore, the inequality $A_p \mu_p \geq d$ *added at step C.8 to* (L, b) *is* $2\mu_1 + 2\mu_2 + \mu_3 + \mu_{2,1} + \mu_{3,1} - \mu_{c,1} - \mu_{c,2} - \mu_{c,3} - \mu_{1,2} \geq 2$.

7.5.6 The Computation of a $\mathcal{T}$-minimal Active Subnet

This section discusses the operations done at step B and at step C.9. The computation of a $\mathcal{T}$-minimal active subnet has been presented in Algorithm 6.2. However, it is not necessary to compute the active subnet that way at every iteration. In fact, it is enough to use Algorithm 6.2 only once, at step B. Then, during the iterations, the computation at step C.9 can be performed using the update algorithm below.

ALGORITHM 7.2   Update of the Active Subnet

Input: $\mathcal{N}^A_{i-1} = (P^A_{i-1}, T^A_{i-1}, F^A_{i-1}, W^A_{i-1})$, $\mathcal{N}_i = (P_i, T_i, F_i, W_i)$, *and the sets* $\Sigma(t)$, *where for all* $t \in T_{i-1}$, $\Sigma(t)$ *is the subset of* $T_i \setminus T_{i-1}$ *containing the transitions resulted from splitting* t.
Output: $\mathcal{N}^A_i = (P^A_i, T^A_i, F^A_i, W^A_i)$.
  1. $T^A_i = T^A_{i-1} \cup \{t \in T_i : \exists t_u \in T^A_{i-1} \text{ and } t \in \Sigma(t_u)\}$.
  2. *The active subnet is* $\mathcal{N}^A_i = (P^A_i, T^A_i, F^A_i, W^A_i)$, $P^A_i = T^A_i \bullet$, $F^A_i = F_i \cap \{(T^A_i \times P^A_i) \cup (P^A_i \times T^A_i)\}$, *and* W^A_i *is the restriction of* W_i *to* F^A_i.

Step C.9 uses Algorithm 7.2. Using Algorithm 7.2 rather than Algorithm 6.2 has computational advantages, as the former is a lot simpler.

7.6 Examples

This section presents examples illustrating the operation of the dp-procedure and of the le-procedure.

EXAMPLE 7.5

In this example the dp-procedure is applied to the Petri net of Fig. 7.6 with the parameter $\mathcal{T} = T_0$. *At the first iteration, the control places* C_1 *and* C_2 *are added with respect to the uncontrolled siphons* $\{p_1, p_2\}$ *and* $\{p_3, p_4\}$,

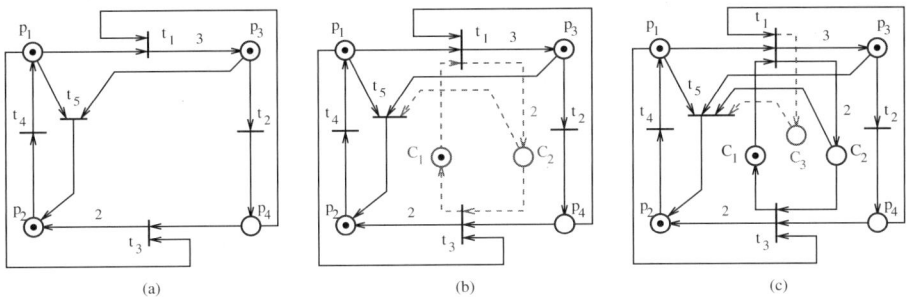

Fig. 7.6. The Petri nets in Example 7.5: (a) $\mathcal{N}_0$, (b) $\mathcal{N}_1$, (c) $\mathcal{N}_2$.

respectively. Consequently, the inequalities $\mu_1 + \mu_2 \geq 1$ and $\mu_3 + \mu_4 \geq 1$ are added to (L, b). At the second iteration, the only uncontrolled siphon is $\{C_1, C_2\}$; the control place C_3 results, and the inequality $\mu(C_1) + \mu(C_2) \geq 1$ is added to (A, d). At the following step C.8, we have $A_p = [0\,0\,0\,0]$, $A_c = [0\,0]$, and

$$L = \begin{bmatrix} 1\,1\,0\,0 \\ 0\,0\,1\,1 \end{bmatrix} \text{ and } b = \begin{bmatrix} 1 \\ 1 \end{bmatrix}.$$

Therefore, $\mu_1 + \mu_2 + \mu_3 + \mu_4 \geq 3$ is added to (L, b). Note that the calculations at step C.8 correspond to the substitution of $\mu(C_1) = \mu_1 + \mu_2 - 1$ and $\mu(C_2) = \mu_3 + \mu_4 - 1$ in $\mu(C_1) + \mu(C_2) \geq 1$. At the third iteration there is no uncontrolled siphon. The procedure terminates with empty constraints (L_0, b_0) and

$$L = \begin{bmatrix} 1\,1\,0\,0 \\ 0\,0\,1\,1 \\ 1\,1\,1\,1 \end{bmatrix} \text{ and } b = \begin{bmatrix} 1 \\ 1 \\ 3 \end{bmatrix}.$$

The supervised Petri net is shown in Fig. 7.9. The supervisor not only prevents deadlock, but is also a least restrictive liveness-enforcing supervisor.

EXAMPLE 7.6

This example concludes our discussion on $\mathcal{T}$-liveness enforcing on the Petri net of Fig. 7.5(a). As discussed in Examples 7.3 and 7.4, two constraints are added during the iterations: $\mu_1 + \mu_2 + \mu_3 \geq 1$ and $\mu_1 + \mu_2 + \mu_{2,1} + \mu_{3,1} + \mu_{2,2} + \mu_{3,2} + \mu(C_1) \geq 1$. At the third iteration the procedure terminates, as there is no uncontrolled minimal active siphon with respect to the active subnet. Recall that the active subnet is given by the set of transitions $T^A = \{t_4, t_5\}$. Again, as shown in Example 7.4, the constraints in (L, b) at the third iteration are $\mu_1 + \mu_2 + \mu_3 - \mu_{1,2} - \mu_{2,2} - \mu_{3,2} \geq 1$ and $2\mu_1 + 2\mu_2 + \mu_3 + \mu_{2,1} + \mu_{3,1} - \mu_{1,2} - \mu_{c,1} - \mu_{c,2} - \mu_{c,3} \geq 2$, while (L_0, b_0) is empty. Consequently, the constraints (L, b) after step D are:

$$\mu_1 + \mu_2 + \mu_3 \geq 1, \tag{7.15}$$
$$2\mu_1 + 2\mu_2 + \mu_3 \geq 2. \tag{7.16}$$

The first constraint is redundant, so only the last constraint remains after step E. The procedure ends with

$$L = \begin{bmatrix} 2 & 2 & 1 \end{bmatrix} \text{ and } b = \begin{bmatrix} 2 \end{bmatrix}$$

and empty constraints (L_0, b_0). *This means that the target Petri net is* $\{t_4, t_5\}$-*live for all initial markings* μ_0 *satisfying* $2\mu_{01} + 2\mu_{02} + \mu_{03} \geq 2$, *when supervised according to* $L\mu \geq b$. *The supervisor enforcing* $L\mu \geq b$ *is shown in Fig. 7.5(g). It is the least restrictive* $\{t_4, t_5\}$-*liveness-enforcing supervisor.*

EXAMPLE 7.7

In this example the le-procedure is applied to the Petri net of Fig. 7.7(a) for full liveness enforcement. The intermediary Petri nets $\mathcal{N}_1$, $\mathcal{N}_2$, $\mathcal{N}_3$, *and* $\mathcal{N}_4$ *are represented in Figures 7.7 and 7.8, where the control places added to* $\mathcal{N}_1$, $\mathcal{N}_2$, *and* $\mathcal{N}_3$ *are connected with dashed lines.*

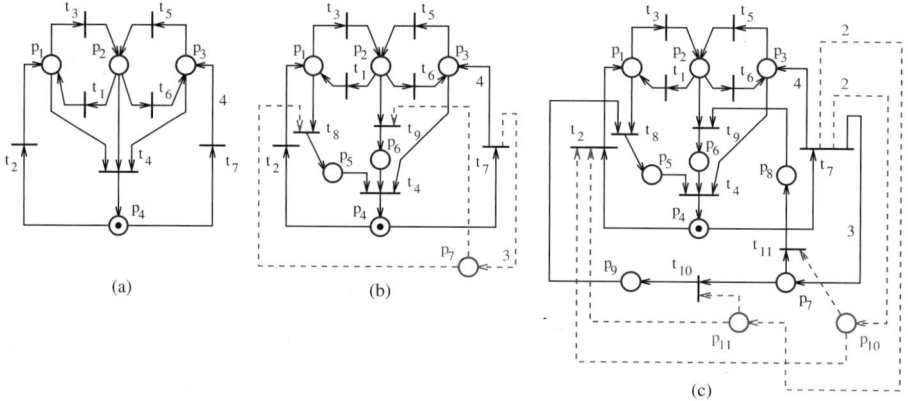

Fig. 7.7. Petri nets in Example 7.7: (a) $\mathcal{N}_0$, (b) $\mathcal{N}_1$, (c) $\mathcal{N}_2$.

In the first iteration there is a single minimal siphon, $\{p_1, p_2, p_3, p_4\}$, *and the control place* p_7 *is added. In the second iteration there are two new minimal siphons:* $\{p_4, p_5, p_7, p_8\}$ *and* $\{p_4, p_6, p_7, p_9\}$, *and two control places* p_{10} *and* p_{11}, *respectively, are thus added. In the third iteration there are two new minimal siphons:* $\{p_4, p_6, p_9, p_{10}, p_{15}\}$ *and* $\{p_4, p_5, p_8, p_{11}, p_{14}\}$, *and so the control places* p_{16} *and* p_{17}, *respectively, are added. At the fourth iteration no new minimal siphons are found, and so the procedure terminates. Note*

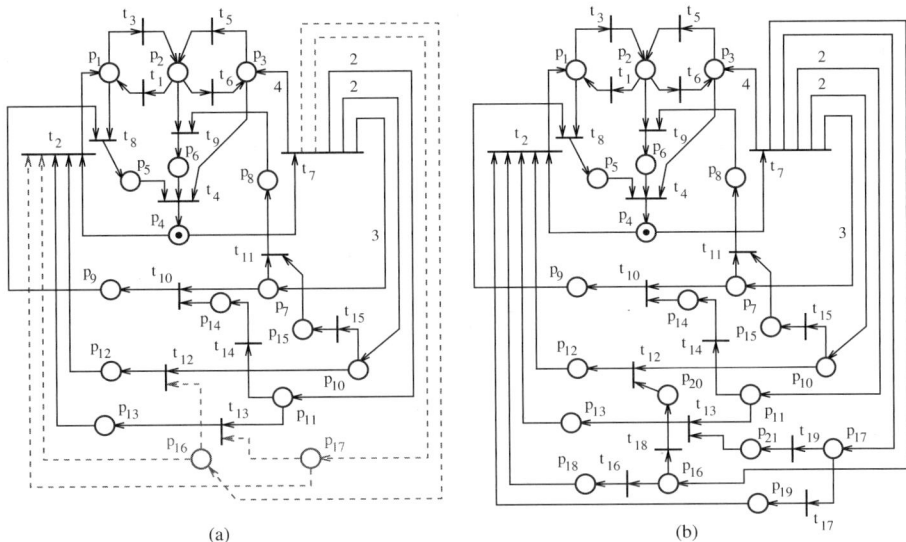

Fig. 7.8. Petri nets in Example 7.7: (a) $\mathcal{N}_3$, (b) $\mathcal{N}_4$.

that the places p_5, p_6, p_8, p_9, p_{12}, ..., p_{15}, p_{18}, ..., p_{21} and the transitions t_8, t_9, ..., t_{19} result through the transition splits of the AC-transformations. At the fourth iteration there are no constraints in (L_0, b_0); the constraints $L\mu \geq b$ are

$$\mu_1 + \mu_2 + \mu_3 + \mu_4 - \mu_8 - \mu_9 \geq 1,$$
$$\mu_1 + \mu_2 + \mu_3 + 2\mu_4 + \mu_5 - \mu_9 - \mu_{12} - \mu_{15} \geq 2,$$
$$\mu_1 + \mu_2 + \mu_3 + 2\mu_4 + \mu_6 - \mu_8 - \mu_{13} - \mu_{14} \geq 2,$$
$$\mu_1 + \mu_2 + \mu_3 + 3\mu_4 + \mu_5 + \mu_6 - \mu_{12} - \mu_{18} - \mu_{20} \geq 3,$$
$$\mu_1 + \mu_2 + \mu_3 + 3\mu_4 + \mu_5 + \mu_6 - \mu_{13} - \mu_{19} - \mu_{21} \geq 3.$$

The constraints above are enforced by p_7, p_{10}, p_{11}, p_{16}, and p_{17}, respectively. After removing the redundant constraints, the supervisor of $\mathcal{N}_0$ is defined by $L = [1, 1, 1, 3]$ and $b = 3$ and is the least restrictive liveness-enforcing supervisor. There are no constraints (L_0, b_0). The supervised Petri net is shown in Fig. 7.9.

Finally, note that the Petri net of this example serves also as an illustration of the difference that may exist between the supervisors designed by the le-procedure and the dp-procedure. Indeed, when the dp-procedure is applied to this net, it terminates in two iterations. At the first iteration we have that $\mathcal{N}_1$ is identical to $\mathcal{N}_0$, as the dp-procedure does not use AC-transformations. $\mathcal{N}_1$ has only one siphon: $S = P_0$, and so the inequality $\mu_1 + \mu_2 + \mu_3 + \mu_4 \geq 1$ is added to (L_0, b_0). At the second iteration the net *is unchanged, and so no new siphon is found. Therefore the dp-procedure terminates with $\mu_1 + \mu_2 + \mu_3 + \mu_4 \geq 1$ in (L_0, b_0) and empty constraints (L, b). This means that the Petri net is deadlock-free for all initial markings μ_0 satisfying $L_0\mu_0 \geq b_0$. However, as can easily be noticed, the Petri net is not live. The le-procedure has the advantage that the supervisors it designs are guaranteed to enforce $\mathcal{T}$-liveness. The supervisors of the dp-procedure are only guaranteed for deadlock prevention, however the dp-procedure is faster and more likely to converge. Nonetheless, note that the dp-procedure may often generate $\mathcal{T}$-liveness-enforcing supervisors. For instance, the dp-procedure generates a liveness-enforcing supervisor in Example 7.5. Another such instance appears when the dp-procedure is applied to the Petri net of Fig. 7.5(a), as it generates the same supervisor as the le-procedure.*

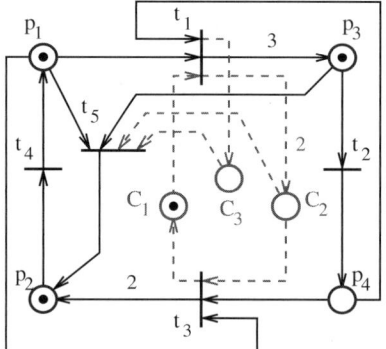 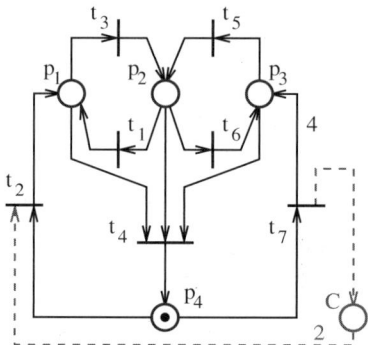

Fig. 7.9. The target Petri nets in Example 7.5 (left) and Example 7.7 (right) with their liveness-enforcing supervisors.

7.7 Properties

This section proves that the procedure is correct, and derives permissiveness results for the supervisors generated by the procedure. First, additional notation and definitions are introduced in section 7.7.1. Then, the correctness proofs are presented in section 7.7.2. Finally, section 7.7.3 proves permissiveness results for the supervisors generated by the procedure.

7.7.1 Preliminaries

In principle, the intermediary Petri nets $\mathcal{N}_i$ generated in the iterations of the procedure could be arbitrarily marked. However, there is a special class of markings of interest in our analysis, that we define below.

Definition 7.2. *A marking μ of $\mathcal{N}_i$ is* **valid** *if it satisfies the following:*

1. *For all control places added in the iterations $1 \ldots i-1$ the invariant equations of the form (7.12) hold true.*
2. *$\mu(p) = 0$ for all places p that are not control places or places of $\mathcal{N}_0$.*

When a marking is valid, the control places in the net are marked in accordance with the supervision based on invariants method, which is used to generate the control places. This is the requirement (a) of the definition. The requirement (b) introduces the additional restriction that only control places and places of the target net may have nonzero markings. This allows comparing markings of $\mathcal{N}_0$ with markings of $\mathcal{N}_i$. As an example, consider the Petri net $\mathcal{N}_3$ in Fig. 7.5(f). The control places are C_1 and C_2, while the places of $\mathcal{N}_0$ are p_1, p_2, p_3. The invariants (7.12) of C_1 and C_2 are (7.13) and (7.14), respectively (see Example 7.4). Let μ the marking of $\mathcal{N}_3$ in Fig. 7.5(f). Then it can easily be checked that μ is valid. However, μ' reached by firing $t_{1,1}$ from μ is not valid, as $\mu'_{1,1} \neq 0$. Furthermore, a marking μ'' differing from μ only in $\mu''_1 = 2$ is not valid, as μ'' does not satisfy (7.13) and (7.14).

Definition 7.3. *Two valid markings μ_i and μ_j of $\mathcal{N}_i$ and $\mathcal{N}_j$ are* **equivalent** *if $\mu_i(p) = \mu_j(p)$ for all places p of $\mathcal{N}_0$.*

Recall that the procedure generates Petri nets $\mathcal{N}_i$ such that $P_0 \subseteq P_1 \subseteq P_2 \subseteq \ldots P_i \ldots$. The equivalence relation defined above allows us to compare markings of different nets $\mathcal{N}_i$ and $\mathcal{N}_j$.

Finally, a notation for the sequences of split transitions is introduced. Both the PT- and AC-transformations (section 7.5.4) perform transition splits. A transition t_i may be split in more than just one iteration, and the transitions $t_{i,k}$ that result from splitting t_i may also be split in subsequent iterations. Given a transition t of $\mathcal{N}_0$ and an iteration j, let $\sigma_{0,j}(t)$ denote an arbitrary transition sequence of $\mathcal{N}_j$ such that (a) $\sigma_{0,j}(t)$ enumerates the transitions (including t itself) in which t of $\mathcal{N}_0$ is successively split until (and including) the iteration $j-1$, and (b) valid markings μ of $\mathcal{N}_j$ exist such that μ enables $\sigma_{0,j}(t)$. In this way firing $\sigma_{0,j}(t)$ in $\mathcal{N}_j$ corresponds to firing t in $\mathcal{N}_0$. If t is not split, we let $\sigma_{0,j}(t) = t$. The notation $\sigma_{i,j}(t)$ for $i < j$ and t in $\mathcal{N}_i$, is similarly defined by taking $\mathcal{N}_i$ instead of $\mathcal{N}_0$. If $\sigma = t_1 t_2 t_3 \ldots$, we let $\sigma_{i,j}(\sigma) = \sigma_{i,j}(t_1)\sigma_{i,j}(t_2)\sigma_{i,j}(t_3)\ldots$. For instance, in Fig. 7.5, $\sigma_{0,2}(t_2) = t_{2,1}t_2$, and in Fig. 7.7, $\sigma_{0,1}(t_4)$ is any of $t_8 t_9 t_4$ and $t_9 t_8 t_4$ and $\sigma_{2,3}(t_{10}) = t_{14}t_{10}$.

The technical result below is used in the correctness proofs that follow next.

Proposition 7.4. *Let μ be a valid marking of $\mathcal{N}_k$, σ an enabled firing sequence, and $t \in T_0$. Assume that t appears in σ. Then each transition $t_i \neq t$ of $\sigma_{0,k}(t)$ appears in σ before the first occurrence of t in σ; let s be the sequence in which these transitions appear in σ before the first occurrence of t in σ. There is a subsequence s_0 of s such that the sequence $s_0 t$ equals a $\sigma_{0,k}(t)$.*

Proof. Let P_R be the set of places that resulted through split operations in the iterations $1 \ldots k - 1$. The marking μ is valid, so t cannot be fired unless the places $\bullet t \cap P_R$ are marked, which cannot become marked unless the transitions in $\bullet(\bullet t \cap P_R)$ are fired. Next, let $T_{x1} = \bullet(\bullet t \cap P_R)$. The transitions of T_{x1} cannot fire unless the places $\bullet T_{x1} \cap P_R$ are marked, which cannot happen unless the transitions in $\bullet(\bullet T_{x1} \cap P_R)$ fire before. Let $T_{x2} = \bullet(\bullet T_{x1} \cap P_R)$. We continue in the same way until we get $T_{xk} = \emptyset$. This proves the first part of the proposition, as the transitions of $\sigma_{0,k}(t)$ are $\{t\} \cup T_{x1} \cup \ldots \cup T_{xk-1}$.

Given a transition t_i, let $T_x(t_i) = \bullet(\bullet t_i \cap P_R)$. Let t_1 be the last transition from $T_x(t)$ appearing in s before t. Let t_2 be the last transition from $(T_x(t) \cup T_x(t_1)) \setminus \{t_1\}$ appearing in s before t_1. Let t_3 be the last transition from $(T_x(t) \cup T_x(t_1) \cup T_x(t_2)) \setminus \{t_1, t_2\}$ appearing in s before t_2. We continue this way until t_m such that $(T_x(t) \cup \bigcup_{i=1}^{m} T_x(t_i)) \setminus \{t_1, t_2, \ldots, t_m\} = \emptyset$. Let s_0 be the sequence $t_m, t_{m-1}, \ldots, t_1, t$. By construction, s_0 is a sequence $\sigma_{0,k}(t)$. □

7.7.2 Proof of Correctness

The next result proves that the supervisors generated by the dp-procedure prevent deadlock and that the supervisors generated by the le-procedure enforce $\mathcal{T}$-liveness. The assumptions are that a $\mathcal{T}$-liveness enforcement is possible for some initial marking and that the procedure terminates. In view of Definition 6.11 and Lemma 6.7, the first assumption ensures that a $\mathcal{T}$-minimal active subnet exists. When no $\mathcal{T}$-minimal active subnet exists, the procedure terminates at step B and declares that there is no initial marking for which $\mathcal{T}$-liveness can be enforced.

Theorem 7.5. *Assume that a $\mathcal{T}$-liveness-enforcing supervisor exists for some initial marking of $\mathcal{N}_0$. Then:*

(a) *If the dp-procedure terminates, $(\mathcal{N}_0, \mu_0)$ supervised according to $L\mu \geq b$ is deadlock-free for all initial markings μ_0 satisfying $L\mu_0 \geq b$ and $L_0\mu_0 \geq b_0$.*
(b) *If the le-procedure terminates, $(\mathcal{N}_0, \mu_0)$ supervised according to $L\mu \geq b$ is $\mathcal{T}$-live for all initial markings μ_0 satisfying $L\mu_0 \geq b$ and $L_0\mu_0 \geq b_0$.*

Proof. (a) The proof is organized as follows. Let k be the number of the last iteration. First it is proved that for any marking μ of $\mathcal{N}_0$ satisfying $L\mu \geq b$ and $L_0\mu \geq b_0$, the equivalent marking μ_k of $\mathcal{N}_k$ exists, and $(\mathcal{N}_k, \mu_k)$ is deadlock-free. Then it is proved that assuming $(\mathcal{N}_0, \mu)$ in deadlock contradicts that $(\mathcal{N}_k, \mu_k)$ is deadlock-free.

Let (L_k, b_k) and $(L_{0,k}, b_{0,k})$ be the sets of constraints (L, b) and (L_0, b_0) at the end of iteration $k - 1$. The final set of constraints (L, b) and (L_0, b_0) is obtained from $(L_k|_{\mathcal{N}_0}, b_k)$[8] and $(L_{0,k}|_{\mathcal{N}_0}, b_{0,k})$, after removing redundant constraints at step E. Let μ be a marking of $\mathcal{N}_0$, μ_k a marking of $\mathcal{N}_k$, $\mu_{k,p} =$

---

[8] $L_k|_{\mathcal{N}_0}$ is L_k restricted to the columns corresponding to the places of $\mathcal{N}_0$.

$\mu_k|_{P_k \setminus C}$, and $\mu_{k,c} = \mu_k|_C$. Assume that $\mu_k|_{P_0} = \mu$ and $\mu_k(p) = 0 \; \forall p \in P_k \setminus (P_0 \cup C)$. Then $L\mu \geq b$ and $L_0\mu \geq b_0$ imply $L_k\mu_{p,k} \geq b_k$ and $L_{0,k}\mu_{p,k} \geq b_{0,k}$. Furthermore, $L_k\mu_{p,k} \geq b_k$ implies that we can define $\mu_{k,c} = L_k\mu_{p,k} - b_k$. Thus μ_k is by construction valid and equivalent to μ. Since the procedure terminates at iteration k, $\mathcal{N}_k$ contains no uncontrolled active siphons, and so $(\mathcal{N}_k, \mu_k)$ is deadlock-free by Proposition 6.15.

Let $\mathcal{N}_S$ be the closed loop of $\mathcal{N}_0$ and the supervisor enforcing $L\mu \geq b$ (Theorem 3.1). Assume that from an initial marking μ_0 of $\mathcal{N}_0$ satisfying $L\mu_0 \geq b$ and $L_0\mu_0 \geq b_0$, the supervised net can reach a marking μ_S of total deadlock. We show that this leads to contradiction. Let $\mu = \mu_S|_{\mathcal{N}_0}$, and let $\mu_{0,k}$ and μ_k be the equivalent markings of μ_0 and μ in $\mathcal{N}_k$. Since $(\mathcal{N}_k, \mu_k)$ is deadlock-free, μ_k enables an infinite firing sequence σ. Let $T_R = T_k \setminus T_0$. Thus, T_R is the set of transitions that appeared by transition split operations in all iterations. Firing any transition $t_x \in T_R$ always reduces the marking of some places in $P_0 \cup C$ (Proposition 7.1), while firing $t_x \in T_0$ may increase the marking of some places in $P_0 \cup C$. Because the total marking of $P_0 \cup C$ is finite, σ must include transitions $t_x \in T_0$. Let t_1 be the first transition in T_0 that appears in σ. Then we can write σ as $\sigma = \sigma_1\sigma_1'$, where t_1 appears only once in σ_1. By Proposition 7.4, σ_1 contains a subsequence $\sigma_{0,k}(t_1)$. Since all transitions of σ before t_1 are in T_R, and firing them only decreases markings of $P_0 \cup C$, $\sigma_{0,k}(t_1)$ is enabled by μ_k. But this implies that μ enables t_1 in $\mathcal{N}_S$, which contradicts that $(\mathcal{N}_S, \mu_S)$ is in deadlock.

(b) The proof is similar to that of part (a) and shares the notation of part (a). The proof is organized as follows. As shown in part (a), for any marking μ of $\mathcal{N}_0$ satisfying $L\mu \geq b$ and $L_0\mu \geq b_0$, the equivalent marking μ_k of $\mathcal{N}_k$ exists; here it is shown that $(\mathcal{N}_k, \mu_k)$ is $\mathcal{T}$-live. Then it is proved that assuming $(\mathcal{N}_0, \mu)$ not $\mathcal{T}$-live contradicts that $(\mathcal{N}_k, \mu_k)$ is $\mathcal{T}$-live.

Let μ_k be the valid marking constructed at the beginning of the proof of part (a) to be equivalent to an arbitrary marking μ of $\mathcal{N}_0$ satisfying $L\mu \geq b$ and $L_0\mu \geq b_0$. Since the procedure terminates at iteration k, $\mathcal{N}_k$ contains no uncontrolled active siphons, and so $(\mathcal{N}_k, \mu_k)$ is $\mathcal{T}$-live by Theorem 6.19.

Assume that from an initial marking μ_0 of $\mathcal{N}_0$ satisfying $L\mu_0 \geq b$ and $L_0\mu_0 \geq b_0$, the supervised net $\mathcal{N}_S$ can reach a marking μ_S for which a transition $t \in \mathcal{T}$ is dead. We show that this leads to contradiction. Let $\mu = \mu_S|_{\mathcal{N}_0}$, and let $\mu_{0,k}$ and μ_k be the equivalent markings of μ_0 and μ in $\mathcal{N}_k$. Since $(\mathcal{N}_k, \mu_k)$ is $\mathcal{T}$-live, μ_k enables a transition sequence σ in $\mathcal{N}_k$ which includes t. Let t_1 be the first transition in T_0 that appears in σ. Then we can write σ as $\sigma = \sigma_1\sigma_1'$, where t_1 appears only once in σ_1. By Proposition 7.4, σ_1 contains a subsequence $\sigma_{0,k}(t_1)$. Since all transitions of σ before t_1 are in T_R, and firing them only decreases markings of $P_0 \cup C$ (Proposition 7.1), $\sigma_{0,k}(t_1)$ is enabled by μ_k. Let t_2 be the next transition of σ in T_0. Similarly, $\sigma_{0,k}(t_1)\sigma_{0,k}(t_2)$ is enabled by μ_k. We continue this way and eventually find t_j in σ and in T_0 such that $t_j = t$. We have that μ_k enables $\sigma_{0,k}(t_1)\sigma_{0,k}(t_2)\ldots\sigma_{0,k}(t_j)$. But this implies that μ enables $t_1t_2\ldots t_j$ in $\mathcal{N}_S$, and since $t_j = t$, t is not dead in $(\mathcal{N}_S, \mu_S)$, which is a contradiction. $\qquad \square$

7.7.3 Permissiveness Properties

The supervisors generated by the procedure are at least as permissive as the least restrictive $\mathcal{T}$-liveness-enforcing supervisor for a large class of Petri nets. This class of Petri nets includes the Petri nets with a single $\mathcal{T}$-minimal active subnet, as shown in the next theorem. Before stating the formal result, the notion of a least restrictive supervisor needs to be clarified, as the supervisors generated by our procedure are defined on a set of initial markings rather than on a single initial marking. We say that a supervisor generated by the procedure is at least as permissive as the least restrictive $\mathcal{T}$-liveness-enforcing supervisor when for all initial markings μ_0 of $\mathcal{N}_0$ the following are satisfied:

- If $L\mu_0 \not\geq b$ or $L_0\mu_0 \not\geq b_0$, no $\mathcal{T}$-liveness-enforcing supervisor of $(\mathcal{N}_0, \mu_0)$ exists.
- If $L\mu_0 \geq b$ and $L_0\mu_0 \geq b_0$, the supervisor enforcing $L\mu \geq b$ is at least as permissive as the least restrictive $\mathcal{T}$-liveness-enforcing supervisor of $(\mathcal{N}_0, \mu_0)$.

We say that a supervisor generated by the procedure is least restrictive when for all initial markings μ_0 of $\mathcal{N}_0$ the following are satisfied:

- If $L\mu_0 \not\geq b$ or $L_0\mu_0 \not\geq b_0$, no $\mathcal{T}$-liveness-enforcing supervisor of $(\mathcal{N}_0, \mu_0)$ exists.
- If $L\mu_0 \geq b$ and $L_0\mu_0 \geq b_0$, the supervisor enforcing $L\mu \geq b$ is the least restrictive $\mathcal{T}$-liveness-enforcing supervisor of $(\mathcal{N}_0, \mu_0)$.

As usual, in the next theorem *the procedure* can be either the dp-procedure or the le-procedure, depending on which operation is selected by the input arguments. Thus, for the case in which the procedure is used as the dp-procedure, the theorem gives a sufficient condition for the generated supervisor to be at least as permissive as the least restrictive $\mathcal{T}$-liveness-enforcing supervisor. This does not mean that the generated supervisor enforces $\mathcal{T}$-liveness, but only that the supervisor is not more restrictive than the least restrictive $\mathcal{T}$-liveness-enforcing supervisor. Finally, in the case of the le-procedure, the theorem gives a sufficient condition for the generated supervisor to be the least restrictive $\mathcal{T}$-liveness-enforcing supervisor.

Theorem 7.6. *Assume that the procedure terminates and $\mathcal{N}_1$ has a single $\mathcal{T}$-minimal active subnet. Then the procedure provides a supervisor at least as permissive as the least restrictive $\mathcal{T}$-liveness-enforcing supervisor.*

Proof. The proof is organized as follows. Let μ_0 be a marking of $\mathcal{N}_0$ and $\mu_{0,i}$ an equivalent marking of $\mathcal{N}_i$. We prove that $(\mathcal{N}_0, \mu_0)$ cannot be made $\mathcal{T}$-live if $(\mathcal{N}_i, \mu_{0,i})$ cannot be made $\mathcal{T}$-live. Then we use this fact to prove that no $\mathcal{T}$-liveness supervisors exist for the initial markings μ_0 which do not satisfy $L\mu_0 \geq b$ and $L_0\mu_0 \geq b_0$. Finally, given μ_0 satisfying $L\mu_0 \geq b$ and $L_0\mu_0 \geq b_0$, we prove that the supervisor enforcing $L\mu \geq b$ is at least as permissive as the least restrictive $\mathcal{T}$-liveness-enforcing supervisor of $\mathcal{N}_0$. Note that the existence of a (least restrictive) $\mathcal{T}$-liveness-enforcing supervisor is guaranteed by the fact that $\mathcal{N}_1$ has a $\mathcal{T}$-minimal active subnet; see Theorem 6.10 and Definition 6.18.

To prove our first claim, we prove by contradiction that $(\mathcal{N}_i, \mu_{0,i})$ cannot be made $\mathcal{T}$-live if $(\mathcal{N}_{i+1}, \mu_{0,i+1})$ cannot be made $\mathcal{T}$-live, where $i \geq 0$ and $\mu_{0,i+1}$ is the equivalent marking of $\mu_{0,i}$. For $i = 0$, assume that $(\mathcal{N}_0, \mu_0)$ can be made $\mathcal{T}$-live when $(\mathcal{N}_1, \mu_{0,1})$ cannot be made $\mathcal{T}$-live. Then μ_0 enables an infinite transition sequence σ in which all transitions of $\mathcal{T}$ appear infinitely often. But this implies that $\sigma_{0,1}(\sigma)$ is also enabled by $\mu_{0,1}$, contradicting the assumption that $(\mathcal{N}_1, \mu_{0,1})$ cannot be made $\mathcal{T}$-live. For $i \geq 1$, assume that $(\mathcal{N}_i, \mu_{0,i})$ can be made $\mathcal{T}$-live when $(\mathcal{N}_{i+1}, \mu_{0,i+1})$ cannot be made $\mathcal{T}$-live. Let σ be an infinite firing sequence enabled by $\mu_{0,i}$ such that all transitions of $\mathcal{T}$ occur infinitely often in σ. Since $(\mathcal{N}_{i+1}, \mu_{0,i+1})$ cannot be made $\mathcal{T}$-live, $\sigma' = \sigma_{i,i+1}(\sigma)$ is not enabled in $\mathcal{N}_{i+1}$. Then $\sigma = \sigma_1 t_1 \sigma_2$, $\mu_{0,i} \xrightarrow{\sigma_1} \mu_1$, $\mu_{0,i+1} \xrightarrow{\sigma_{i,i+1}(\sigma_1)} \mu_1'$, μ_1 enables t_1, but μ_1' does not enable $\sigma_{i,i+1}(t_1)$. This corresponds to the following: $\mathcal{N}_i$ has an active siphon S_1 which is controlled in $\mathcal{N}_{i+1}$ with C_1, and $\mu_1'(C_1)$ does not allow $\sigma_{i,i+1}(t_1)$ to fire. Hence $t_1 \in C_1\bullet$ was satisfied when C_1 was added to $\mathcal{N}_i$. This implies $t_1 \in S_1\bullet$. Firing $\sigma_{i,i+1}(t_1)$ in $\mathcal{N}_{i+1}$ produces the same marking change for the places in P_i as firing t_1 in $\mathcal{N}_i$. Since $\sigma_{i,i+1}(t_1)$ is not allowed by $\mu_1'(C_1)$ to fire, firing t_1 from μ_1 empties S_1. Since t_1 is fired in the sequence $\sigma = \sigma_1 t_1 \sigma_2$, S_1 is an empty active siphon of $(\mathcal{N}_i, \mu_1)$. An empty active siphon implies a set T_x of dead transitions from the active subnet. Therefore the transitions in T_x do not appear infinitely often in σ. Let $T_{x1} = \{t \in T_1^A : \exists t_u \in \sigma_{1,i}(t) \text{ and } t_u \in T_x\}$. The active subnets $\mathcal{N}_i^A$ for $i > 1$ are computed using the update algorithm of section 7.5.6, so $T_{x1} \neq \emptyset$. Using the same construction as in the proof of Theorem 7.5(b), the projection of σ on T_1 (let it be σ^1) is enabled by $\mu_{1,0}$, where $\mu_{1,0}$ is the restriction of $\mu_{i,0}$ to the places of P_1. Note that the transitions of T_{x1} do not appear infinitely often in σ^1. We apply Lemma 6.7 for $\mathcal{N}_1$ and σ^1, and using the notation of Lemma 6.7, we let $T_x^A = \|x\|$; T_x^A defines an active subnet and $\mathcal{T} \subseteq T_x^A$, as all transitions of $\mathcal{T}$ appear infinitely often in σ^1. However T_1^A is not a subset of T_x^A, for $T_{x1} \subseteq T_1^A \setminus T_x^A$. Therefore $\mathcal{N}_1^A$ is not the single $\mathcal{T}$-minimal subnet, which contradicts the theorem assumptions.

The second part of the proof, showing that all $\mathcal{T}$-liveness-enforcing supervisors forbid the markings such that $L\mu \not\geq b$ or $L_0\mu \not\geq b_0$, is also by contradiction. Assume that $\mathcal{N}_0$ can be made $\mathcal{T}$-live for a marking μ_0 which does not satisfy all constraints $L\mu \geq b$ and $L_0\mu \geq b_0$. Let (L_d, b_d) and $(L_{0,d}, b_{0,d})$ be the constraints (L, b) and (L_0, b_0) before step D. Since step D only removes redundant constraints, μ_0 does not satisfy all constraints of $L_d\mu \geq b_d$ and $L_{0,d}\mu \geq b_{0,d}$. Let i be the first iteration in which an inequality $l_1'\mu \geq b_1$ is added such that its restriction $l_1\mu \geq b_1$ to P_0 is one of the inequalities of $L_d\mu \geq b_d$ and $L_{0,d}\mu \geq b_{0,d}$ not satisfied by μ_0. The markings forbidden at every iteration i are those for which there are empty active siphons. Therefore $\mathcal{N}_i$ has an empty active siphon for $\mu_{0,i}$, where $\mu_{0,i}$ is the equivalent marking of μ_0 in $\mathcal{N}_i$. As shown in the previous paragraph, this implies that $(\mathcal{N}_i, \mu_{0,i})$

cannot be made $\mathcal{T}$-live. Then $(\mathcal{N}_0, \mu_0)$ cannot be made $\mathcal{T}$-live, which is a contradiction.

Finally, let μ_0 be a marking satisfying $L\mu_0 \geq b$ and $L_0\mu_0 \geq b_0$. Let Ξ_0 be the supervisor enforcing $L\mu \geq b$ on $(\mathcal{N}_0, \mu_0)$. Assume there is a $\mathcal{T}$-liveness-enforcing supervisor Ξ less restrictive than Ξ_0. We show that this leads to contradiction. Let $(\mathcal{N}_0, \mu_0, \Xi_0)$ and $(\mathcal{N}_0, \mu_0, \Xi)$ be the closed loops of $(\mathcal{N}_0, \mu_0)$ with Ξ_0 and Ξ, respectively. Then there is a (possibly empty) firing sequence σ enabled from μ_0 in both $(\mathcal{N}_0, \mu_0, \Xi_0)$ and $(\mathcal{N}_0, \mu_0, \Xi)$, such that $\mu_0 \xrightarrow{\sigma} \mu$ and $\exists t \in T_0$, t is enabled by μ, t is allowed to fire at μ by Ξ, and t is not allowed to fire at μ by Ξ_0. Then the marking μ' such that $\mu \xrightarrow{t} \mu'$ satisfies $L\mu' \not\geq b$. Therefore, by the previous part of the proof, $\mathcal{T}$-liveness cannot be enforced in $(\mathcal{N}_0, \mu')$. Then Ξ is not a $\mathcal{T}$-liveness-enforcing supervisor of $(\mathcal{N}_0, \mu_0)$, which is a contradiction. $\square$

Note that in case of liveness enforcement, $\mathcal{T}$ equals the whole set of transitions T_0 of $\mathcal{N}_0$. Then the only possible $\mathcal{T}$-minimal active subnet is the whole net. Consequently, in view of Theorem 7.5(b), Theorem 7.6 has the following corollary.

Corollary 7.7. *Assume that liveness is enforcible in $\mathcal{N}_0$ for some initial marking and that the le-procedure terminates. If $\mathcal{T} = T_0$, the le-procedure provides the least restrictive liveness-enforcing supervisor.*

The previous corollary can be stated in a more general form, allowing one to characterize the permissiveness of the supervisors in the case when $\mathcal{N}_1$ has more than a single $\mathcal{T}$-minimal active subnet. Let $T^A = T_1^A \cap T_0$, where T_1^A is the set of transitions of the $\mathcal{T}$-minimal active subnet $\mathcal{N}_1^A$ of $\mathcal{N}_1$ that is computed at step B. Then, even though $\mathcal{N}_1^A$ may not be the only $\mathcal{T}$-minimal active subnet of $\mathcal{N}_1$, it is still the only T_1^A-minimal active subnet, and also the only T^A-minimal active subnet. Thence the next corollary follows. (Note that $\mathcal{T} \subseteq T^A$.)

Corollary 7.8. *Assume that $\mathcal{T}$-liveness is enforcible in $\mathcal{N}_0$ for some initial marking and that the procedure terminates. The generated supervisor is at least as permissive as the least restrictive T^A-liveness-enforcing supervisor.*

An important consequence of Theorem 7.6 and Theorem 7.5(b) is that the le-procedure will not terminate for a Petri net $\mathcal{N}_0$ with a single $\mathcal{T}$-minimal subnet when the set of markings for which $\mathcal{T}$-liveness can be enforced cannot be represented as a conjunction of linear marking inequalities.

Corollary 7.9. *Assume that $\mathcal{N}_1$ has a single $\mathcal{T}$-minimal active subnet and that the markings of $\mathcal{N}_0$ for which $\mathcal{T}$-liveness enforcement is possible cannot be represented as the markings μ satisfying a set of linear inequalities $M\mu \geq g$. Then the le-procedure diverges.*

Note that the divergence condition of the previous corollary is only sufficient. It will be shown later that the le-procedure can diverge in other conditions as well. The dp-procedure can also diverge. For instance, for Petri nets for
which it can be proved that the dp-procedure generates $\mathcal{T}$-liveness-enforcing
supervisors if it terminates, the dp-procedure diverges when the assumptions
of Corollary 7.9 apply. The discussion on the procedure convergence will be
carried out in more detail in section 8.8.

Finally, note that two classes of Petri nets for which the supervisors generated by the dp-procedure are guaranteed to enforce $\mathcal{T}$-liveness are characterized in Theorem 6.8 and Theorem 6.9. Identifying such Petri nets is of
interest, as using the dp-procedure instead of the le-procedure has computational advantages.

7.8 Extending the Permissiveness of the Procedure

Theorem 7.6 and also Corollary 7.7 show that for a large class of Petri nets
and liveness specifications, the procedure generates supervisors at least as
permissive as the least restrictive $\mathcal{T}$-liveness-enforcing supervisor. The natural
question of whether we can use our procedure to ensure this property for an
even larger class of Petri nets has a positive answer, as we show in this section.
We consider the case when $\mathcal{N}_1$ has the $\mathcal{T}$-minimal active subnets $\mathcal{N}_1^{A,1}$, $\mathcal{N}_1^{A,2}$,
..., $\mathcal{N}_1^{A,p}$. Let $\mathcal{N}_0^{A,1}$, $\mathcal{N}_0^{A,2}$, ..., $\mathcal{N}_0^{A,p}$ be the corresponding $\mathcal{T}$-minimal active
subnets in $\mathcal{N}_0$. Theorem 7.6 does not apply, as we have p $(p > 1)$ $\mathcal{T}$-minimal
subnets. However, if $\mathcal{T}$ were equal to any of $T_0^{A,i}$, Theorem 7.6 would apply,
as there is a single $T_0^{A,i}$-minimal active subnet: $\mathcal{N}_1^{A,i}$. $(T_0^{A,i}$ denotes the set
of transitions of $\mathcal{N}_0^{A,i}$ for $i = 1 \ldots p$.) Assume that the procedure terminates
for all $i = 1 \ldots p$ when the input argument $\mathcal{T}$ is set to $T_0^{A,i}$. Let $L^{(i)}\mu \geq b^{(i)}$
and $L_0^{(i)}\mu \geq b_0^{(i)}$ be the generated constraints for each $i = 1 \ldots p$. Let $\varXi$ be
the supervisor defined as follows. $\varXi$ requires the initial marking μ_0 to be in
the set $\mathcal{M}$, where

$$\mathcal{M} = \bigcup_{i=1}^{p} \left\{ \mu : L^{(i)}\mu \geq b^{(i)} \text{ and } L_0^{(i)}\mu \geq b_0^{(i)} \right\}.$$

Furthermore, $\varXi$ allows a transition to fire only if the next reached marking
is in $\mathcal{M}$. Clearly, regardless of which of the dp- or le-procedure was used for
$i = 1 \ldots p$, $\varXi$ prevents deadlock. Furthermore, $\varXi$ enforces $\mathcal{T}$-liveness if the
le-procedure was used for all $i = 1 \ldots p$.

Theorem 7.10. *Assume that for each $i = 1 \ldots p$ the procedure terminates.
Then $\varXi$ is at least as permissive as the least restrictive $\mathcal{T}$-liveness-enforcing
supervisor.*

Proof. The proof is by contradiction. Assume that there is a $\mu_0 \notin \mathcal{M}$ such
that μ_0 enables a firing sequence σ which includes all transitions in $\mathcal{T}$ infinitely

often. In the notation of Lemma 6.7, let $T^A = \|x\|$. Then T^A defines an active subnet, and note that $\mathcal{T} \subseteq T^A$. Since $\mathcal{N}_0^{A,i}$, $i = 1 \ldots p$, are all the $\mathcal{T}$-minimal active subnets of $\mathcal{N}_0$, there is a j, $1 \leq j \leq p$, such that $T_0^{A,j} \subseteq T^A$. This leads to contradiction, since by Theorem 7.6 not all transitions of $T_0^{A,j}$ can be made live for $\mu_0 \notin \mathcal{M}$, and so not all of them can appear in σ infinitely often.   □

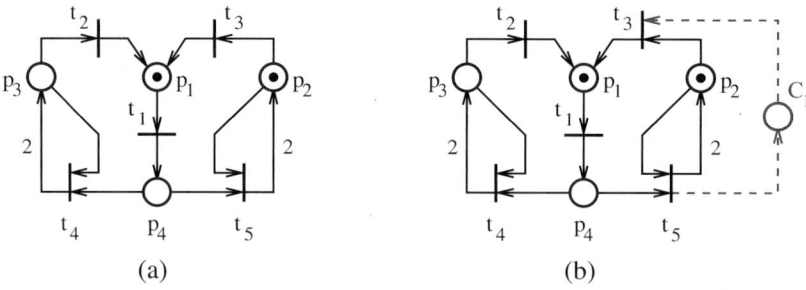

Fig. 7.10. (a) A Petri net with two $\{t_1\}$-minimal active subnets. (b) Resulting supervisor after selecting one of the two active subnets.

EXAMPLE 7.8

Here we consider an example in which the supervisor generated by the le-procedure is not least restrictive due to the existence of several $\mathcal{T}$-minimal active subnets. The Petri net of Fig. 7.10(a) has two $\{t_1\}$-minimal active subnets, $\mathcal{N}_1^A$ and $\mathcal{N}_2^A$. They are given by $T_1^A = \{t_1, t_3, t_5\}$ and $T_2^A = \{t_1, t_2, t_4\}$, respectively. There are two cases, depending on which of $\mathcal{N}_1^A$ or $\mathcal{N}_2^A$ is selected by the procedure at step B.
If the procedure selects $\mathcal{N}_1^A$ as the $\mathcal{T}$-minimal active subnet, we have the following constraints:

$$\mu_2 \geq 1, \tag{7.17}$$

which is implemented by C_1 in Fig. 7.10(b), and

$$\mu_1 + \mu_2 + \mu_3 + \mu_4 \geq 2, \tag{7.18}$$

which is an (L_0, b_0) type constraint.
If the procedure selects $\mathcal{N}_2^A$ as the $\mathcal{T}$-minimal active subnet, the outcome is similar, except that instead of (7.17) we have

$$\mu_3 \geq 1. \tag{7.19}$$

Thus it can be seen that neither case produces the least restrictive supervisor. Nonetheless, the least restrictive supervisor can be obtained from the disjunction of the two supervisors corresponding to the two $\mathcal{T}$-minimal active subnets. Thus, the least restrictive $\{t_1\}$-liveness-enforcing supervisor enforces

$$\mu_2 \geq 1 \vee \mu_3 \geq 1 \tag{7.20}$$

and requires the initial marking μ_0 to satisfy (7.18) and (7.20).

8

Liveness Enforcement in Petri Nets: A Structural Approach. Part II

8.1 Introduction

The previous chapter has introduced the simplified dp- and le-procedures. This chapter presents the dp- and le-procedures in their general form. As discussed in the introduction of the previous chapter, the procedures in their general form are able to deal with partially controllable and observable Petri nets, and they accept initial constraints and initial-marking constraints. The setting of partial controllability and observability considered here is that in which controllable (observable) transitions can be individually controlled (observed).

The chapter is organized as follows. Section 8.2 presents the problem statement. Section 8.3 presents a motivation for the procedure features that are present only in the general form of the procedures. Section 8.4 defines the dp-procedure and the le-procedure. Section 8.5 provides examples illustrating the operation of the procedures. Section 8.6 contains theoretical results characterizing the performance of the procedures. Section 8.8 describes convergence issues.

8.2 Problem Statement

This section presents the purpose of the dp- and le-procedures defined in this chapter. Section 8.2.1 presents the problem statement for deadlock prevention, while section 8.2.2 presents the problem statement for $\mathcal{T}$-liveness enforcement.

8.2.1 Deadlock Prevention

In the deadlock prevention problem of this chapter, the input is a Petri net $\mathcal{N}$, and optionally any of the following: a set of transitions $\mathcal{T}$, initial constraints $L_I\mu \geq b_I$, and initial-marking constraints $L_{I0}\mu \geq b_{I0}$. The output consists of the two sets of marking constraints $L\mu \geq b$ and $L_0\mu \geq b_0$, and a set $\mathcal{T}'$. They are to satisfy that:

1. The constraints $L\mu \geq b$ are admissible with respect to the uncontrollable and unobservable transitions of the net.
2. The supervisor enforcing $L\mu \geq b$ via SBPI prevents deadlock for all initial markings μ_0 that satisfy $L\mu_0 \geq b$, $L_0\mu_0 \geq b_0$, $L_I\mu_0 \geq b_I$, and $L_{I0}\mu_0 \geq b_{I0}$.
3. The supervisor is not overly restrictive.
4. The supervisor is a good approximation of a $\mathcal{T}$-liveness enforcement supervisor.
5. The set $\mathcal{T}' \subseteq \mathcal{T}$ excludes any transitions of $\mathcal{T}$ that have been detected as having the property that they cannot be live in the closed loop.

Finally, the case when $\mathcal{T}$-liveness enforcement is impossible at all initial markings should be identified.

Note that the first and the fifth requirements are in addition to the requirements given in the problem statement of the previous chapter. The first requirement is necessary in order for the supervision to be feasible in partially controllable and observable Petri nets. The fifth requirement means the following. During the supervisor design process, it may be possible to identify transitions of $\mathcal{T}$ that the supervisor would cause to be dead in the closed loop. There are four factors that may cause such situations to arise: the partial controllability and observability of the net, the structure of the net, the initial constraints, and the initial-marking constraints. For instance, examples can easily be found to show that inappropriate initial constraints and initial-marking constraints may not allow $\mathcal{T}$-liveness enforcement in a Petri net.

8.2.2 $\mathcal{T}$-liveness Enforcement

In the $\mathcal{T}$-liveness enforcement problem of this chapter, the input is a Petri net $\mathcal{N}$, and optionally any of the following: a set of transitions $\mathcal{T}$, initial constraints $L_I\mu \geq b_I$, and initial-marking constraints $L_{I0}\mu \geq b_{I0}$. The output consists of the two sets of marking constraints $L\mu \geq b$ and $L_0\mu \geq b_0$, and a set $\mathcal{T}'$. They are to satisfy that:

1. The constraints $L\mu \geq b$ are admissible with respect to the uncontrollable and unobservable transitions of the net.
2. The supervisor enforcing $L\mu \geq b$ via SBPI enforces $\mathcal{T}'$-liveness for all initial markings μ_0 that satisfy $L\mu_0 \geq b$, $L_0\mu_0 \geq b_0$, $L_I\mu_0 \geq b_I$, and $L_{I0}\mu_0 \geq b_{I0}$.
3. The supervisor is not overly restrictive.
4. The set $\mathcal{T}' \subseteq \mathcal{T}$ should be as large as possible.

Finally, the case when $\mathcal{T}$-liveness enforcement is impossible at all initial markings should be identified.

Compared to the problem statements presented so far, the second and the last requirements are different. The second requirement mentions $\mathcal{T}'$-liveness instead of $\mathcal{T}$-liveness, as the latter may not be enforcible. Four factors that may cause $\mathcal{T}$-liveness enforcement to be impossible are: the partial controllability and observability of the net, the structure of the net, the initial constraints, and the initial-marking constraints. The last requirement is not to be taken in a strict sense; the procedure of this chapter is not designed to find the largest subset $\mathcal{T}'$ for which $\mathcal{T}'$-liveness enforcement is possible.

8.3 Motivation

Section 7.4 has motivated our approach for the case of fully controllable and observable Petri nets. Here, a motivation of the additional features of the procedures in the general case is presented. Section 8.3.1 illustrates the need for constraint transformations in the case of partially controllable and observable Petri nets. Section 8.3.2 shows that constraint transformations can be a source of deadlock. Section 8.3.3 illustrates the use of the set $\mathcal{T}'$. Finally, section 8.3.4 illustrates the use of initial constraints.

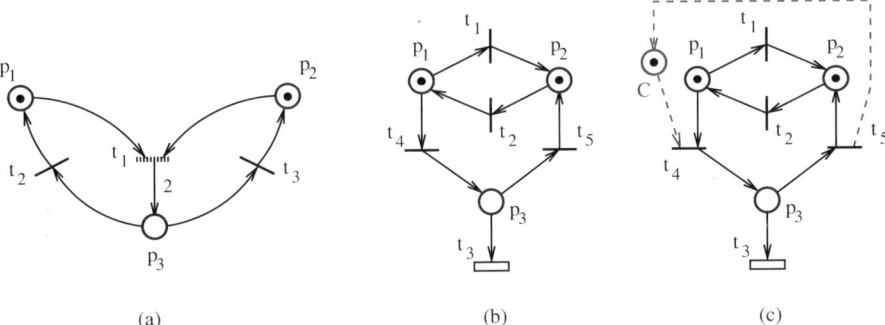

(a)   (b)   (c)

Fig. 8.1. Partially controllable and observable Petri nets.

8.3.1 Partially Controllable and Observable Petri Nets

This section shows that constraint transformations are necessary in the procedures when the target Petri net is partially controllable and observable. Consider the Petri net of Figure 8.1(a), where the transition t_1 is unobservable. As shown in section 7.4.1, liveness is enforced by ensuring that (7.1) and (7.2) are satisfied at all reachable markings. However, the constraints (7.1) and (7.2) are inadmissible, because enforcing them via SBPI (see Theorem 3.1) requires observing the firing of t_1. Nonetheless they can be transformed without loss of permissiveness to the admissible constraints

$$2\mu_1 + \mu_3 \geq 1, \tag{8.1}$$

$$2\mu_2 + \mu_3 \geq 1. \tag{8.2}$$

This example has shown that for partially controllable and observable Petri nets, inadmissible constraints may arise. For this reason the procedures include a function that checks the admissibility of the constraints and that transforms the constraints to an admissible form when they are inadmissible.

8.3.2 Constraint Transformations and Deadlock

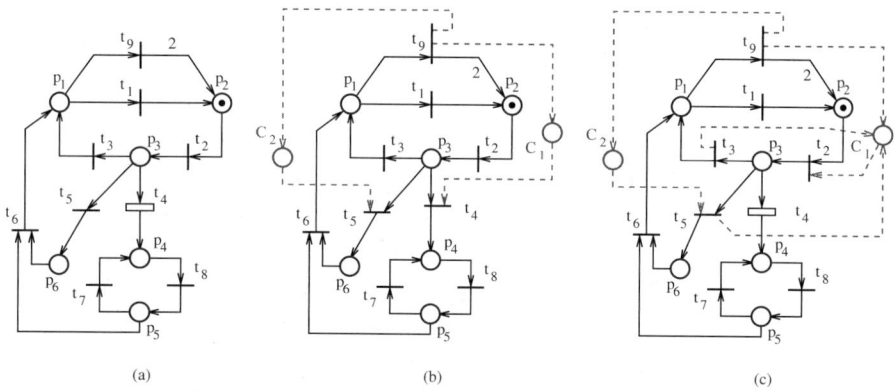

(a)   (b)   (c)

Fig. 8.2. Constraint transformations may lead to deadlock.

The previous chapter has presented procedures for deadlock prevention and $\mathcal{T}$-liveness enforcement for fully controllable and observable Petri nets. On the other hand, transformations to admissible constraints are readily available in the literature [141, 142]. This may suggest that the procedures of the previous chapter can be extended as follows to partially controllable and observable Petri nets: apply first the dp-/le-procedures, and then transform the resulting constraints $L\mu \geq b$ to be admissible. This section shows that such an approach can be inappropriate, as constraint transformations may cause new deadlock possibilities.

The Petri net of Figure 8.2(a) has all transitions controllable and observable, except for t_4, which is uncontrollable. Assume that a liveness-enforcing supervisor is to be designed. If t_4 were controllable, it would be enough to control the siphons $\{p_1, p_2, p_3, p_6\}$ and $\{p_1, p_2, p_3, p_4, p_5\}$. The resulting control places would be C_1 and C_2 shown in Figure 8.2(b), enforcing:

$$\mu_1 + \mu_2 + \mu_3 + \mu_6 \geq 1 \tag{8.3}$$

and

$$\mu_1 + \mu_2 + \mu_3 + \mu_4 + \mu_5 \geq 1. \tag{8.4}$$

As t_4 is uncontrollable, we have that (8.3) is inadmissible, while (8.4) is admissible. The constraint (8.3) can be transformed to the admissible constraint

$$\mu_1 + \mu_2 + \mu_6 \geq 1. \tag{8.5}$$

The control place C_1 enforcing the transformed constraint is shown in Figure 8.2(c). However, note that enforcing (8.4) and (8.5) fails to enforce liveness. Indeed, let μ be the marking shown in Figure 8.2(a); μ satisfies both (8.4) and (8.5). As expected, the closed loop of Figure 8.2(b) is live at the marking μ. However, the closed loop of Figure 8.2(c) is in deadlock at the same marking! This shows that the constraint transformation of (8.3) to the admissible form (8.5) creates a deadlock possibility that did not exist for the original constraint (8.3). To cope with this problem, the procedures presented in this chapter perform the constraint transformation during the iterations that remove the deadlock possibilities, rather than after completing the iterations.

8.3.3 The Set $\mathcal{T}'$

This section illustrates the role of the set $\mathcal{T}'$ generated by the procedures together with the constraints $L\mu \geq b$ and $L_0\mu \geq b_0$. In the case of the fully controllable and observable Petri nets with no initial constraints and no initial-marking constraints, the set of transitions that can be made live depends only on the structure of the Petri net. However, in the general case, removing deadlock possibilities may come at the cost of accepting some (less important) deadlocks in the system. The Petri net of Figure 8.1(b) is an example. All transitions are controllable and observable except for t_3, which is uncontrollable and observable. Assume that the supervisory purpose is $\mathcal{T}$-liveness with $\mathcal{T} = \{t_1, t_2, t_4, t_5\}$. Deadlock occurs in the system when the siphon $S = \{p_1, p_2, p_3\}$ is emptied of tokens. To prevent it, the constraint

$$\mu_1 + \mu_2 + \mu_3 \geq 1 \tag{8.6}$$

is to be enforced. However, this constraint is inadmissible, and it can be transformed to the admissible constraint

$$\mu_1 + \mu_2 \geq 1. \tag{8.7}$$

Figure 8.1(c) shows the control place C enforcing (8.7). Note that C only enforces $\{t_1, t_2\}$-liveness. In fact, it can be seen that this is the most that can be done to prevent deadlocks in our Petri net. Indeed, the only way in which the number of tokens of the system can be preserved is by avoiding the states which enable the uncontrollable transition t_3. However, this implies avoiding the firing of t_4, and hence of t_5. So we can say that C enforces $\mathcal{T}'$-liveness instead of $\mathcal{T}$-liveness, for $\mathcal{T}' = \{t_1, t_2\}$.

The way in which a set $\mathcal{T}'$ is estimated by the procedures will be discussed in more detail in section 8.4. In this example, the procedures will declare a *siphon control failure* when attempting to control the siphon $\{p_3, C\}$. A siphon control failure is an instance in which the procedures cannot or will not control a siphon S. This means that the siphon S is allowed to be emptied, and so the transitions $t \in S\bullet$ are allowed to die. Then, $\mathcal{T}'$ can be estimated as $\mathcal{T} \setminus \{p_3, C\}\bullet$, and hence $\mathcal{T}' = \{t_1, t_2\}$ results.

8.3.4 The Use of Initial Constraints

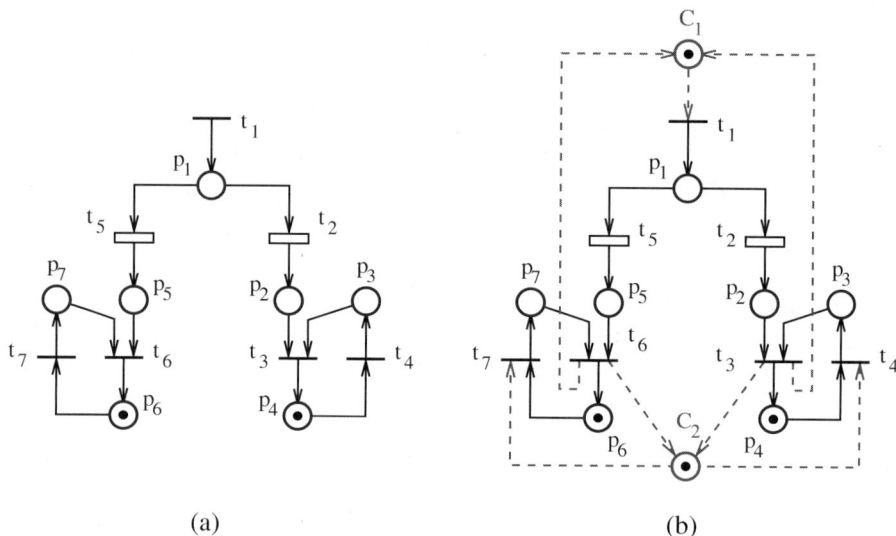

(a)   (b)

Fig. 8.3. Using initial constraints.

This section illustrates the use of initial constraints. Consider the Petri net of Figure 8.3(a), adapted from [141], pp. 122–129. The Petri net represents the model of an unreliable machine [45, 141]. The transitions t_2 and t_5 are uncontrollable. The Petri net is live for all initial markings μ which satisfy

$$\mu_3 + \mu_4 \geq 1, \tag{8.8}$$
$$\mu_6 + \mu_7 \geq 1. \tag{8.9}$$

Assume that we desire to enforce the following constraints

$$\mu_1 + \mu_2 + \mu_5 \leq 1, \tag{8.10}$$
$$\mu_3 + \mu_7 \leq 1. \tag{8.11}$$

The supervisor enforcing these two constraints is shown in Figure 8.3(b). Note that starting from the initial marking shown in the figure, the firing sequence t_1, t_2, t_7 leads to deadlock. This shows that enforcing constraints on a live Petri net may introduce deadlock possibilities. Therefore, once a supervisor is generated that enforces the desired safety specification, it is useful to run a liveness-enforcing procedure to add constraints ensuring liveness properties. When our dp- and le-procedures have as input the closed loop of a plant Petri net and a Petri net supervisor, we can have them take advantage of the fact that not all markings are possible by using *initial constraints*. For instance, in the example of Figure 8.3(b) for all reachable markings

$$\mu_{C_1} = 1 - \mu_1 - \mu_2 - \mu_5, \tag{8.12}$$
$$\mu_{C_2} = 1 - \mu_3 - \mu_7. \tag{8.13}$$

These can be used as initial constraints for our procedure when applied to the closed-loop Petri net of Figure 8.3(b).

Initial-marking constraints can also be used to restrict the initial markings considered in the procedures. For instance, if the system shown in Figure 8.3(a) is always started with a nonzero marking in p_6, then the initial-marking constraint $\mu_6 \geq 1$ can be used.

8.4 Procedure Definition

8.4.1 Description

This section describes the usage of the dp- and le-procedures. As in the case of the simplified dp- and le-procedures defined in section 7.5, the two procedures are defined as a single procedure in which an input argument *type* selects between deadlock prevention design and $\mathcal{T}$-liveness enforcement design. Thus *type* $= DP$ requests deadlock prevention design, and *type* $= LE$ requests $\mathcal{T}$-liveness enforcement design.

The input of the procedure consists of: *type*, the target Petri net $\mathcal{N}_0$, the set of uncontrollable transitions of $\mathcal{N}_0$, the set of unobservable transitions of $\mathcal{N}_0$, the set $\mathcal{T}$, and optionally a set of initial constraints $L_I \mu \geq b_I$ and a set of initial-marking constraints $L_{I0} \mu \geq b_{I0}$.

The output of the procedure consists of a set of transitions $\mathcal{T}'$, and two sets of constraints $L\mu \geq b$ and $L_0 \mu \geq b_0$. The procedure generates this output if it terminates (converges) and if it does not declare a failure. The output of the procedure satisfies the following. Let $\mathcal{M}$ be the set of initial markings of $\mathcal{N}_0$ for which all reachable markings satisfy the initial constraints, i.e., $\mathcal{M} = \{\mu_0 : \forall \mu \in \mathcal{R}(\mathcal{N}_0, \mu_0) : L_I \mu \geq b_I\}$. Then:

- If *type* $= LE$, $\mathcal{T}'$-liveness is enforced for all initial markings $\mu_0 \in \mathcal{M}$ such that $L\mu_0 \geq b$ and $L_0 \mu_0 \geq b_0$, when $(\mathcal{N}_0, \mu_0)$ is supervised according to $L\mu \geq b$.

- If $type = DP$, deadlock is prevented for all initial markings $\mu_0 \in \mathcal{M}$ such that $L\mu_0 \geq b$ and $L_0\mu_0 \geq b_0$, when $(\mathcal{N}_0, \mu_0)$ is supervised according to $L\mu \geq b$.
- Regardless of $type$, markings μ_0 satisfying $L_{I0}\mu_0 \geq b_{I0}$, $L\mu_0 \geq b$, and $L_0\mu_0 \geq b_0$ exist.
- $L\mu \geq b$ are admissible constraints.

If initial constraints are given to the procedure, note that unless the feasible set of the initial-marking constraints is a subset of $\mathcal{M}$, there may be no markings μ_0 satisfying all of $\mu_0 \in \mathcal{M}$, $L_{I0}\mu_0 \geq b_{I0}$, $L\mu_0 \geq b$, and $L_0\mu_0 \geq b_0$. This is normal, as $\mathcal{M}$ is not known in the procedure.

The set $\mathcal{T}'$ is a subset of $\mathcal{T}$. Ideally, $\mathcal{T}' = \mathcal{T}$. The following four factors may cause $\mathcal{T}' \neq \mathcal{T}$:

- the partial controllability and observability of the net,
- the initial constraints being overrestrictive,
- the initial-marking constraints being overrestrictive,
- the Petri net structure $\mathcal{N}_0$ not allowing $\mathcal{T}$-liveness enforcement.

To enforce liveness for as many transitions as possible, the argument $\mathcal{T}$ may be set to T_0, the total set of transitions. Unlike the simplified procedure of the previous chapter, this procedure won't declare a failure when it detects that $\mathcal{T}$-liveness enforcement is impossible. Instead, it will attempt $\mathcal{T}'$-liveness enforcement for $\mathcal{T}' \subset \mathcal{T}$ and report the set $\mathcal{T}'$ for which the generated supervisor is designed. Failure is declared when the procedure terminates without finding such a nonempty set $\mathcal{T}'$.

8.4.2 Definition

This section defines the dp- and le-procedures in their general form. As previously mentioned, the two procedures are defined as a single procedure in which an input argument selects between deadlock prevention design and $\mathcal{T}$-liveness enforcement design. This procedure is similar to the procedure of section 7.5. Naturally, the same notations are used. The following changes appear in the procedure defined in this section:

- During any iteration i and given any uncontrolled minimal active siphon S of $\mathcal{N}_i$, the procedure enforces

$$\sum_{p \in S} \mu(p) \geq 1 \tag{8.14}$$

only if (8.14) is admissible with respect to $\mathcal{N}_0$, the target Petri net. A constraint (8.14) is *admissible with respect to* $\mathcal{N}_0$ if its contribution to the final constraints $L\mu \geq b$ is an admissible constraint with respect to the uncontrollable and unobservable transitions of $\mathcal{N}_0$. This concept is discussed in more detail in section 8.4.3. When (8.14) is not admissible with respect to $\mathcal{N}_0$, (8.14) is transformed, if possible, to be so.

- The procedure stores in a variable X the transitions in the postset of any active siphon that it cannot control. An instance in which the procedure cannot control a siphon is said to be a *siphon control failure*.
- The simplified procedure converges when it reaches an intermediary net $\mathcal{N}_k$ in which all active siphons are controlled. This is also true of the procedure of this section. However, note that rather than terminating when a siphon control failure occurs, the procedure shrinks the active subnet such that the active siphons causing failures are no longer active. This is done by recomputing the active subnet such that it excludes the transitions in X.
- Siphon control failures, when they occur, cause the le-procedure to generate supervisors enforcing $\mathcal{T}'$-liveness instead of $\mathcal{T}$-liveness. The subset $\mathcal{T}'$ of $\mathcal{T}$ is an additional output of the procedure. In the case of the dp-procedure, the output $\mathcal{T}'$ indicates that the designed supervisor approximates or enforces $\mathcal{T}'$-liveness.
- To avoid generating overrestrictive $\mathcal{T}'$-liveness-enforcing supervisors, the procedure may restart itself with an adjusted argument $\mathcal{T}$ when certain siphon control failures occur.

Note that siphon control failures cannot occur in the simplified procedure. They may occur in the general procedure when a constraint (8.14) cannot be transformed to an admissible form, or when (8.14) conflicts with the initial constraints or the initial-marking constraints. Thus, siphon control failures may occur due to any of the following three factors: the net is only partially controllable and/or observable, the initial constraints are overrestrictive, and the initial-marking constraints are overrestrictive.

PROCEDURE 8.1   Deadlock Prevention/$\mathcal{T}$-liveness Enforcement

Input: The target Petri net $\mathcal{N}_0$, a nonempty set of transitions $\mathcal{T}$, the set of uncontrollable transitions T_{uc}, the set of unobservable transitions T_{uo}, type $\in \{DP, LE\}$, and optionally a set of initial constraints $L_I \mu \geq b_I$ and a set of initial-marking constraints $L_{I0}\mu_0 \geq b_{I0}$.

Output: Two sets of constraints $L\mu \geq b$ and $L_0\mu \geq b_0$, and a set of transitions $\mathcal{T}'$.

A. (L_0, b_0) is initialized to (L_I, b_I) and (L, b) to be empty. $\mathcal{N}_0$ is transformed to be PT-ordinary if type $= DP$, and asymmetric-choice PT-ordinary if type $= LE$ (the transformations appear in section 6.5.2 and in Algorithm 6.3).[1] Let the transformed net be $\mathcal{N}_1$. The initial constraints (L_0, b_0), if any, are updated according to the transformations (refer to equation (7.10)). If not previously defined, let $X = \emptyset$. Let $i = 1$, $P = P_1$, and $\mathcal{C} = \emptyset$.

---

[1] The transformation to PT-ordinary Petri nets has no effect if the Petri net is already PT-ordinary; the same is true of the transformation to asymmetric-choice nets.

B. *A $\mathcal{T}$-minimal active subnet $\mathcal{N}_1^A$ is computed for $\mathcal{N}_1$ such that the transitions in X are not included (Algorithm 6.2). When none exists, Algorithm 6.2 computes a T_x-minimal active subnet such that $T_x \subset \mathcal{T}$ and the transitions in X are not included. If no such $T_x \neq \emptyset$ exists, the procedure terminates and declares failure. Let $\mathcal{T}' = \mathcal{T} \cap T_1^A$.*

C. While *true* **do**

1. *Let (A, d) and (A_0, d_0) be empty sets of marking constraints.*

2. *If no uncontrolled minimal active siphon is found (section 7.5.2), the next step is D.*[2]

3. *For every uncontrolled minimal active siphon S:*

   (a) *Test whether (8.14) needs control place enforcement (section 7.5.2).*

   (b) *If (8.14) does not need control place enforcement, include (8.14) in (A_0, d_0).*

   (c) *If (8.14) needs control place enforcement*

      (i) *Transform*[3] *(8.14) to an inequality $l\mu \geq c$ which is admissible with respect to $\mathcal{N}_0$, T_{uc}, and T_{uo} (section 8.4.3).*[4]

      (ii) *If the procedure could not transform (8.14) to an admissible constraint (l, c), let $X = X \cup S\bullet$ and continue with the next siphon at step C.3.a.*

      (iii) *Else include (l, c) to (A, d).*

   (d) *Let*[5] *$\mu_p = \mu|_P$ and $\mu_c = \mu|_C$. Check whether the system $\mu_c = L\mu_p - b$, $L_0\mu_p \geq b_0$, $A\mu \geq d$, $A_0\mu \geq d_0$, $L_{I0}\mu|_{P_0} \geq b_{I0}$, and $\mu_p|_{P_i \setminus P_0} = 0$ is feasible. If not feasible, let $X = X \cup S\bullet$, remove (l, c) from (A, d) if the last step was C.3.c.iii, and remove (8.14) from (A_0, d_0) if the last step was C.3.b.*

4. *Let $\mathcal{N}_i' = (P_i', T_i', F_i', W_i')$ be the Petri net structure obtained by enforcing $A\mu \geq d$ in $\mathcal{N}_i$ via SBPI (see Theorem 3.1), and let $A^I\mu' = d$ be the corresponding place invariant equations (see equation (7.8)).*

5. *If type $= DP$, $\mathcal{N}_i'$ is transformed to be PT-ordinary; if type $= LE$, the Petri net is transformed to be PT-ordinary and with asymmetric choice*

---

[2]In the worst case, the number of uncontrolled minimal siphons depends exponentially on the size of the net. Checking whether a siphon is uncontrolled may involve solving a linear integer program.

[3]In our implementation, this operation may involve integer programming.

[4]$l\mu \geq c$ is the same as (8.14) if the latter is admissible; in particular, (8.14) is always admissible for fully controllable and observable Petri nets, i.e., for $T_{uc} = T_{uo} = \emptyset$. Furthermore, note that we are only interested in having final constraints (L, b) admissible in $\mathcal{N}_0$; this is why the observability or controllability of the transitions of $\mathcal{N}_i$ for $i \geq 1$ does not matter in our approach.

[5]Given a set of places X, $\mu|_X$ is the restriction of μ to the places of X.

(section 6.5.2 and Algorithm 6.3)[1]; note that the argument M of the Algorithm 6.3 is set to $M = P_i' \setminus P_i$. Let $\mathcal{N}_{i+1}$ be the transformed net.

6. Update A^I according to the net transformations performed at step 5 (section 7.5.5). Let A^u be the updated A^I (this means that $A^I \mu' = d$ in $\mathcal{N}_i'$ corresponds to $A^u \mu = d$ in $\mathcal{N}_{i+1}$, where μ' and μ are markings of $\mathcal{N}_i'$ and $\mathcal{N}_{i+1}$).

7. Let $P = P \cup (P_{i+1} \setminus P_i')$, $\mathcal{C}^o = \mathcal{C}$, and $\mathcal{C} = \mathcal{C}^o \cup (P_i' \setminus P_i)$. Let $\mu_p = \mu|_P$ and $\mu_c = \mu|_{\mathcal{C}}$, for any marking μ of $\mathcal{N}_{i+1}$. For each place in $P_{i+1} \setminus P_i'$ add a null column to L and L_0, to match the size of μ. Similarly, add null columns to A_0 to match the size of μ. Let[6] $A_p = A^u|_P$, $A_{p0} = A_0|_P$, $A_c = A^u|_{\mathcal{C}^o}$, and $A_{c0} = A_0|_{\mathcal{C}^o}$.

8. If (L, b) is empty, include $A_p \mu_p \geq d$ in (L, b) and $A_{p0} \mu_p \geq d_0$ in (L_0, b_0).

   Else, do the following:

   (a) If (A_0, d_0) is not empty, include $(A_{p0} + A_{c0}L)\mu_{p0} \geq d_0 + A_{c0}b$ in (L_0, b_0).

   (b) If (A, d) is not empty, include $(A_p + A_c L)\mu_p \geq d + A_c b$ in (L, b).

9. Compute the new active subnet $\mathcal{N}_{i+1}^A$ such that it does not contain the transitions in X (section 8.4.4). Let $\mathcal{T}' = \mathcal{T} \cap T_{i+1}^A$. If $\mathcal{T}' = \emptyset$, exit and declare failure.

10. If an infeasibility occurred at a step C.3.d of the current iteration, let $X = T_0 \setminus T_{i+1}^A$, and the procedure is restarted at step A with this value of X.

11. Else let $i = i + 1$. The next step is C.1.

D. The constraints (L, b) and (L_0, b_0) are restricted to the columns corresponding to the places in $\mathcal{N}_0$.

E. Optionally, the redundant constraints of (L, b) and (L_0, b_0) are removed. The constraints of (L_I, b_I) appearing in (L_0, b_0) are removed from (L_0, b_0).

We proceed by describing the specific operations involved in the procedure that have not yet been (completely) defined.

8.4.3 Transforming Constraints to Admissible Constraints

This section describes step C.3.c.i of the procedure. In this step, (8.14) is transformed to a constraint $l\mu \geq c$ admissible with respect to $\mathcal{N}_0$, where l is an integer row vector. The admissibility requirement is that the constraint $l\mu \geq c$ is admissible in $\mathcal{N}_0$ when written in terms of the places of $\mathcal{N}_0$. That is,

---

[6] $A^u|_P$ is the restriction of A^u to the columns corresponding to places in P; $A_0|_P$, $A^u|_{\mathcal{C}^o}$, ..., have a similar meaning.

$((l_p + l_c L)\mu_p)|_{\mathcal{N}_0} \geq c + l_c b$ is to be admissible in $\mathcal{N}_0$, for $l_c = l|_C$ and $l_p = l|_P$. This section presents an algorithm to generate the constraints $l\mu \geq c$ such that $l\mu \geq c$ is admissible and satisfies additional constraints. Let i denote the iteration number of the algorithm.

The admissibility requirement can be written as follows. Let D_{uc} and D_{uo} be the restrictions of the incidence matrix of $\mathcal{N}_0$ to the uncontrollable transitions and unobservable transitions, respectively. Let N be the matrix defined by the relation $lN = (l_p + l_c L)|_{\mathcal{N}_0}$. Then, the sufficient admissibility conditions in terms of D_{uc} and D_{uo} [141, 142] can be written as

$$lND_{uc} \geq 0, \tag{8.15}$$
$$lND_{uo} = 0. \tag{8.16}$$

The requirements R1, R2, and R3 stated in section 7.5.4 for the simplified procedure are to be satisfied by the general procedure as well. The only requirement that can be affected by constraint transformations is the requirement R1. The requirement R1 states that no control place is in the postset of a transition t generated by transition splits, that is, a transition $t \notin T_0$. Let C be the control place enforcing $l\mu \geq c$ in $\mathcal{N}'_i$. The requirement R1 for C can be written as $C \notin (T_i \setminus T_0)\bullet$, which corresponds to

$$lD_s \leq 0, \tag{8.17}$$

where D_s is the restriction of the incidence matrix D_i of $\mathcal{N}_i$ to the columns corresponding to the transitions of $T_i \setminus T_0$. Ensuring that the control places enforcing transformed constraints satisfy R1 is enough to guarantee that the control places enforcing constraints that do not need admissibility transformations satisfy also R1. To see this, the reader is referred to the proof of Proposition 7.1.

To ensure that (8.14) is satisfied for all markings satisfying $l\mu \geq c$, we impose:

$$l(p) \geq 0 \quad \forall p \in S, \tag{8.18}$$
$$l(p) \leq 0 \quad \forall p \in P_i \setminus S, \tag{8.19}$$
$$\sum_{p \in S} l(p) \geq 1, \tag{8.20}$$
$$c = 1. \tag{8.21}$$

One situation which may cause the $\mathcal{T}$-liveness procedure to diverge is when l ends up with a single nonzero entry; that entry is positive, in view of (8.20). To avoid this, the algorithm declares failure if l contains a single nonzero entry.

The following algorithm finds the constraints $l\mu \geq c$ satisfying the constraints above.

ALGORITHM 8.2   Constraint Transformation

Input: $\mathcal{N}_0 = (P_0, T_0, F_0, W_0)$; T_{uc}, *the set of uncontrollable transitions of* $\mathcal{N}_0$; T_{uo}, *the set of unobservable transitions of* $\mathcal{N}_0$; P_i, *the set of places at the current iteration* i; *the current constraints* $L\mu \geq b$ *and* $L_0\mu \geq b_0$, *and the siphon* S.

Output: *A constraint* $l\mu \geq c$ *admissible with respect to* $\mathcal{N}_0$.
1. *Let* $c = 1$, $l(p) = 1$ $\forall p \in S$, *and* $l(p) = 0$ $\forall p \notin S$.

2. **If** *(8.15) and (8.16) are satisfied* **then** *exit and return* l *and* c.

3. *Let* $f = TRUE$ *and* $A = S$.

4. **While** f *is* $TRUE$

   (a) *Check*[7] *the feasibility of* $\sum_{p \in A} l(p) \geq 1$ *with the additional constraints (8.15–8.20).*

   (b) **If** *infeasible, set* $f = FALSE$.

   (c) **Else** *let* $A = A \setminus \{p \in S : l(p) \neq 0\}$; **if** $A = \emptyset$, *set* $f = FALSE$.

5. **If** $|S \setminus A| < 2$ **then** *declare siphon control failure and exit.*[8]

6. *Solve the linear integer program* $\min_l (\sum_{p \in S} l(p) - \sum_{p \notin S} l(p))$ *subject to* $l(p) \geq 1$ $\forall p \in S \setminus A$ *and (8.15)–(8.20).*

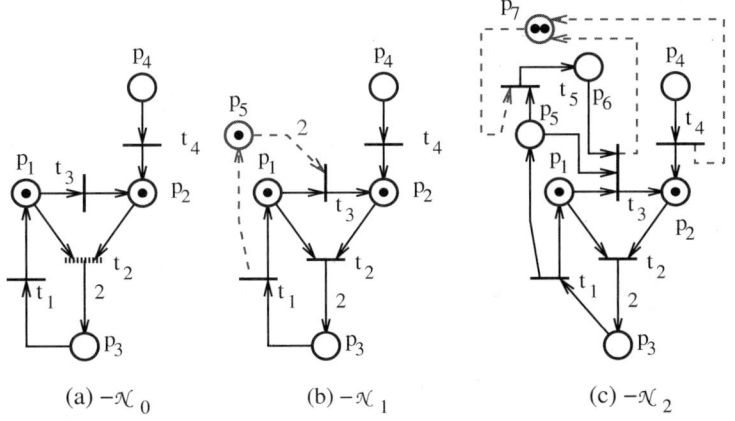

(a) $-\mathcal{N}_0$ (b) $-\mathcal{N}_1$ (c) $-\mathcal{N}_2$

Fig. 8.4. Example of the proposed constraint transformations.

---

[7]The feasibility check involves solving a linear program.
[8]$|S \setminus A|$ denotes the number of elements of $S \setminus A$.

EXAMPLE 8.1

To illustrate the transformation, consider the Petri net of Fig. 8.4(a), in which all transitions but t_2 are controllable and observable, and t_2 is controllable and unobservable. When the procedure is applied for $\{t_1\}$-liveness, the control place p_5 is added at the first iteration, to enforce the admissible constraint $2\mu_1 + \mu_3 \geq 1$ (Fig. 8.4(b)). Then, as $W(p_5, t_3) = 2$, t_3 is split, and so the place p_6 is generated (Fig. 8.4(c)). We illustrate the transformation to admissible constraints on the constraint (8.14) for the active siphon $S = \{p_2, p_3, p_4, p_5\}$ obtained at the second iteration. At the second iteration the matrices L_0 and b_0 are empty, while

$$L = \begin{bmatrix} 2 & 0 & 1 & 0 & -1 \end{bmatrix}, \qquad b = \begin{bmatrix} 1 \end{bmatrix}.$$

At step 1 of the transformation, $l = [0, 1, 1, 1, 1, 0]$ and $c = 1$. At step 2, $l_p = [l_1, l_2, l_3, l_4, l_6]$ (i.e., $l_p = [0, 1, 1, 1, 0]$), $l_c = l_5$ (i.e., $l_c = 1$). Let L_x be L restricted to the first four columns. Then $N = [I_4, L_x^T, 0_{4 \times 1}]^T$ and $lN = [2, 1, 2, 1]$. There are no inequalities (8.15) to check, as there are no uncontrollable transitions. The inequality of (8.16) is not satisfied, as $D_{uo} = [-1, -1, 2, 0]^T$. Therefore (8.14) is not admissible with respect to $\mathcal{N}_0$. Further, at step 4, the constraints (8.15)–(8.20) are $-l_1 - l_2 + 2l_3 = 0$ as (8.16), $-l_5 + l_6 \leq 0$ as (8.17), $l_i \geq 0$ for $i = 2 \ldots 5$ as (8.18), $l_i \leq 0$ for $i = 1, 6$ as (8.19), and $l_2 + l_3 + l_4 + l_5 \geq 1$ as (8.20). In constraint (8.17) $D_s = [0, 0, 0, 0, -1, 1]^T$ is the restriction of the incidence matrix to the transition t_5, which is the only transition of the net generated by transition splits. Thus step 4 generates $A = \emptyset$, and $l\mu \geq b$ at step 6 has $l = [0, 2, 1, 1, 1, 0]$ and $c = 1$. The constraint $l\mu \geq c$ in $\mathcal{N}_2$ corresponds to $((l_p + l_c L)\mu_p)|_{\mathcal{N}_0} \geq c + l_c b$ in $\mathcal{N}_0$, that is, $2\mu_1 + 2\mu_2 + 2\mu_3 + \mu_4 \geq 2$, which is indeed admissible. When $l\mu \geq c$ is enforced, the control place p_7 of Fig. 8.4(c) is generated.

8.4.4 The Computation of the Active Subnet

This section discusses the computation of the active subnet at step B and at step C.9. The computation at step B is similar to that of the simplified procedure, as the same Algorithm 6.2 is used. However, there are two differences. Recall that Algorithm 6.2 finds a $\mathcal{T}$-minimal active subnet if one exists, or else it computes a $\mathcal{T}'$-minimal active subnet for the largest subset $\mathcal{T}'$ of $\mathcal{T}$ for which a $\mathcal{T}'$-minimal active subnet exists. The first difference occurs when there is no $\mathcal{T}$-minimal active subnet; this situation is detected by checking whether all transitions in $\mathcal{T}$ are contained in the computed active subnet. In this situation, the simplified procedure terminates, declaring that $\mathcal{T}$-liveness enforcement is impossible. In contrast, the general procedure defined in this chapter continues its computations with the $\mathcal{T}'$-minimal active subnet if $\mathcal{T}' \neq \emptyset$, and terminates otherwise. The second difference is that the general procedure uses the optional argument X of Algorithm 6.2. The argument X is used to ensure that all transitions in X are excluded from the active subnet. Note that X at step B is nonempty if and only if step B has been reached after a restart at step C.10.

The computation at step C.9 is similar to that presented in section 7.5.6. However, there is a difference when siphon control failures occur during the iteration. Siphon control failures change the set X. When no siphon control failures occur in the iteration, X is unchanged, and so the update algorithm 7.2 can be used. However, when siphon control failures occur, the active subnet needs to be recomputed using Algorithm 6.2, to insure that the newly added transitions to X are excluded from the active subnet. Note that the simplified procedure uses exclusively the update Algorithm 7.2 at step C.9.

8.5 Examples

This section illustrates the operation of the procedure on Petri nets with uncontrollable and/or unobservable transitions and in the presence of initial constraints.

EXAMPLE 8.2

Consider the repetitive Petri net of Fig. 8.5(a), where t_1 is unobservable. In the first iteration we have that $\mathcal{N}_1$ is identical to $\mathcal{N}_0$, since $\mathcal{N}_0$ is a PT-ordinary asymmetric-choice Petri net. At the first iteration there are two minimal siphons: $\{p_1, p_3\}$ and $\{p_2, p_3\}$. The marking constraint for $\{p_1, p_3\}$ is $\mu_1 + \mu_3 \geq 1$. The constraint needs control place enforcement and is not admissible. Therefore it is transformed to the admissible constraint $2\mu_1 + \mu_3 \geq 1$, which is added to (L, b). The control place enforcing this constraint in $\mathcal{N}_1'$ is C_1, and its corresponding place invariant is $\mu(C_1) = 2\mu_1 + \mu_3 - 1$. Similarly, for the siphon $\{p_2, p_3\}$ the control place C_2 is added to enforce $2\mu(p_2) + \mu(p_3) \geq 1$, which is the other constraint added to (L, b); the invariant of C_2 is $\mu(C_2) = 2\mu_2 + \mu_3 - 1$. Since $\mathcal{N}_1'$ is PT-ordinary and with asymmetric choice, $\mathcal{N}_2$ is the same as $\mathcal{N}_1'$, and is shown in Fig. 8.5(b).

At the second iteration there is a single new minimal siphon: $\{C_1, C_2\}$. The control place which would result by enforcing $\mu(C_1) + \mu(C_2) \geq 1$ is C_3 such that $C_3\bullet = \emptyset$. Therefore, $\{C_1, C_2\}$ does not need control, and so $\mu(C_1) + \mu(C_2) \geq 1$ is added to (A_0, d_0) in step C.3.b. Hence, $2\mu(p_1) + 2\mu(p_2) + 2\mu(p_3) \geq 3$ is added to (L_0, b_0) at step C.7.a.

The procedure terminates at the third iteration, since there is no uncontrolled siphon. The final matrices (L, b) and (L_0, b_0) are:

$$L = \begin{bmatrix} 2 & 0 & 1 \\ 0 & 2 & 1 \end{bmatrix}, \quad b = \begin{bmatrix} 1 \\ 1 \end{bmatrix}, \quad L_0 = \begin{bmatrix} 2 & 2 & 2 \end{bmatrix}, \quad b_0 = \begin{bmatrix} 3 \end{bmatrix}.$$

The supervised net is shown in Fig. 8.5(b). For all initial markings μ_0 such that $L\mu_0 \geq b$ and $L_0\mu_0 \geq b_0$, liveness is enforced in a least restrictive manner.

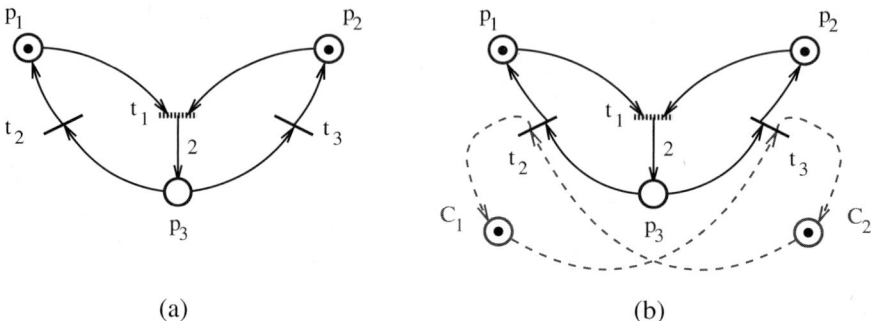

(a)   (b)

Fig. 8.5. The Petri nets of Example 8.2: (a) $\mathcal{N}_0$; (b) the final Petri net supervised for liveness.

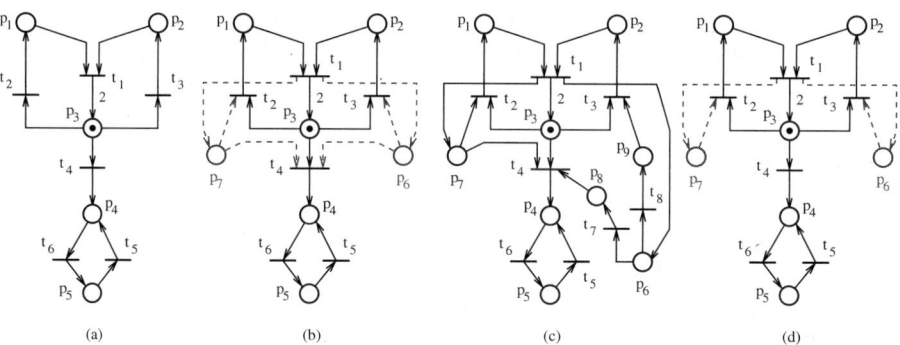

(a)   (b)   (c)   (d)

Fig. 8.6. The Petri nets of Example 8.3: (a) $\mathcal{N}_0$; (b) the final Petri net supervised for liveness; (c) $\mathcal{N}_2$; (d) $\mathcal{N}_1$ with the control places added in the first iteration after the procedure is restarted.

EXAMPLE 8.3

Consider the Petri net of Fig. 8.6(a), and assume that the procedure is started with the initial constraint $\mu_1 + \mu_2 + \mu_3 + \mu_4 + \mu_5 \leq 1$. Recall that an initial constraint specifies a constraint satisfied by all markings reached from the set of initial markings that are of interest in a specific application. In this example $\mathcal{T}$ is set to equal the whole set of transitions. At the first iteration $\mathcal{N}_1$ equals $\mathcal{N}_0$. At step B, no active subnet including $\mathcal{T}$ exists, since the structure of the Petri net does not allow t_4 to be live. Therefore the largest T_x-minimal active subnet is computed such that $T_x \subseteq \mathcal{T}$; thus $\mathcal{N}_1^A$ is defined by $T_1^A = \{t_1, t_2, t_3, t_5, t_6\}$ and $T_x = T_1^A$.

At the first iteration there are two minimal active siphons: $\{p_1, p_3\}$ and $\{p_2, p_3\}$. Therefore the control places p_6 and p_7 are added; their place invariants in $\mathcal{N}_1'$ are $\mu_6 = \mu_1 + \mu_3 - 1$ and $\mu_7 = \mu_2 + \mu_3 - 1$; $\mathcal{N}_1'$ is shown in Fig. 8.6(b). Then, after transforming the Petri net to be with asymmetric choice, $\mathcal{N}_2$ is obtained and shown in Fig. 8.6(c). The invariants of $A^u \mu = d$ in $\mathcal{N}_2$ are $\mu_6 = \mu_1 + \mu_3 - \mu_8 - \mu_9 - 1$ and $\mu_7 = \mu_2 + \mu_3 - 1$.

At the second iteration there are two new minimal siphons: $\{p_1, p_7\}$ and $\{p_2, p_6, p_9\}$. When the procedure considers $\{p_1, p_7\}$, the system at step C.3.d is: $\mu_6 = \mu_1 + \mu_3 - \mu_8 - \mu_9 - 1$, $\mu_7 = \mu_2 + \mu_3 - 1$, $\mu_1 + \mu_2 + \mu_3 + \mu_4 + \mu_5 \leq 1$, $\mu_1 + \mu_7 \geq 1$, $\mu_8 = 0$, and $\mu_9 = 0$. This system is infeasible, so X is set to $X = \{p_1, p_7\}\bullet$, that is, $X = \{t_1, t_2, t_4\}$. Thus, the attempt to control $\{p_1, p_7\}$ results in a siphon control failure due to the initial constraint. The same happens for the siphon $\{p_2, p_6, p_9\}$: at step C.3.d we have the system $\mu_6 = \mu_1 + \mu_3 - \mu_8 - \mu_9 - 1$, $\mu_7 = \mu_2 + \mu_3 - 1$, $\mu_1 + \mu_2 + \mu_3 + \mu_4 + \mu_5 \leq 1$, $\mu_2 + \mu_6 + \mu_9 \geq 1$, $\mu_8 = 0$, and $\mu_9 = 0$, which also is infeasible. Then X is set to $X = X \cup \{p_2, p_6, p_9\}\bullet$, so $X = \{t_1, t_2, t_3, t_4, t_7, t_8\}$. Since no other siphons appear, $\mathcal{N}_3$ is the same as $\mathcal{N}_2$. At step C.9 $\mathcal{N}_{i+1}^A$ is computed to exclude the transitions of X, and so $T_{i+1}^A = \{t_5, t_6\}$ and $\mathcal{T}' = \{t_5, t_6\}$. At step C.10 X is set to $\{t_1, t_2, t_3, t_4\}$, and the procedure is restarted with this value of X.

As the procedure is restarted with $X = \{t_1, t_2, t_3, t_4\}$, the active subnet $\mathcal{N}_1^A$ is given by $T_1^A = \{t_5, t_6\}$. At the first iteration there are only two minimal active siphons with respect to $\mathcal{N}_1^A$: $\{p_1, p_3, p_4, p_5\}$ and $\{p_2, p_3, p_4, p_5\}$. The control places p_6 and p_7 are added as in Fig. 8.6(d). At the second iteration $\mathcal{N}_2$ is the same as $\mathcal{N}_1'$ (Fig. 8.6(d)), and there are no new minimal active siphons. (Indeed, even though adding p_6 and p_7 generate the minimal siphons $\{p_1, p_7\}$ and $\{p_2, p_6\}$, they do not generate new minimal active siphons.) Thus the procedure terminates with $\mathcal{T}' = \{t_5, t_6\}$, empty (L_0, b_0), and

$$L = \begin{bmatrix} 1 & 0 & 1 & 1 & 1 \\ 0 & 1 & 1 & 1 & 1 \end{bmatrix}, \qquad b = \begin{bmatrix} 1 \\ 1 \end{bmatrix}. \qquad (8.22)$$

The supervisor enforcing $L\mu \geq b$ enforces $\{t_5, t_6\}$-liveness and is least restrictive.

EXAMPLE 8.4

Here we consider liveness enforcement for the example of Fig. 8.1(b)–(c). Thus we start with the Petri net $\mathcal{N}_0$ shown in Fig. 8.7(a) and the initial

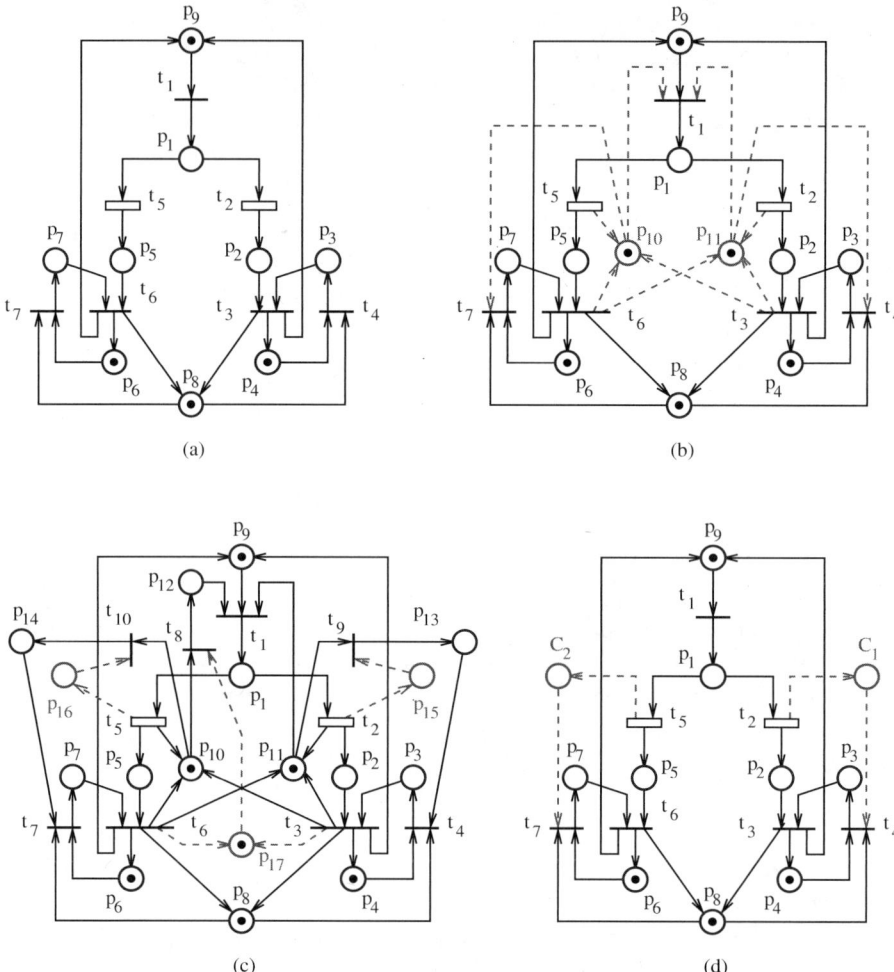

Fig. 8.7. The Petri nets of Example 8.4: (a) $\mathcal{N}_0$; (b) $\mathcal{N}_1$; (c) $\mathcal{N}_2$; (d) the final Petri net supervised for liveness.

constraints $\mu_1 + \mu_2 + \mu_5 + \mu_9 = 1$ and $\mu_3 + \mu_7 + \mu_8 = 1$. The transitions t_2 and t_5 are uncontrollable. The operations performed by the procedure are summarized in Table 8.1, where the iteration number appears on the first column, the uncontrolled minimal active siphons on the second column, the associated constraint on the third column, and the control places on the fifth column. At the first iteration $\mathcal{N}_1$ is the same as $\mathcal{N}_0$ and two control places p_{10} and p_{11} are added (Fig. 8.7(b)). At the second iteration the control places p_{15}, p_{16}, and p_{17} are added (Fig. 8.7(c)). No control places are added at the third iteration, and so the procedure terminates.

Table 8.1. Summary of operations in Example 8.4

i	Min. siphon	Constraint	Added to	C. place
1	$\{p_3, p_4\}$	$\mu_3 + \mu_4 \geq 1$	(A_0, d_0)	—
	$\{p_1, p_3, p_5, p_8, p_9\}$	$\mu_3 + \mu_5 + \mu_8 + \mu_9 \geq 1$	(A, d)	p_{10}
	$\{p_6, p_7\}$	$\mu_6 + \mu_7 \geq 1$	(A_0, d_0)	—
	$\{p_1, p_2, p_7, p_8, p_9\}$	$\mu_2 + \mu_7 + \mu_8 + \mu_9 \geq 1$	(A, d)	p_{11}
2	$\{p_1, p_2, p_5, p_{11}\}$	$\mu_1 + \mu_2 + \mu_5 + \mu_{11} \geq 1$	(A, d)	p_{15}
	$\{p_1, p_2, p_5, p_{10}, p_{12}\}$	$\mu_1 + \mu_2 + \mu_5 + \mu_{10} + \mu_{12} \geq 1$	(A, d)	p_{16}
	$\{p_1, p_2, p_7, p_9, p_{10}, p_{14}\}$	$\mu_1 + \mu_2 + \mu_7 + \mu_9 + \mu_{10} + \mu_{14} \geq 1$	(A, d)	p_{17}

In the first iteration there are two siphons for which the transformation to admissible constraints is required: $\{p_1, p_3, p_5, p_8, p_9\}$ and $\{p_1, p_2, p_7, p_8, p_9\}$; their transformed constraints appear in the third column of Table 8.1. The initial constraints help the procedure converge. Indeed, some siphons are identified as not uncontrolled due to the initial constraints. For instance, at the first iteration the minimal siphon $\{p_1, p_2, p_5, p_9\}$ is not uncontrolled, since one initial constraint is $\mu_1 + \mu_2 + \mu_5 + \mu_9 = 1$.

After removing the redundant constraints, the procedure terminates with the constraints

$$\mu_1 + 2\mu_2 + \mu_5 + \mu_7 + \mu_8 + \mu_9 \geq 2, \tag{8.23}$$
$$\mu_1 + \mu_2 + \mu_3 + 2\mu_5 + \mu_8 + \mu_9 \geq 2 \tag{8.24}$$

in (L, b), and the constraints

$$\mu_3 + \mu_4 \geq 1, \tag{8.25}$$
$$\mu_6 + \mu_7 \geq 1 \tag{8.26}$$

in (L_0, b_0). Note that by substituting μ_8 and μ_9 from the initial constraints, (8.23) and (8.24) become

$$\mu_2 - \mu_3 \geq 0, \tag{8.27}$$
$$\mu_5 - \mu_7 \geq 0. \tag{8.28}$$

The supervisor enforces liveness. The supervised Petri net is shown in Fig. 8.7(d).

8.6 Properties

This section proves that the procedure is correct and derives permissiveness results for the supervisors generated by the procedure. The majority of the results presented here can be seen as extensions to the general procedure of the results presented in section 7.7 for the simplified procedure. The notation is described in section 8.6.1. The correctness proofs are presented in section 8.6.2. Section 8.6.3 proves permissiveness results for the supervisors generated by the procedure.

8.6.1 Preliminaries

The notation and definitions assumed in section 7.7.1 for the simplified procedure are maintained for the general procedure of this chapter. In addition the following notation is introduced.

In our results we will refer to the sets $\mathcal{M}_0$ and $\mathcal{M}_I$, which we define as follows. Let $\mathcal{M}_I$ be the set of initial markings of interest if initial constraints $L_I\mu \geq b_I$ are given to the procedure (recall that this means that $\mathcal{M}_I$ satisfies that $\forall \mu_0 \in \mathcal{M}_I \ \forall \mu \in \mathcal{R}(\mathcal{N}_0, \mu_0)$: $L_I\mu \geq b_I$), or $\mathcal{M}_I = \mathbb{N}^{|P_0|}$ otherwise. Then we define $\mathcal{M}_0 = \mathcal{M}_I$ if no initial-marking constraints are given, and $\mathcal{M}_0 = \mathcal{M}_I \cap \{\mu : L_{I0}\mu \geq b_{I0}\}$ otherwise. The meaning of the set $\mathcal{M}_0$ is that the procedure generates supervisors such that they can be used for some initial markings $\mu_0 \in \mathcal{M}_0$; recall that the supervisors are defined for all initial markings μ_0 satisfying $\mu_0 \in \mathcal{M}_I$, $L\mu_0 \geq b$, and $L_0\mu_0 \geq b_0$.

Finally, recall that a *siphon control failure* is the situation in which a siphon cannot be controlled. This happens when no admissible constraint is found at step C.3.c.i or when an infeasibility occurs at step C.3.d.

8.6.2 Proof of Correctness

As mentioned in the procedure description, if the procedure detects that it fails to generate a $\mathcal{T}$-liveness-enforcing supervisor, it attempts to generate a $\mathcal{T}'$-liveness-enforcing supervisor, for $\mathcal{T}' \subseteq \mathcal{T}$ (and $\mathcal{T}' \neq \emptyset$). Thus the procedure returns the parameter $\mathcal{T}'$ together with the constraints (L, b) and (L_0, b_0). Furthermore, note that the procedure may terminate at any of steps B, C.9, or E. The procedure terminates at step B or C.9 if it fails, and it terminates at step E if it is successful. The next result shows that if the procedure terminates at step E, it generates a $\mathcal{T}'$-liveness-enforcing supervisor when the procedure is used as the le-procedure, and a deadlock prevention supervisor when the procedure is used as the dp-procedure. From the procedure it can easily be seen that when no siphon control failures occur in the steps C.3.c.i and C.3.d, the value of $\mathcal{T}'$ is as follows: $\mathcal{T}' = \mathcal{T}$ if the problem is well formulated (if there is a $\mathcal{T}$-liveness-enforcing supervisor for some initial marking) or $\mathcal{T}' = \mathcal{T} \cap T_1^A$ otherwise.

Theorem 8.1. (a) *If the dp-procedure terminates at step* E, $(\mathcal{N}_0, \mu_0)$ *supervised according to* $L\mu \geq b$ *is deadlock-free for all initial markings* $\mu_0 \in \mathcal{M}_I$ *satisfying* $L\mu_0 \geq b$ *and* $L_0\mu_0 \geq b_0$.

(b) *If the le-procedure terminates at step* E, $(\mathcal{N}_0, \mu_0)$ *supervised according to* $L\mu \geq b$ *is* $\mathcal{T}'$-*live for all initial markings* $\mu_0 \in \mathcal{M}_I$ *satisfying* $L\mu_0 \geq b$ *and* $L_0\mu_0 \geq b_0$.

Proof. Let $\mathcal{N}_k$ be the Petri net of the last iteration and T_k^A the set of transitions of its active subnet $\mathcal{N}_k^A$. Then the same proof as that of Theorem 7.5 can be used to prove $\mathcal{T}'$-liveness enforcement for $\mathcal{T}' = T_k^A \cap \mathcal{T}$.   □

As an example, Theorem 8.1 applies in Examples 8.2–8.4, as the procedure terminates at step E.

8.6.3 Permissiveness

The simplified procedure has been proved to generate supervisors that are at least as permissive as the least restrictive $\mathcal{T}$-liveness-enforcing supervisor when $\mathcal{N}_1$ has a single $\mathcal{T}$-minimal active subnet. This result is not affected by the initial/initial-marking constraint extension. However, when uncontrollable and unobservable transitions exist, the result can easily be proved only for a particular situation. This situation is characterized by the following conditions:

C1 The procedure terminates.
C2 The procedure does not terminate by declaring failure.
C3 No siphon control failures occur in the procedure.
C4 For every siphon S considered at step C.3.c.i., the inequality $l\mu \geq c$ at that step is such that $l(p) > 0 \; \forall p \in S$, and $l(p) = 0 \; \forall p \notin S$.
C5 $\mathcal{N}_1^A$ has a single $\mathcal{T}$-minimal active subnet.

For fully controllable and observable Petri nets, condition C4 is always satisfied, while the conditions C2 and C3 are also satisfied if the problem is well formulated, that is, if $\mathcal{T}$-liveness enforcement does not conflict with the initial/initial-marking constraints. Condition C.5 can be violated if there are more than one $\mathcal{T}$-minimal active subnets or if there is no $\mathcal{T}$-minimal active subnet. In the latter case the problem is not well formulated, since $\mathcal{T}$-liveness cannot be enforced for any initial marking.

Before presenting our results, we need a new definition. First, note that a supervisor enforcing $\mathcal{T}$-liveness in a Petri net under partial controllability and observability assumptions will tend to be more restrictive than a $\mathcal{T}$-liveness-enforcing supervisor of the same Petri net under a full controllability and observability assumption. Furthermore, while a least restrictive $\mathcal{T}$-liveness-enforcing supervisor always exists for a fully observable Petri net allowing $\mathcal{T}$-liveness enforcement, a least restrictive $\mathcal{T}$-liveness-enforcing supervisor may not exist for a partially observable Petri net allowing $\mathcal{T}$-liveness enforcement. The following permissiveness results will compare the supervisors generated by the procedure to the supervisor that enforces $\mathcal{T}$-liveness in the net in a least restrictive fashion under the full controllability and observability assumption. For ease of notation, we denote the latter as the least restrictive $\mathcal{T}$-liveness-enforcing *fco-supervisor*.

Definition 8.2. *A supervisor relying on the assumption that the Petri net is fully controllable and observable is said to be a* **fco-supervisor***.*

The next result states that under the assumptions C1–C5 the procedure generates a supervisor that is at least as permissive as the least restrictive $\mathcal{T}$-liveness-enforcing fco-supervisor. As discussed in section 7.7.3, this means that no $\mathcal{T}$-liveness-enforcing fco-supervisor exists for any of the markings μ_0 which do not satisfy $L_0\mu_0 \geq b_0$ or $L\mu_0 \geq b$, and that for all initial markings μ_0 satisfying $L_0\mu_0 \geq b_0$ and $L\mu_0 \geq b$, the supervisor enforcing $L\mu \geq b$ is at least as permissive as the least restrictive $\mathcal{T}$-liveness-enforcing fco-supervisor of $(\mathcal{N}_0, \mu_0)$.

Theorem 8.3. *Assuming conditions* C1–C5 *are satisfied, the procedure generates a supervisor at least as permissive as the least restrictive $\mathcal{T}$-liveness-enforcing fco-supervisor.*

Proof. The proof is identical with the proof of Theorem 7.6, once it is noticed that all minimal active siphons are controlled in a least restrictive way. Indeed, the transformation to admissible constraints generates constraints with $c = 1$, and so C4 implies that $l\mu \geq c$ is not satisfied only if S is empty.   □

Note that under the assumptions C1–C5, Theorem 8.1(b) guarantees that the le-procedure generates a supervisor enforcing $\mathcal{T}$-liveness for all initial markings $\mu_0 \in \mathcal{M}_I$ satisfying $L_0\mu_0 \geq b_0$ and $L\mu \geq b$. Recall that $\mathcal{M}_I$ is the set of initial markings for which the initial constraints are satisfied. However, $\mathcal{M}_I$ does not appear in Theorem 8.3 or its proof. The reason is that under the assumptions C1–C5, the le-procedure generates a supervisor satisfying the following: (a) $\mathcal{T}$-liveness enforcement (even under the full controllability and observability assumption) is impossible for all initial markings μ_0 that do not satisfy $L_0\mu_0 \geq b_0$ and $L\mu_0 \geq b$; (b) for all initial markings allowing $\mathcal{T}$-liveness enforcement, the le-procedure-generated supervisor is at least as permissive as any of the $\mathcal{T}$-liveness-enforcing supervisors. Moreover, the set of initial markings $\mathcal{M}_T$ for which $\mathcal{T}$-liveness enforcement is possible is guaranteed to satisfy $\mathcal{M}_L \subseteq \mathcal{M}_T \subseteq \mathcal{M}_U$, for $\mathcal{M}_U = \{\mu_0 : L\mu_0 \geq b \text{ and } L_0\mu_0 \geq b_0\}$ and $\mathcal{M}_L = \mathcal{M}_I \cap \mathcal{M}_U$.

To illustrate the application of Theorem 8.3, note that in Example 8.2 the assumptions C1–C5 hold true. However, Theorem 8.3 does not apply in Examples 8.3 and 8.4. It does not apply to Example 8.3 due to assumption C.3, as siphon control failures occur when the procedure attempts to control the siphons $\{p_1, p_7\}$ and $\{p_2, p_6, p_9\}$. It also does not apply to Example 8.4, as the constraints enforced for the siphons $\{p_1, p_3, p_5, p_8, p_9\}$ and $\{p_1, p_2, p_7, p_8, p_9\}$ (see Table 8.1) do not satisfy assumption C4.

The next result is a consequence of Theorem 8.3, and it deals with the question of whether there is some supervisor enforcing $\mathcal{T}$-liveness when the supervisor generated by our procedure enforces $\mathcal{T}'$-liveness with $\mathcal{T}' \subset \mathcal{T}$, or when the procedure declares failure. (Recall that $\mathcal{T}' \neq \mathcal{T}$ if and only if siphon control failures occur or the Petri net structure does not allow $\mathcal{T}$-liveness enforcement.)

Proposition 8.4. *Consider T_1^A at the first step* B *(i.e., not a step* B *reached by the procedure restarting itself at step* C.10*).* $\mathcal{T}$*-liveness is not enforcible in* $\mathcal{N}_0$ *for any initial marking if* $\mathcal{T} \not\subseteq T_1^A$. *Assume that the condition* C5 *is satisfied, siphon control failures occur, their first occurrence is at a step* C.3.d, *and the condition* C4 *is satisfied in all iterations previous to this occurrence. Then, for no initial marking* $\mu_0 \in \mathcal{M}_0$ *is* $\mathcal{T}$*-liveness enforcible in* $\mathcal{N}_0$.

Proof. The first part is a consequence of Theorem 6.10, as the Algorithm 6.2 for the computation of the active subnets does not fail to find a $\mathcal{T}$-minimal active subnet if such a subnet exists. For the second part, as C5 is satisfied, there is a $\mathcal{T}$-minimal active subnet, and so $\mathcal{T} \subseteq T_1^A$. Let j be the iteration in which the first siphon control failure occurs. Since conditions C3–C5 apply at the iterations $1, 2, \ldots, j - 1$, we can show as in Theorem 7.6 that if μ_0 is a marking of $\mathcal{N}_0$, $i \in \{1, 2, \ldots, j - 1\}$, the equivalent marking $\mu_{0,i}$ of μ_0 in $\mathcal{N}_i$ exists, and $(\mathcal{N}_i, \mu_{0,i})$ cannot be made $\mathcal{T}$-live, then $(\mathcal{N}_0, \mu_0)$ cannot be made $\mathcal{T}$-live. The fact that C4 has been satisfied before the first siphon control failure and that the failure takes place at a step C.3.d. indicate that the procedure has been given initial constraints $L_I\mu \geq b_I$ and/or initial-marking constraints $L_{I0}\mu_0 \geq b_{I0}$. Let $\mu_0 \in \mathcal{M}_0$ be an initial marking of $\mathcal{N}_0$. If no equivalent marking $\mu_{0,j}$ of $\mathcal{N}_j$ exists, there is $i \in \{1, 2, \ldots, j-1\}$ such that the equivalent marking $\mu_{0,i}$ of $\mathcal{N}_i$ exists and an active siphon of $(\mathcal{N}_i, \mu_{0,i})$ is empty. This implies that $(\mathcal{N}_i, \mu_{0,i})$ cannot be made $\mathcal{T}$-live, and so $(\mathcal{N}_0, \mu_0)$ cannot be made $\mathcal{T}$-live. If an equivalent marking $\mu_{0,j}$ of $\mathcal{N}_j$ exists, we reach the same conclusion as follows. The first siphon control failure occurs at step C.3.d; this failure implies that there is an active siphon S_x of $\mathcal{N}_j$ which, due to (L_{I0}, b_{I0}) and (L_I, b_I), must be empty for all valid markings μ_j such that $L_I\mu_j|_{\mathcal{N}_0} \geq b_I$ and $L_{I0}\mu_j|_{\mathcal{N}_0} \geq b_{I0}$. Therefore S_x is empty at $\mu_{0,j}$, so $(\mathcal{N}_j, \mu_{0,j})$ cannot be made $\mathcal{T}$-live, which implies that $(\mathcal{N}_0, \mu_0)$ cannot be made $\mathcal{T}$-live. This concludes our proof. $\square$

In view of Theorem 8.3 and Proposition 8.4, corollaries similar to the Corollaries 7.7–7.9 (pages 182–183) of the simplified procedure can be defined. The corollaries mention the set $\mathcal{M}_0$, which, as mentioned in the preliminaries, is the set of markings in $\mathcal{M}_I$ that satisfy the initial-marking constraints.

Corollary 8.5. *Assume that liveness is enforcible in* $\mathcal{N}_0$ *for some initial marking* $\mu_0 \in \mathcal{M}_0$, $\mathcal{N}_0$ *is fully controllable and observable, and the le-procedure terminates. If* $\mathcal{T} = T_0$, *the le-procedure provides the least restrictive liveness-enforcing supervisor.*

The previous corollary extends Corollary 7.7 to the case of initial constraints and initial marking constraints. The next corollary extends Corollary 7.8 to the same case. As in Corollary 7.8, we denote by T^A the set of transitions $T^A = T_1^A \cap T_0$, where T_1^A is the set of transitions of the active subnet $\mathcal{N}_1^A$ of $\mathcal{N}_1$.

Corollary 8.6. *Assume that T-liveness is enforcible in $\mathcal{N}_0$ for some initial marking $\mu_0 \in \mathcal{M}_0$, $\mathcal{N}_0$ is fully controllable and observable, and the procedure terminates. The generated supervisor is at least as permissive as the least restrictive T^A-liveness-enforcing supervisor.*

The corollary stating a sufficient condition for divergence to occur is stated next. In the corollary, an **integer convex set** is a set of integer vectors that can be expressed as the set of integer vectors x satisfying a matrix inequality $Mx \geq g$, where M and g have rational elements.

Corollary 8.7. *Assume that $\mathcal{N}_1$ has a single T-minimal active subnet, $\mathcal{N}_0$ is fully controllable and observable, T-liveness enforcement is possible for some initial marking $\mu_0 \in \mathcal{M}_0$, and the markings $\mu_0 \in \mathcal{M}_I$ for which T-liveness enforcement is possible cannot be represented as the intersection of $\mathcal{M}_I$ with an integer convex set. Then the le-procedure diverges.*

Proof. Assume the contrary, that the le-procedure terminates. In view of Proposition 8.4, no siphon control failures can occur, and so $T' = T$. Then, by Theorem 8.1(b), the supervisor enforcing $L\mu \geq b$ is a T-liveness-enforcing supervisor for all initial markings $\mu_0 \in \mathcal{M}_L$, for $\mathcal{M}_L = \mathcal{M}_I \cap \mathcal{M}_U$ and $\mathcal{M}_U = \{\mu : L\mu \geq b \wedge L_0\mu \geq b_0\}$. The corollary statement explicitly guarantees that the assumptions C1 and C5 are satisfied. Because of Proposition 8.4, it can be seen that the corollary statement implies that C2–C4 are also satisfied. Therefore, Theorem 8.3 applies. By Theorem 8.3, T-liveness enforcement is impossible for all initial markings $\mu_0 \notin \mathcal{M}_U$. Then, it follows that the set of markings of $\mathcal{M}_I$ for which T-liveness is enforcible is the set $\mathcal{M}_L = \mathcal{M}_I \cap \mathcal{M}_U$. This contradicts the last corollary assumption, and so concludes the proof. $\square$

The conditions of the previous corollary are sufficient only. Compared to Corollary 7.9 of the simplified procedure, this corollary suggests that initial constraints may be used to help convergence for a Petri net for which divergence is guaranteed when no initial constraints are used. A more detailed discussion of the termination issues follows in section 8.8.

Finally, note that checking the conditions C2–C5 is trivial on a computer. The procedure can be easily extended to check these conditions in order to report in the end which of the results of this section apply.

8.7 Extending Permissiveness

It was shown in section 7.8 that the simplified procedure can be extended to improve permissiveness when $\mathcal{N}_1$ has more than one T-minimal active subnet. This section considers a similar extension for the general procedure. Recall that the assumption that $\mathcal{N}_1$ has a single T-minimal active subnet has been denoted as the condition C5 in the list of conditions C1–C5 (page 207) for Theorem 8.3. This section considers the case when the condition C5 is not satisfied, and shows how the procedure can be extended to guarantee the permissiveness property of Theorem 8.3 in this case.

Let $\mathcal{N}_1^{A,1}$, $\mathcal{N}_1^{A,2}$, ..., $\mathcal{N}_1^{A,p}$ be the $\mathcal{T}$-minimal active subnets of $\mathcal{N}_1$, and let $\mathcal{N}_0^{A,1}$, $\mathcal{N}_0^{A,2}$, ..., $\mathcal{N}_0^{A,p}$ be the corresponding $\mathcal{T}$-minimal active subnets of $\mathcal{N}_0$. Theorem 8.3 does not apply, as we have p $(p > 1)$ $\mathcal{T}$-minimal subnets. However, it may apply for $T_0^{A,i}$-liveness, as there is a single $T_0^{A,i}$-minimal active subnet: $\mathcal{N}_1^{A,i}$ (we denote by $T_0^{A,i}$ the set of transitions of $\mathcal{N}_0^{A,i}$ and $i = 1 \ldots p$). Assume that the procedure terminates for all $i = 1 \ldots p$ when used to enforce $T_0^{A,i}$-liveness. Let $u \in \{0, 1, \ldots, p\}$, and assume that the $\mathcal{T}$-minimal active subnets are ordered such that for $i \in \{j \in \mathbb{N} : 1 \leq j \leq u\}$ the procedure has no siphon control failures when used for $T_0^{A,i}$-liveness, but for each $i \in \{j \in \mathbb{N} : u + 1 \leq j \leq p\}$ it has some siphon control failures. (In particular, if $u = 0$ siphon control failures occur for all $i = 1 \ldots p$, and if $u = p$ no siphon control failure occurs for any i.) Let $L^{(i)}\mu \geq b^{(i)}$ and $L_0^{(i)}\mu \geq b_0^{(i)}$ be the generated constraints for all $i = 1 \ldots u$. Let Ξ be the supervisor defined as follows. Ξ requires the initial marking μ_0 to be in the set $\mathcal{M}$, where $\mathcal{M} = \emptyset$ for $u = 0$ and

$$\mathcal{M} = \bigcup_{i=1}^{u} \left\{ \mu : L^{(i)}\mu \geq b^{(i)} \text{ and } L_0^{(i)}\mu \geq b_0^{(i)} \right\} \qquad (8.29)$$

for $u \geq 1$. Furthermore, Ξ allows a transition to fire only if the next reached marking is in $\mathcal{M}$. By construction and Theorem 8.1, Ξ is a $\mathcal{T}$-liveness-enforcing supervisor for all initial markings $\mu_0 \in \mathcal{M}_I \cap \mathcal{M}$. (Of course, for $u = 0$ we have $\mathcal{M} = \emptyset$, in which case no such initial markings μ_0 exist.) The next theorem states that Ξ is at least as permissive as the least restrictive fco-supervisor with respect to initial markings in $\mathcal{M}_0$. This means that no $\mathcal{T}$-liveness-enforcing supervisors exist for initial markings $\mu_0 \in \mathcal{M}_0 \setminus \mathcal{M}$ and that Ξ is at least as permissive as any $\mathcal{T}$-liveness-enforcing supervisor for initial markings $\mu_0 \in \mathcal{M}_0 \cap \mathcal{M}$.

Theorem 8.8. *Assume that for each $i = 1 \ldots u$ the procedure satisfies C1–C4. Assume also that for each $i = u + 1 \ldots p$ C1 is satisfied, the first siphon control failure occurs at step C.3.d, and C4 is satisfied in all iterations previous to the first siphon control failure. If $u > 0$, Ξ is at least as permissive as the least restrictive $\mathcal{T}$-liveness-enforcing fco-supervisor for initial markings in $\mathcal{M}_0$. Otherwise, if $u = 0$, there is no $\mathcal{T}$-liveness-enforcing supervisor for any initial marking $\mu_0 \in \mathcal{M}_0$.*

Proof. The assumptions of the theorem ensure that for any $i = u + 1 \ldots p$, the first failure at step C.3.d is only possible when initial and/or initial-marking constraints are given. Then, by Proposition 8.4, $T_0^{A,i}$-liveness cannot be enforced for all $i = u + 1 \ldots p$ for the given constraints. The proof of the theorem is by contradiction. Let $\mu_0 \in \mathcal{M}_0$, and assume there is a $\mathcal{T}$-liveness-enforcing supervisor allowing a marking $\mu \notin \mathcal{M}$ to be reached; then μ enables a firing sequence σ which includes all transitions in $\mathcal{T}$ infinitely often. Let U be the set of transitions appearing infinitely often in σ, and D the incidence matrix of $\mathcal{N}_0$. By Lemma 6.7, there is $x \geq 0$ such that $Dx \geq 0$, $x(i) > 0$ $\forall t_i \in U$ and $x(i) = 0$ $\forall t_i \in T_0 \setminus U$; let $T^A = \|x\|$. Then T^A defines an active subnet, and note that $\mathcal{T} \subseteq T^A$. Since $\mathcal{N}_0^{A,i}$, $i = 1 \ldots p$, are all the $\mathcal{T}$-minimal active subnets, there is j, $1 \leq j \leq p$, such that $T_0^{A,j} \subseteq T^A$. If $j \leq u$, we have a contradiction, since by Theorem 8.3 not all transitions of $T_0^{A,j}$ can be made live for $\mu \notin \mathcal{M}$, and so not all of them can appear in σ. If $j > u$ we again have a contradiction, since for all initial markings $\mu_0 \in \mathcal{M}_0$ not all transitions of $T^{A,j}$ can be made live. $\square$

While the construction of Ξ yields a supervisor more permissive than the supervisors generated by the procedure itself, it is not so clear whether Ξ is admissible. The problem arises as follows. Let D_{uc} and D_{uo} be the restrictions of the incidence matrix D of $\mathcal{N}_0$ to the uncontrollable and unobservable transitions, respectively. The admissibility transformation used by the procedure (see section 8.4.3) guarantees that $L^{(i)} D_{uc} \geq 0$ and $L^{(i)} D_{uo} = 0$ (see equations (8.15) and (8.16)) for all $i = 1 \ldots u$. A consequence of $L^{(i)} D_{uo} = 0$ is that given the initial state, it is possible to monitor at all times the value of $x_i = L^{(i)} \mu - b^{(i)}$, as its value is only affected by the firings of observable transitions. However, there is no guarantee that $L_0^{(i)} D_{uo} = 0$. Therefore, it may not be possible to monitor the value of $x_{i,0} = L_0^{(i)} \mu - b_0^{(i)}$. Note that $x_{i,0} \geq 0$ is guaranteed at a given state μ_k if there is a prior state μ_{k-j} in which $x_{i,0} \geq 0$, $x_{i,0} \geq 0$, and all subsequent states $\mu_{k-j+1} \ldots \mu_k$ satisfy that $x_i \geq 0$. As this condition may not be satisfied in the context of (8.29), it may be necessary to monitor $x_{i,0}$, in order to know whether $x_{i,0} \geq 0$ or not. However, when $L_0^{(i)} D_{uo} \neq 0$, firing some unobservable transitions may cause uncertainty with regard to whether $x_{i,0} \geq 0$ is satisfied or not. For this reason, the exact implementation of Ξ may not always be possible.

Note that the admissibility requirement with regard to the uncontrollable transitions is always satisfied. Indeed, $L^{(i)} D_{uc} \geq 0$ guarantees that only controllable transitions need to be disabled in order to ensure that $L^{(i)} \mu \geq b^{(i)}$ is not violated. Therefore Ξ is always admissible if the net has no unobservable transitions. To guarantee in general that Ξ is admissible, step C.3.a of the procedure can be changed to prevent an inequality from being added to (A_0, d_0) when that inequality written in terms of $\mathcal{N}_0$ is[9] $l\mu \leq c$ such that $l D_{uo} \neq 0$. (This change will cause such an inequality to be transformed to be admissible (step C.3.c.i), added to (A, d), and enforced with a control place.) This will ensure that both $L^{(i)} D_{uo} = 0$ and $L_0^{(i)} D_{uo} = 0$, and so that Ξ is admissible. Then Ξ could be implemented as the disjunction of the supervisors Ξ_i enforcing $[L^{(i)T}, L_0^{(i)T}]^T \mu \geq [b^{(i)T}, b_0^{(i)T}]^T$.

---

[9]Recall that an inequality $l_i \mu \geq c_i$ of $\mathcal{N}_i$ is $l\mu \geq c$ when written in terms of $\mathcal{N}_0$ if $l = l_i|_{P_0} + l_i|_C L|_{P_0}$ and $c = c_i + l_i|_C b$.

8.8 Convergence Issues

This section considers issues regarding whether the procedure can provide a supervisor in a reasonable amount of time. From a theoretical viewpoint, the procedure does not have a guaranteed termination. From a practical viewpoint, computational complexity may further limit the usage of the procedure. We discuss the first issue in section 8.8.1 and the second one in section 8.8.2.

8.8.1 Termination Issues

The procedure does not have guaranteed termination. In the cases when the procedure does not terminate, it may be possible to help it terminate by using initial constraints. This topic is discussed first. Then, an example is used to illustrate how initial constraints can help termination. Finally, other termination results are discussed.

Initial constraints can help the procedure terminate, as they may cause some of the siphons generated during the iterations to be controlled (as the reachable markings are restricted to the markings satisfying the initial constraints). Intuitively, less uncontrolled siphons implies fewer control places, fewer control places implies that fewer new possibilities of deadlock are introduced in the system, which in turn implies faster termination. Example 8.4 illustrates this point. In Example 8.4, the same control places are added in the first iteration, regardless of whether initial constraints are given or not. However, at the second iteration, five additional control places are generated when no initial constraints are given; they correspond to the siphons $\{p_1, p_3, p_5, p_8, p_{11}\}$, $\{p_1, p_3, p_5, p_8, p_{10}, p_{12}\}$, $\{p_1, p_2, p_7, p_8, p_{11}\}$, $\{p_1, p_2, p_7, p_8, p_{10}, p_{12}\}$, and $\{p_1, p_2, p_7, p_{10}, p_{11}, p_{14}\}$. Note that the additional control places cause 167 additional minimal active siphons at the next iteration.

A particular case is when we are only interested in a finite set of initial markings and the target Petri net is bounded. Then initial constraints can be chosen to define a bounded set including all markings reachable from the initial markings of interest. Then, if the procedure is started with these initial constraints, assuming that no transition splits occur during the iterations (which in practice is often the case for the dp-procedure), the procedure terminates. Termination occurs because each time a new constraint is added to (L, b) or (L_0, b_0) in the procedure, at least one new marking is forbidden, and the number of markings which can be forbidden is finite due to the initial constraints. To summarize, given a target Petri net $\mathcal{N}$:

- Find a set of constraints $L_I \mu \geq b_I$ with bounded feasible set $\mathcal{F}$ such that for all initial markings μ_0 of interest for $\mathcal{N}$: $\mathcal{R}(\mathcal{N}, \mu_0) \subseteq \mathcal{F}$. Let $\mathcal{M}_I$ be the set of initial markings of interest.
- Apply the procedure on $\mathcal{N}$ with initial constraints (L_I, b_I).

- The resulting supervisor can be used for the initial markings $\mu_0 \in \mathcal{M}_I$ satisfying $L\mu_0 \geq b$ and $L_0\mu_0 \geq b_0$, where (L, b) and (L_0, b_0) are the two sets of constraints generated by the procedure.

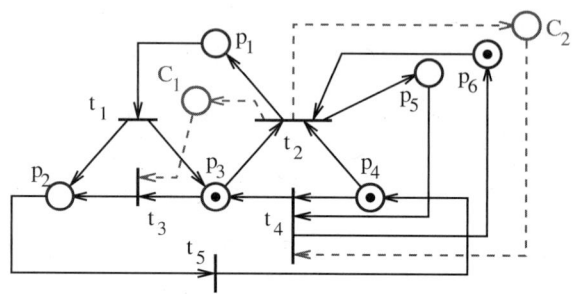

Fig. 8.8. A Petri net for which the dp-procedure does not terminate unless appropriate initial constraints are given.

EXAMPLE 8.5

Consider the Petri net of Fig. 8.8 for deadlock prevention with $\mathcal{T} = T_0$. At the first iteration the uncontrolled siphons are: $S_1 = \{p_1, p_3, p_5\}$, $S_2 = \{p_1, p_2, p_3, p_4\}$, and $S_3 = \{p_5, p_6\}$. The control place C_1 is added to control S_1; the inequality $\mu_1 + \mu_3 + \mu_5 \geq 1$ is added to (L, b). However, S_2 and S_3 do not need a control place, so $\mu_1 + \mu_2 + \mu_3 + \mu_4 \geq 1$ and $\mu_5 + \mu_6 \geq 1$ are added to (L_0, b_0). At the second iteration, there is a single uncontrolled siphon, $\{p_1, p_2, C_1, p_4\}$, and the control place C_2 results. At the third iteration the uncontrolled siphons are $\{p_1, p_3, C_2\}$ and $\{C_2, p_6\}$. Note that C_2 has the same connections as p_5, and so the siphon $\{p_1, p_3, C_2\}$ corresponds to $S_1 = \{p_1, p_3, p_5\}$, and $\{C_2, p_6\}$ to $S_3 = \{p_5, p_6\}$. The procedure diverges. At each iteration it adds a control place as follows: (a) at an iteration $n = 2k$, the control place C_n is added to control the siphon $\{p_1, p_2, C_{n-1}, p_4\}$, and (b) at an iteration $n = 2k + 1$, the control place C_n is added to control the siphon $\{p_1, p_3, C_{n-1}\}$. Then it can be noticed that C_n, for $n = 1, 2, \ldots$, enforces

$$n\mu_1 + \left\lfloor \frac{n}{2} \right\rfloor \mu_2 + \left\lceil \frac{n}{2} \right\rceil \mu_3 + \left\lfloor \frac{n}{2} \right\rfloor \mu_4 + \mu_5 \geq n. \tag{8.30}$$

It can be shown that the system of inequalities (8.30) for $n = 1$ and $n = n_1$ implies (8.30) for $n = n_1 - 1$, for all $n_1 \geq 3$. Furthermore, it can also easily be shown that the new markings forbidden by adding (8.30) at the iteration n are as follows: (a) for $n = 2k$, $\mu_1 = \mu_2 = \mu_4 = 0$, $\mu_3 = 1$, and $\mu_5 = k - 1$; (b) for $n = 2k + 1$, $\mu_1 = \mu_3 = 0$, $\mu_2 + \mu_4 = 1$, and $\mu_5 = k$. Now assume that we start with the initial constraints $\mu_i \leq 4$ for all $i = 1 \ldots 6$. Recall that the usage of initial constraints assumes that for all initial markings of interest $\mu_0 \in \mathcal{M}_I$, all reachable markings μ satisfy the initial constraints. For instance, $\mathcal{M}_I$ could be $\{[1,0,0,0,0,1]^T, [0,0,1,1,0,1]^T\}$ for our initial constraints. Note that the markings forbidden at the iteration $n = 11$ if (8.30) would be enforced are $\mu_1 = \mu_3 = 0$, $\mu_2 + \mu_4 = 1$, and $\mu_5 = 5$. However, according to the initial constraints, these markings can never be reached, so the siphon $\{p_1, p_3, C_{10}\}$ is controlled. Therefore, no control place is added at the iteration $n = 11$, and so the procedure terminates. After removing the redundant constraints, the procedure terminates with:

$$L = \begin{bmatrix} 1 & 0 & 1 & 0 & 1 & 0 \\ 10 & 5 & 5 & 5 & 1 & 0 \end{bmatrix}, \quad b = \begin{bmatrix} 1 \\ 10 \end{bmatrix}, \tag{8.31}$$

and (L_0, b_0) containing $\mu_5 + \mu_6 \geq 1$. Deadlock prevention is guaranteed for all initial markings of interest $\mu_0 \in \mathcal{M}_I$ that satisfy $L\mu_0 \geq b$ and $L_0\mu_0 \geq b_0$.

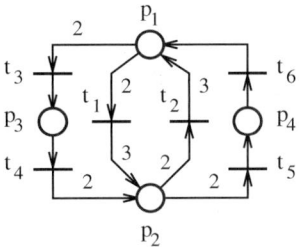

Fig. 8.9. A Petri net for which the least restrictive liveness-enforcing supervisor cannot be expressed by linear marking inequalities.

There are situations in which it is clear that the procedure will not terminate. Such a situation has been characterized by Corollary 7.9 for the simplified procedure, and by Corollary 8.7 for the general procedure. This situation may arise when the set of initial markings for which the Petri net can be made $\mathcal{T}$-live is not an integer convex set. Then, it is clear that the least restrictive $\mathcal{T}$-liveness-enforcing supervisor cannot be represented by any constraints $L\mu \geq b$ and $L_0\mu \geq b_0$. This is illustrated on the Petri net of Fig. 8.9. The Petri net has the property that a deadlock marking, $\mu_3 = [1, 1, 0, 0]$, is a linear combination of two markings $\mu_1 = [2, 0, 0, 0]$ and $\mu_2 = [0, 2, 0, 0]$ for which liveness can be enforced: $\mu_3 = 0.5\mu_1 + 0.5\mu_2$. This means that the least restrictive liveness-enforcing supervisor cannot be represented by any constraints $L\mu \geq b$ and $L_0\mu \geq b_0$. Note that even if we restrict our attention to the single initial marking $\mu_0 = \mu_1$, the least restrictive liveness-enforcing supervisor cannot be represented by constraints $L\mu \geq b$, as both μ_2 and μ_3 are reachable from μ_1.

8.8.2 Computational Complexity

Due to the great variety of Petri net structures, it is not possible to characterize the computational complexity for all cases as a function of the size of the net. In other words, we can find very large nets for which the procedure converges quickly, and small nets for which the procedure gets itself into serious computational problems. However, we may attempt to characterize the operations of the procedure by a worst case complexity. Furthermore, experiments may be used to obtain average complexities for various classes of Petri nets.

Table 8.2. Complexity of the procedure steps

Computation	Involves	Complexity
T-minimal active subnet	Linear programming	Polynomial
PT-transformation/AC-transformation		Polynomial
Finding the minimal active siphons	Search	Exponential
Checking whether a siphon is uncontrolled	Integer programming	Exponential
Admissibility transformation	Integer programming	Exponential
Feasibility check at step C.3.d	Integer programming	Exponential
Removal of redundant constraints	Integer programming	Exponential

The computational complexity of the major operations that may be performed in an iteration is shown in Table 8.2. Some operations admit a "sub-optimal" solution, and so allow replacing integer programming by linear programming. This is the case of the operation checking whether a siphon is uncontrolled. Using linear programming will cause more siphons to be declared "uncontrolled." This would only affect the convergence of the procedure, as the procedure may have to control more siphons. The removal of redundant constraints is another operation that allows replacing integer programming by linear programming. Of course, this may cause some redundant constraints to remain undetected, and so to have a more complex supervisor. Further research may also lead to an admissibility transformation that employs linear programming instead of integer programming. On the other side, removing the feasibility check of step C.3.d or using linear programming there instead of integer programming may have more serious consequences. Step C.3.d is used to ensure that there are initial markings μ satisfying the constraints $L\mu \geq b$, $L_0\mu \geq b_0$, $L_I\mu \geq b_I$, and $L_{I0}\mu \geq b_{I0}$. Without a proper feasibility check at step C.3.d, it is hard or impossible to guarantee that $L\mu \geq b$ and $L_0\mu \geq b_0$ are feasible, even when no initial/initial-marking constraints are given. This is due to the fact that the final matrices L and L_0 may contain negative elements. Note that the matrices L and L_0 are guaranteed to have only nonnegative elements if the admissibility transformation of step C.3.c.i is changed to generate constraints $l\mu \geq c$ with $l(p) \geq 0$ for all places p.

Experimentally, finding the minimal active siphons tends to be the most time-expensive computation. The computation of this operation obviously depends on the number of minimal siphons. It can be noticed that the number of siphons is exponential in the worst case. To see this, consider a Petri net with m places and n transitions. The maximal number of distinct subsets of places is 2^m, so we know that the number of minimal siphons cannot exceed 2^m. On the other hand, consider the free-choice Petri net constructed as shown in Fig. 8.10. The Petri net has $2k$ places and k transitions, where k may be set to $k = \min\left(2\left\lfloor\frac{m}{2}\right\rfloor, n\right)$. Note that the Petri net can be extended to have m places and n transitions by adding disconnected places and disconnected transitions, as necessary. Note that the Petri net has 2^k minimal siphons of the form $\{p_{1,i_1}, p_{2,i_2}, \dots, p_{k,i_k}\}$, for $i_1 \dots i_k \in \{1, 2\}$. This proves that *the maximal number of minimal siphons that a Petri net with m places and n transitions may have is lower bounded by 2^k and upper bounded by 2^m, for $k = \min\left(2\left\lfloor\frac{m}{2}\right\rfloor, n\right)$.*

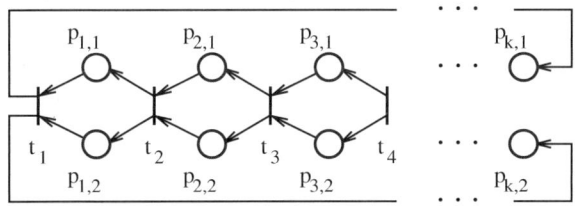

Fig. 8.10. In the worst case, the number of minimal siphons has an exponential dependence on the size of the Petri net.

Finally, note that the experimental results indicate that the dp-procedure is significantly more likely to converge than the le-procedure.

8.9 Applications

8.9.1 Deadlock Prevention in a Manufacturing System

This section illustrates how the procedure can be used for deadlock prevention/liveness enforcement in manufacturing systems.

EXAMPLE 8.6

Consider the target Petri net structure of Fig. 8.12(a) for deadlock prevention ($\mathcal{T} = T_0$). The Petri net may be seen as the representation of the manufacturing system shown in Fig. 8.11. The manufacturing system is described

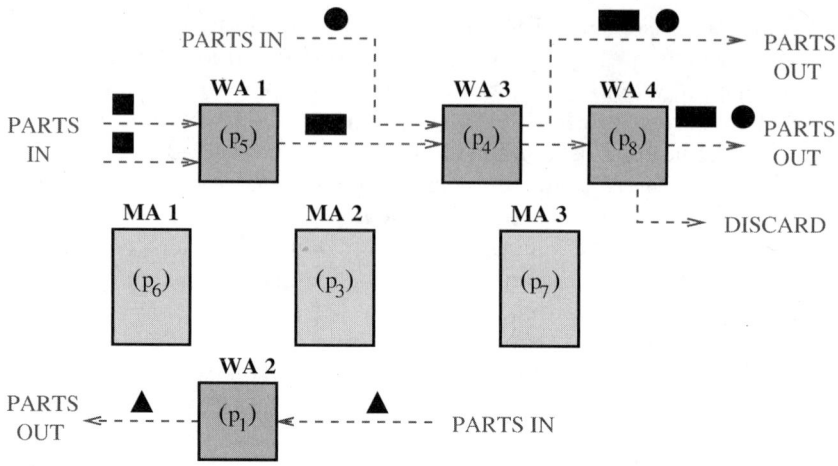

Fig. 8.11. The manufacturing system of Example 8.6.

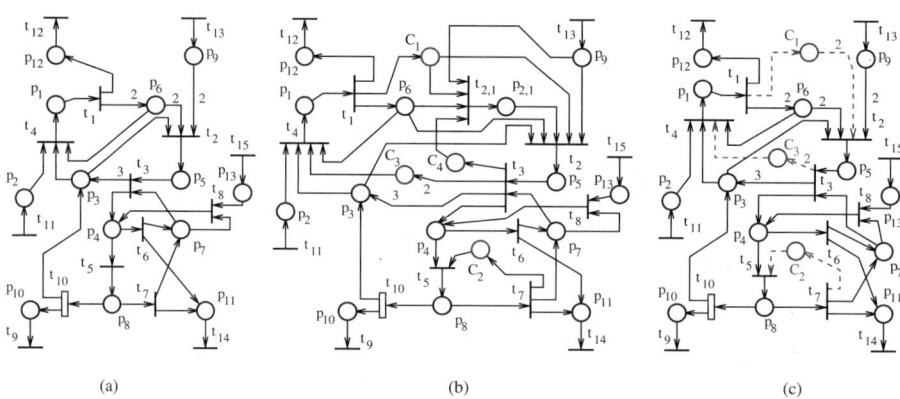

Fig. 8.12. Petri nets in Example 8.6: (a) target Petri net; (b) the Petri net after five iterations; (c) the supervised Petri net.

next. The system has four work areas, $WA1\ldots, WA4$, and three machine areas $MA1$, $MA2$, and $MA3$. In $WA1$ two parts are assembled, and this operation involves two machines from $MA1$ and one from $MA2$; upon completion, all three machines should be in $MA2$. Work in $WA2$ involves one part, one machine from $MA2$, and one from $MA1$; upon completion, both machines should be in $MA1$. Work in $WA3$ involves one part which may be of two different types, and one machine from $MA3$; upon completion, the machine returns to $MA3$. Optionally, the operation in $WA3$ is continued with an additional operation in $WA4$; when this is the case, the machine of $MA\,3$ is released when the process in $WA\,4$ is completed. If no failure occurs in $WA\,4$, the machine returns to $MA\,3$. When a failure occurs, the machine no longer may be used in $MA\,3$, but it can still be used in $MA\,1$ or $MA\,2$, and is moved to $MA\,2$. The marking of the places p_3, p_6, and p_7 corresponds to available machines. The marking of the places p_5, p_1, p_4, and p_8 corresponds to the number of working processes in $WA\,1\ldots,WA\,4$. The markings of p_2, p_9, p_{10}, p_{11}, p_{12}, and p_{13} represent the number of parts in buffer areas. The uncontrollable transition t_{10} models the failure in $WA\,4$.

In the first iteration, the Petri net structure $\mathcal{N}_1 = (P_1, T_1, F_1, W_1)$ is that of Fig. 8.12(b), but without the control places $C_1, \ldots, C_4$ and their transition arcs. The place $p_{2,1}$ and the transition $t_{2,1}$ appear by splitting t_1. The maximal active subnet has the transitions in $T_1 \setminus \{t_9, t_{10}\}$. There are two minimal active siphons: $\{p_1, p_6\}$ and $\{p_4, p_7, p_8\}$. They are controlled with two new control places: C_1 and C_2, respectively, where the constraint of C_2 is transformed to (8.33), which is admissible.

In the second iteration, the maximal active subnet still has the transitions $T_1 \setminus \{t_9, t_{10}\}$ and the only uncontrolled minimal active siphon is $\{C_2, p_8\}$. No admissible constraint is found for the control of $\{C_2, p_8\}$. ($\mu(C_2) \geq 1$ is not acceptable; see step 5 of the constraint transformation Algorithm 8.2.) Therefore X, the set of transitions that should not appear in the active subnets of the following iterations, is set to $X = \{t_5, t_7, t_{10}\}$.

In the third iteration and the remaining iterations the active subnet has the set of transitions $T_1 \setminus \{t_5, t_7, t_9, t_{10}\}$. The only uncontrolled minimal active siphon is $S = \{C_2, p_8, p_3, p_5\}$. The constraint $\sum_{p \in S} \mu(p) \geq 1$ is admissible, and the control place enforcing it is C_3.

In the fourth iteration the only uncontrolled minimal active siphon is $\{p_1, C_1, p_5, C_3\}$, and so the control place C_4 is added.

In the fifth iteration the only uncontrolled minimal active siphon is $S = \{C_4, p_{2,1}, p_5\}$. Since the control place which would control this siphon satisfies $C\bullet \subseteq \bullet S$, no control place is added, and so its constraint is included in (A_0, d_0).

The procedure terminates at the sixth iteration, as there is no uncontrolled minimal active siphon left. The constraints after step D of the procedure are:

$$\mu_1 + \mu_6 \geq 1, \tag{8.32}$$

$$\mu_4 + \mu_7 \geq 1, \tag{8.33}$$

$$\mu_3 + \mu_4 + \mu_5 + \mu_7 + \mu_8 \geq 2, \tag{8.34}$$

$$2\mu_1 + \mu_3 + \mu_4 + 2\mu_5 + \mu_6 + \mu_7 + \mu_8 \geq 4, \tag{8.35}$$

$$2\mu_1 + \mu_3 + \mu_4 + 3\mu_5 + \mu_6 + \mu_7 + \mu_8 \geq 5. \tag{8.36}$$

The inequalities (8.32)–(8.35) are included in $L\mu \geq b$ and correspond to $C_1 \ldots C_4$ in this order, while the inequality (8.36) is written as $L_0\mu \geq b_0$. The inequality (8.35) is redundant, and so it can be omitted. The Petri net supervised for deadlock-freedom is obtained by enforcing the constraints (L, b) on the target net (Fig. 8.12(c)).

8.9.2 Minimization of the Number of Resources

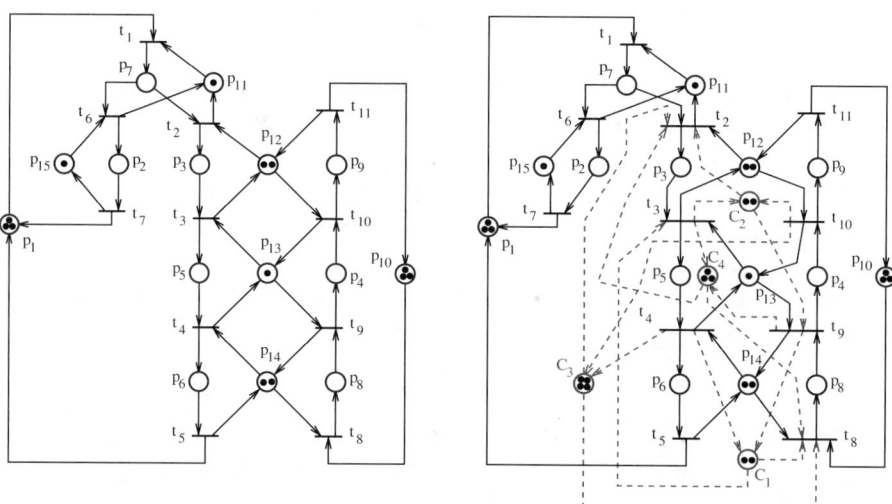

Fig. 8.13. Target Petri net (left) and supervised Petri net (right) in Example 8.7.

In some manufacturing applications, it may be of interest to know the minimal number of resources for which a manufacturing process can operate. In principle, this problem is nontrivial, as even for particular classes of manufacturing systems, the realizability problem is NP-complete [163]. In terms of Petri nets, the realizability problem corresponds to checking whether a given initial marking allows liveness enforcement or not. This section illustrates the fact that the supervisors generated by the procedure of this chapter allow the designer to obtain the minimal number of resources for which the system has no deadlocks. Of course, when the supervisors are not least restrictive, we only obtain an upper bound of the minimal number of resources.

Typically, the Petri net models of manufacturing systems associate a place with each kind of resource. Furthermore, the number of idle resources of a kind is represented by the marking of the corresponding resource place. Assuming that all resources are idle at the initialization of the system, the initial state of the system can be represented by an initial marking in which only the resource places are marked. This constraint on the initial marking together with the constraints $L\mu \geq b$ and $L_0\mu \geq b_0$ generated by the procedure can be used in an integer linear program to find the minimal number of resources. This idea is illustrated in the following example.

EXAMPLE 8.7

The Petri net in Fig. 8.13 is used to model a simple manufacturing system in [48]. In this example the le-procedure is applied to the Petri net model. The resulting supervisor has the control places shown in Fig. 8.13. They correspond to the constraints of $L\mu \geq b$:

$$\mu_4 + \mu_6 + \mu_{13} + \mu_{14} \geq 1, \tag{8.37}$$

$$\mu_5 + \mu_9 + \mu_{12} + \mu_{13} \geq 1, \tag{8.38}$$

$$\mu_6 + \mu_9 + \mu_{12} + \mu_{13} + \mu_{14} \geq 1, \tag{8.39}$$

$$2\mu_4 + 2\mu_5 + \mu_6 + \mu_9 + \mu_{12} + 2\mu_{13} + \mu_{14} \geq 3. \tag{8.40}$$

The constraints of $L_0\mu \geq b_0$ are:

$$\mu_6 + \mu_8 + \mu_{14} \geq 1, \tag{8.41}$$

$$\mu_2 + \mu_{15} \geq 1, \tag{8.42}$$

$$\mu_1 + \mu_2 + \mu_3 + \mu_5 + \mu_6 + \mu_7 \geq 1, \tag{8.43}$$

$$\mu_4 + \mu_5 + \mu_{13} \geq 1, \tag{8.44}$$

$$\mu_3 + \mu_9 + \mu_{12} \geq 1, \tag{8.45}$$

$$\mu_7 + \mu_{11} \geq 1, \tag{8.46}$$

$$\mu_4 + \mu_8 + \mu_9 + \mu_{10} \geq 1. \tag{8.47}$$

The places p_{11}, p_{12}, p_{13}, p_{14}, and p_{15} are resource places. At the initial marking only the resource places are marked. This can be written as

$$\mu_1 + \mu_2 + \cdots + \mu_{10} = 0. \tag{8.48}$$

To find the minimum number of resources for which the system remains live, the following integer program can be solved:

$$\min \mu_{11} + \mu_{12} + \mu_{13} + \mu_{14} + \mu_{15}$$
$$\text{subject to } L\mu \geq b, \ L_0\mu \geq b_0, \text{ and } (8.48).$$

The solution to the integer program is $\mu_{11} = \mu_{12} = \cdots = \mu_{15} = 1$. This shows that the minimal number of resources for which the system can operate is reached when there is a single resource of each kind.

8.9.3 Resource Preallocation

In practice, it may be desirable to automatize the initialization of a manufacturing system. When resources of the same type can be used for various tasks at various locations in the plant, a problem that may arise is to allocate such resources to the different locations in a way that guarantees that the system can operate without deadlocks. For instance, given the manufacturing system of Example 8.6, assume that at the initialization there are n machines that can be (irreversibly) dedicated as machines of either $MA1$, $MA2$, or $MA3$. The problem is to allocate machines to the locations $MA1$, $MA2$, and $MA3$, such that the system can operate. In principle, we could find a solution by solving a linear integer program constrained by the constraints $L\mu \geq b$ and $L_0\mu \geq b_0$ generated by the procedure in Example 8.6. Alternatively, as proposed in this section, the initialization problem can be solved together with the liveness enforcement/deadlock prevention problem, in which case the supervisor will disable any allocation decisions that would lead to deadlock. This solution is illustrated in the following example.

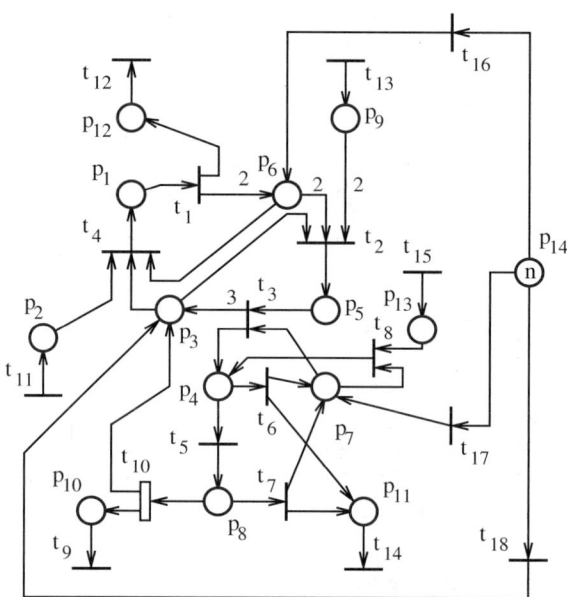

Fig. 8.14. The manufacturing system of Example 8.8.

EXAMPLE 8.8

Consider the Petri net of Fig. 8.14. The initialization problem discussed above can be included in the deadlock prevention problem of Example 8.6 by adding the source place p_{14} of initial marking n. The place p_{14} is connected through the transitions t_{16}, t_{17}, and t_{18} to the places p_6, p_7, and p_3, corresponding to $MA1$, $MA3$, and $MA2$, respectively.

This problem is more complex than that of Example 8.6, as significantly more active siphons appear. After the dp-procedure is run, the following constraints are obtained:

$$\mu_1 + \mu_6 + \mu_{14} \geq 1, \qquad (8.49)$$

$$\mu_4 + \mu_7 + \mu_{14} \geq 1, \qquad (8.50)$$

$$\mu_1 + \mu_4 + \mu_6 + \mu_7 + \mu_{14} \geq 2, \qquad (8.51)$$

$$\mu_3 + \mu_4 + \mu_5 + \mu_7 + \mu_8 + \mu_{14} \geq 2, \qquad (8.52)$$

$$2\mu_1 + \mu_3 + \mu_4 + 3\mu_5 + \mu_6 + \mu_7 + \mu_8 + \mu_{14} \geq 5. \qquad (8.53)$$

The inequalities (8.49)–(8.52) are in $L\mu \geq b$, while the inequality (8.36) is in $L_0\mu \geq b_0$. The closed-loop Petri net has a more involved graphical representation, which is omitted.

9

DES Control of Concurrent Hybrid Systems

9.1 Introduction

This chapter extends the DES framework toward the description of concurrency in hybrid systems. Concurrency could arise in various ways in hybrid systems. A possible paradigm is presented in Figures 9.1 and 9.2. Fig. 9.1 presents n subsystems operating in parallel. Fig. 9.2 presents a two-level control architecture. The lower level consists of the controllers $C1$, $C2$, ..., Cn. The role of the controllers is to govern the hybrid dynamics of the subsystems. The lower level design generates the controllers $C1$, $C2$, ..., Cn and obtains a logical model for each closed loop of a controller and a subsystem. The logical models are composed to form the global logical model of the system. This global model of the system is used when the supervisor is generated at the higher level design step. In Fig. 9.2 the global model is implemented (for instance, as a computer program), and the variables of the global model are updated (according to the rules of the global model) when a subsystem generates an event. The supervisor uses the global model variables in order to decide which controller inputs are appropriate. The control input of a controller Ci dictates the region where the state of the controlled subsystem should be. Given a state of the system, the supervisor may have more than one choice in selecting the inputs of the controllers, and so an operator (for instance, another computer program) can select a specific input according to some optimization criteria.

The material of this chapter corresponds to the upper level design in the architecture of Fig. 9.2. The lower level design is approached in the next chapter.

Fig. 9.1. Concurrent system.

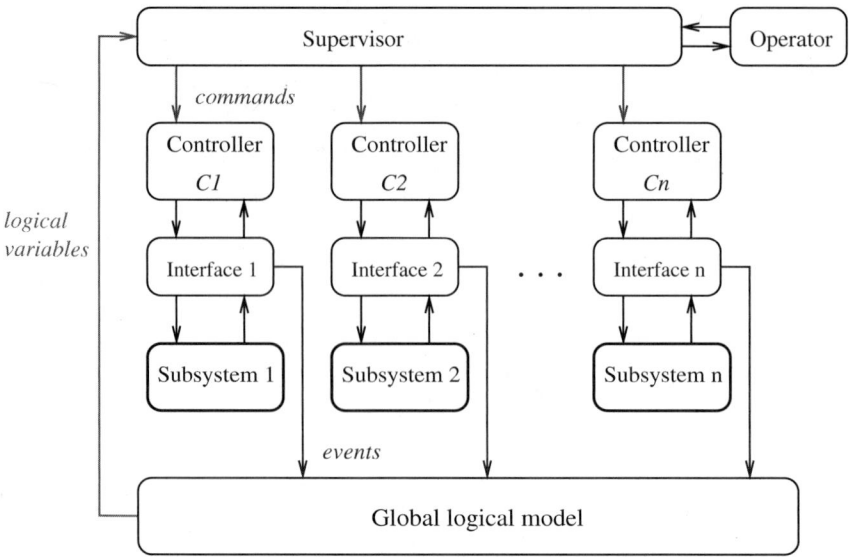

Fig. 9.2. The Control Architecture

Here, an extended DES framework is presented for the representation of the global logical model. As in the rest of the book, Petri net modeling is used. The specifications considered here involve safety and liveness requirements. Safety refers to restricting the space of Petri net markings by linear-marking inequalities. The liveness requirements are given as a set of actions (transitions) which are to be live. The design problem is to synthesize a supervisor enforcing the specifications.

This chapter extends the traditional DES framework in two directions. First, it is no longer assumed that a supervisor always has the ability to keep the system in any of the DES states indefinitely long. DES states with this property are called **unstable**. Thus, when an unstable state is reached, the supervisor is to enable at least one of the transitions moving the system from that state. Otherwise, the system would still move out of the unstable state, but according to unmodeled dynamics. In hybrid system terms, an unstable state corresponds to a mode that does not have a controlled invariant.

A second extension is that the uncontrollability of transitions is refined to distinguish between transitions that cannot be forced to fire and transitions that cannot be prevented from firing (when enabled). Further distinction is made between the uncontrollable transitions that can be guaranteed to fire (eventually or in a timely fashion) when enabled, and the uncontrollable transitions with no such guarantee.

In the traditional DES framework, the SBPI [141, 142, 214] solves the problem of enforcing linear-marking inequalities. Further, liveness can be enforced using the procedure of chapters 7 and 8 or other methods from the literature. However, in the following it is shown that these approaches cannot be applied directly to the extended DES framework.

First, consider the enforcement of safety specifications, such as the linear-marking inequalities. Clearly, we may generate the supervisor as in the traditional framework when we can have the system stay as long as desired in each state. Indeed, for this class of constraints, staying in a state (as long as the state is not forbidden) is always a legal choice. Therefore, a supervisor avoiding forbidden states always has at least a legal choice. However, when there are unstable states, it may be possible to reach a situation with no legal choices. In such a situation the supervisor can do nothing to stop the constraints from being violated. As an example, Fig. 9.3(a) displays the part of a Petri net model describing a hybrid system mode m. The fact that the mode has no self-loop indicates that it represents an unstable state. Thus, once the mode m is entered, the supervisor has to choose one of t_1, t_2, or t_3. If choosing any of t_1, t_2, or t_3 violates the specifications, the supervisor is not well defined, as it does not satisfy the specifications. In traditional supervisory problems it is assumed that we can inhibit all of t_1, t_2, and t_3 until the global state changes so that it is safe to fire one of them. However, in our setting this is not a choice: if no control policy is selected, the hybrid system of mode m will continue to run anyway (we cannot have it freeze its continuous dynamics until circumstances change), but according to some dynamics not modeled in the DES abstraction. When we can keep the system in mode m at will, we may add in Fig. 9.3(a) an action t_6 that is a self-loop transition (starts from m and enters m). When all modes of all component (hybrid) systems have this property, we have a traditional supervisory problem. Of course, we could still use the traditional framework by modeling t_1, t_2, and t_3 as uncontrollable. However, this may create an overly restrictive supervisor.

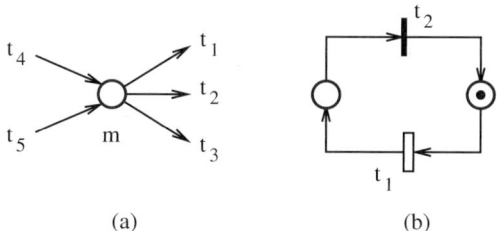

(a)   (b)

Fig. 9.3. Illustrations for two supervisory problems.

The enforcement of liveness specifications also has the same problem. (For instance, recall that the liveness enforcement approach of the previous chapters generates supervisors enforcing linear-marking inequalities.) In addition, note that removing the assumption that an enabled uncontrollable transition eventually fires creates the following problem. Consider the Petri net of Fig. 9.3(b), in which t_1 is uncontrollable and t_2 is controllable. In the traditional DES framework the system is live. However, in this framework, the system may not be live, since t_1 may never fire. Note also that such uncontrollable transitions cannot be deleted from the model either, since their firing, if it occurs, affects the state of the system.

This chapter is organized as follows. A brief overview of the related work is given is section 9.2. The Petri net modeling in the extended DES framework is presented in section 9.3. Approaches for supervisory specifications in this framework are considered in section 9.4.

9.2 Related Work

Petri nets have been used to generalize the hybrid automaton model. The resulting model has been called *programmable timed Petri net* (PTPN) and is described in [123, 65]. PTPNs have been used in a number of other works. Using supervisory control based on place invariants for the control of systems modeled by PTPNs has been considered in [110, 107]. Examples include a power system [108] and a gas storage system [105]. When PTPNs have been used, it has usually been assumed that the state gradually moves from an equilibrium point to another by switching between the subsystems (or modes), where the switching policies are described by the Petri net structure [104]. However, this chapter considers also the case when the state does not always move through equilibrium points.

Significant research effort has been made in the area of *hybrid Petri nets* [69]. Two types of hybrid Petri nets have been considered: the combination of traditional Petri nets and continuous Petri nets [40], and Petri nets enhanced with continuous variables satisfying differential equations (such as PTPNs). The study of such Petri nets requires hybrid systems methodologies. However, the approach in this chapter is different. It attempts to decompose the problem into a pure DES problem and problems requiring hybrid systems methodologies. This approach takes advantage of existing DES methodologies and simplifies the problems to be solved at the hybrid system level.

Other work on the Petri net modeling of continuous systems includes [131]. There, a Petri net model is obtained by composing state machine models, each representing logic sequences between various situations that may occur in the plant. The approach there seems to be aimed for the untimed diagnosis of faults. A related but different approach also appears in [111]. There, a timed Petri net modeling of complex hybrid systems is proposed. The Petri net is used as a reference model for fault diagnosis. The hybrid systems there are complex in that they consist of a large number of components.

In the context of the automata representation of DES, forcible events have been incorporated in [60] and in the timed setting of [28].

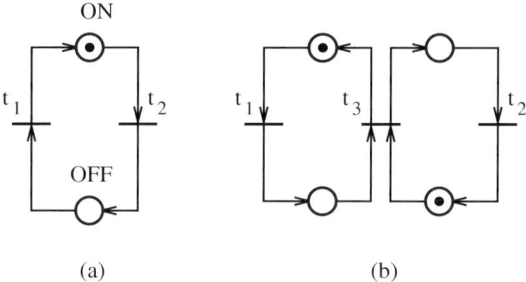

ON

OFF

(a)   (b)

Fig. 9.4. Examples of Petri net DES models.

9.3 Modeling

This section describes the global Petri net model of Fig. 9.2. Several issues are of interest: (a) the type of information stored in the model, (b) the way in which behaviors of interest are modeled, (c) ways in which to reduce the amount of nondeterminism contained by a model, and (d) ways in which to reduce the size of the model.

The global Petri net model is a composition of DES abstractions for each individual hybrid system. The DES models assumed for the individual systems are described in section 9.3.1. The individual systems may be coupled in several ways; this is considered in section 9.3.2. The Petri net transitions are classified in section 9.3.3 according to their controllability attributes. Section 9.3.4 describes how nondeterminism can be incorporated. The use of self-loops is described in section 9.3.5. Note that a place participating in a self-loop with a controllable transition indicates the ability to force the hybrid system to stay in the mode modeled by the place. Synchronization is treated in section 9.3.6. Then, the modeling approach is outlined in Table 9.2. Examples are given in sections 9.3.7 and 9.3.8.

9.3.1 The DES Representation of Hybrid Systems

The transitions between the modes of a hybrid system can be represented by a state machine or a more general Petri net. (The state machines are a particular class of Petri nets [145].) For instance, a system with two modes ON and OFF can be represented by the state machine of Fig. 9.4(a).

In the case of a state machine model, places correspond to modes, the total marking is always 1, and the marked place corresponds to the current mode of operation. However, multiple identical systems can be represented by a single state machine with total marking greater than 1. Then each token would correspond to one of the systems. Thus the number of tokens in a place would be the number of systems in the mode corresponding to that place.

A hybrid system could also be represented by Petri nets that are not state machines, such as the Petri net of Fig. 9.4(b). A typical way in which such Petri net models arise is by composing hybrid systems with state machine representation. Finally, note that in the Petri net representation of a hybrid system, the continuous dynamics of the hybrid system depends on the Petri net marking.

9.3.2 Coupling among Hybrid Systems

Here, coupling refers to the interdependence between hybrid system components. Three types of couplings could be considered: *state*, *mode*, and *transition* couplings. **State coupling** is due to constraints involving the continuous state variables and modes of two or more systems. **Mode couplings** are the state couplings that can be written as constraints involving only the modes of two or more systems. For instance, if the system #1 should not be in mode #3 when the system #2 is in mode #4, the two systems are mode coupled. Two or more systems are **transition coupled** when they share one or more common transitions.

State coupling seems difficult to consider at the DES level. Rather, it may be more convenient to consider state-coupled systems as a single system and deal with the state constraints at the hybrid system level. Thus, only mode and transition couplings are considered in this chapter. Note also that the mode couplings that can be expressed as linear-marking inequalities are of special interest here.

Finally, note that state machine methods may not be the most appropriate approach to the supervision at the DES level. Indeed, the global Petri net model (composing the DES models of the component hybrid systems) may not be a state machine even if all components are state machines. This is due to the transition couplings. Further, even when the global model is a state machine, the Petri net resulting after enforcing some constraints (such as linear-marking constraints) is no longer a state machine.

9.3.3 Transition Types

Here, the concept of uncontrollability is refined to distinguish between inability to disable a transition and inability to force a transition to fire. Table 9.1 outlines the notation. Thus, the *controllable* transitions are the transitions of type FI, that is, the transitions that can be forced/inhibited. The *uncontrollable* transitions are the transitions of the types NFI, FNI, and NFNI. An NFI transition can be inhibited but not forced, an FNI transition can be forced but not inhibited, and an NFNI transition can be neither forced nor inhibited.

It is important to note the various options a supervisor may have depending on the controllability of the transitions that are plant enabled. This is done by means of the example of Fig. 9.5(a). First, recall that contrary to the traditional DES setting, in this setting a supervisor has to ensure that some

Table 9.1. Transition types

	Can inhibit firing	Cannot inhibit firing
Can force firing	FI	FNI
Cannot force firing	NFI	NFNI

transition is fired once a mode is entered. Thus, at every plant state that is reached, a supervisor has to choose one out of several possible actions that keep the plant operation within the specification. This decision is taken randomly or at the hierarchical level above the supervisor. Further, the possible actions available at a certain reachable state are determined at the time of the supervisor design (not online).

In the example of Fig. 9.5(a), t_{FI} has the type FI, t_{NFI} has the type NFI, and so on. When the place p is marked, the supervisor has only two possibilities: it can choose to force t_{FI} or t_{FNI}. Assuming the supervisor chooses to fire t_{FI}, it has to do the following: force t_{FI} and inhibit t_{NFI}. This does not guarantee t_{FI} will occur, but does ensure that if t_{NFNI} and t_{FNI} do not occur during the process of firing t_{FI}, t_{FI} will occur. The other possibility is to force t_{FNI}. In this case, t_{FNI} is to be forced and t_{FI} and t_{NFI} are to be inhibited. This does not guarantee t_{FNI} will occur, but does ensure that if t_{NFNI} does not occur during the process of firing t_{FNI}, t_{FNI} will occur. Of course, the firing would always be guaranteed if t_{NFNI} were missing from the model.

Note also the following. The transitions of the type NFI can be excluded from a DES model, since they can be disabled at will, but they cannot be forced to fire. Consequently, *the DES models are assumed to have no transitions of the type NFI*. Referring back to the example of Fig. 9.3(b), notice that a transition whose firing cannot be guaranteed is a transition of the type NFNI. Finally, as in the traditional DES framework, the transitions can be classified as *observable* or *unobservable*, depending on whether their firing can be detected or not.

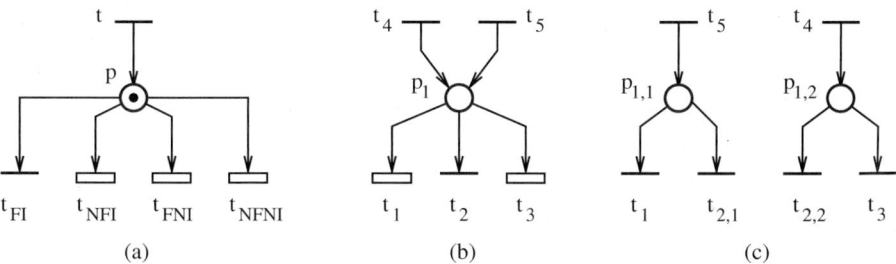

Fig. 9.5. Illustration of the transition types.

Sometimes there is a better way to deal with the NFI transitions rather than just deleting them from the net. For instance, consider the Petri net of Fig. 9.5(b) and assume that all of t_1, t_2, and t_3 can be inhibited; t_2 can always be fired; t_1 may not always be fireable if the mode p_1 was entered through t_4; and t_3 may not always be fireable if the mode was entered through t_5. Then, t_1 and t_3 are of type NFI. However, we can refine the Petri net as in Fig. 9.5(c), where both $p_{1,1}$ and $p_{1,2}$ correspond to the mode p_1, both $t_{2,1}$ and $t_{2,2}$ correspond to t_2, and all of t_1, $t_{2,1}$, $t_{2,2}$, and t_3 are FI. Another similar transformation is illustrated in Fig. 9.7, in which $t_{i,1} \ldots t_{i,k}$ correspond to a subset of t_1, t_2, $\ldots$, t_n which can be fired after firing t_0, and $t_{i,0}$ corresponds to t_0.

9.3.4 Modeling Nondeterminism

It may be possible that a mode of a hybrid system may switch nondeterministically to a mode within a set of modes. Such nondeterministic switching is represented by a number of uncontrollable transitions. This corresponds to Fig. 9.6(a), which illustrates a nondeterministic switch from p_1 to either of p_2 or p_3; the transitions t_1 and t_2 are uncontrollable. However, if we can inhibit the nondeterministic switch, it is better to use the model of Fig. 9.6(b). In Fig. 9.6(b) we have modeled through the controllable transition t_3 that we can inhibit the nondeterministic switch. (t_3 is of type FI, as in the case NFI we do not need to include t_1 and t_2 in the DES model, as discussed in section 9.3.3.) A similar example is given in Fig. 9.7, in which the nondeterminism is caused by t_0.

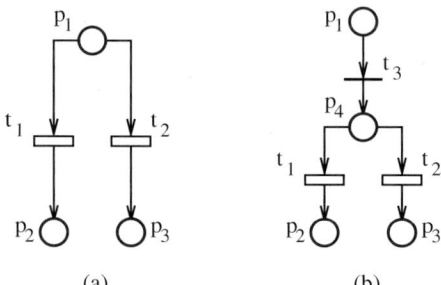

(a)   (b)

Fig. 9.6. Modeling nondeterminism.

9.3.5 Self-Loops

We may or may not have the ability to have a hybrid system stay in a certain mode for an unlimited amount of time. Recall that a *self-loop* is a pair of a place and a transition t such that $p \in \bullet t \cap t \bullet$. However, we restrict the definition of a **self-loop** to pairs p and t such that $t \bullet = \bullet t = \{p\}$. As an example, consider Fig. 9.7(a), in which the transition t_0 is a self-loop transition. A FI or FNI self-loop transition indicates that we can have the hybrid system stay infinitely long in the mode represented by p. The possibility that we may not be able to have a system leave a mode at will can be modeled by a self-loop transition of type FNI or NFNI.

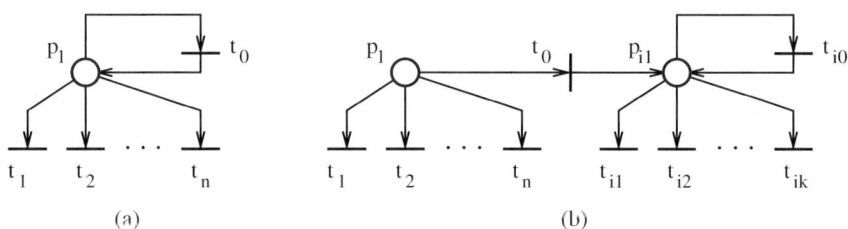

Fig. 9.7. Self-loop examples.

When a place p does not have a self-loop, we do not have the ability to keep the system in the mode corresponding to p for an unlimited amount of time. This means that the supervisor cannot disable all transitions in $p \bullet$ at the same time. The simplest way to incorporate this constraint in the global Petri net model is to consider all transitions in $p \bullet$ as uncontrollable (the type FNI or NFNI). However, in section 9.4.2 a better solution is suggested.

9.3.6 Synchronization

Consider the case in which a place p has a self-loop involving a controllable transition and there is a transition $t \in p \bullet$ such that $|\bullet t| > 1$. When the marking μ is reached such that $\mu(p)$ becomes 1 but t is not enabled, the supervisor has the choice of staying in the mode p by firing the self-loop transition, until t is enabled; then it can choose to fire t. However, if the place p does not participate in a self-loop with a controllable transition, we may include in the Petri net model the case when the supervisor has the option to wait until t is enabled. This corresponds to Fig. 9.8. In Fig. 9.8, the transition t' models the decision to wait until t is enabled; the controllable self-loop transition of p' indicates the ability of the system to wait until t is enabled.

9.3.7 Modeling Example

This section illustrates the proposed modeling technique on an automotive example from [12]. The example involves a four-stroke combustion engine.

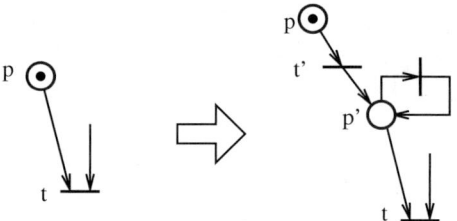

Fig. 9.8. Modeling synchronization.

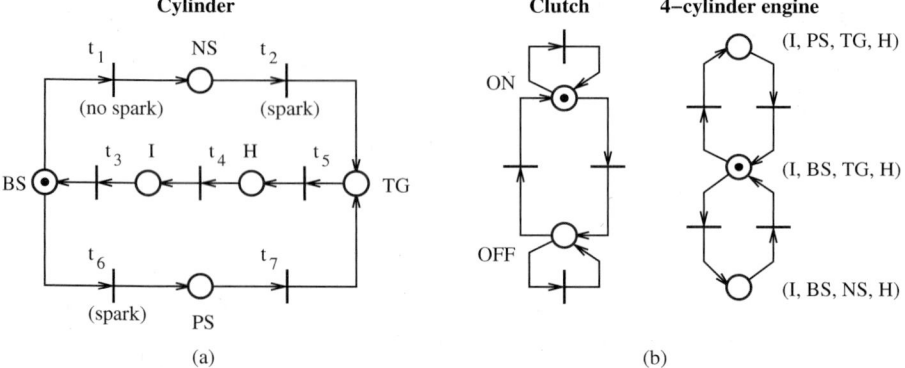

Fig. 9.9. Modeling of (a) cylinder dynamics in a four-stroke engine; (b) clutch and four-cylinder engine.

Every cylinder of the engine cycles through the following runs: intake (I), combustion, expansion, and exhaust (H). The compression phase preceding the ignition spark is denoted by BS (before spark). When the spark is generated before the piston reaches the top of the cylinder, the next phase is denoted by PS (positive spark). If the spark is generated after the piston reaches the top, the next phase is denoted by NS (negative spark). After the fuel is ignited, a torque-generating phase (TG) is entered during the expansion. A Petri net model of the cylinder dynamics is shown in Fig. 9.9(a).

In a four-cylinder engine, there is always one cylinder in each of the four runs. Therefore, because of the symmetry, the discrete dynamics of the engine can be described by three discrete states, corresponding to the three possible combinations of cylinder phases. (A fourth state (I, PS, NS, H) cannot occur, because two cylinders cannot be in PS and NS, respectively, at the same time: PS occurs at the end of the compression phase, while NS at the beginning of the expansion phase.) The discrete model is shown in Fig. 9.9(b). Further, Fig. 9.9(b) also displays the model of the clutch.

Table 9.2. Modeling summary

A hybrid system mode is represented by a place. A switching action is represented by a transition. A mode is active when it has a token. A place may have more tokens when we model several identical systems with the same abstraction.	
Controllable transitions exiting a place indicate that the controller always has the ability to select between them, no matter how the mode has been entered.	
An uncontrollable transition exiting a place indicates that the controller may not have the ability to select between output transitions. Uncontrollable transitions can be used to model nondeterminism.	
A controllable self-loop transition indicates the ability of the controller to keep the hybrid system (indefinitely long) in the current mode.	
Synchronization (which we call *transition coupling*) reflects the situation in which an action cannot be taken unless another system (or systems) is in a certain mode (or modes).	
We denote by *mode coupling* desirable constraints which forbid certain combinations of modes of different hybrid systems to be active at the same time. Mode coupling corresponds to marking constraints.	

Note that the systems of Fig. 9.9(b) are state coupled, as the engagement or disengagement of the clutch affects the mode dynamics of the engine. Further, if the lower level controller has the freedom to engage or disengage the clutch, self-loops can be added in the clutch model, as in Fig. 9.9(b).

The model of Fig. 9.9(a) can be used to illustrate that a supervisor may not be able to keep a system in a certain mode, at will. Consider the mode BS. Note that the transitions t_1 and t_6 are controllable, in the sense that a controller can chose which should be the next state: NS or PS. However, if

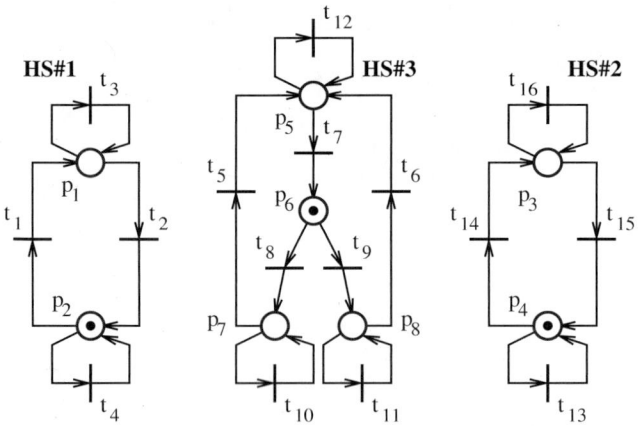

Fig. 9.10. Petri net abstraction of three subsystems.

none of t_1 or t_6 is enabled, that is, no spark is ignited, the mode will move anyway to NS (since the crankshaft keeps rotating), and then to H (exhaust of the ... unburned fuel). Note that the transition from NS to H is unmodeled dynamics!

9.3.8 Supervision Example

Fig. 9.10 shows the Petri net models of three systems, in which all transitions are controllable. Note that the place p_6 does not have a self-loop, and therefore a supervisor cannot force HS#3 to stay in the mode p_6. This complicates the synthesis of the supervisor, as we show next. Let's assume the specification

$$\mu_2 + \mu_7 \leq 1, \tag{9.1}$$

$$\mu_4 + \mu_8 \leq 1. \tag{9.2}$$

A supervisor enforcing them by SBPI is shown in Fig. 9.11. However, note that for the marking displayed in the figure none of the transitions t_8 and t_9 can fire, even though p_6 is marked. Therefore, the supervision is inadequate, as it attempts to keep HS#2 in the mode p_6 (until either of HS#1 or HS#2 change their state). It can be verified that this problem can be avoided by enforcing the additional constraint

$$\mu_2 + \mu_4 + \mu_6 + \mu_7 + \mu_8 \leq 2. \tag{9.3}$$

9.4 DES-Level Supervision

This section presents several types of supervisory specifications for the extended DES framework of this chapter. Recall that a system is in *deadlock*

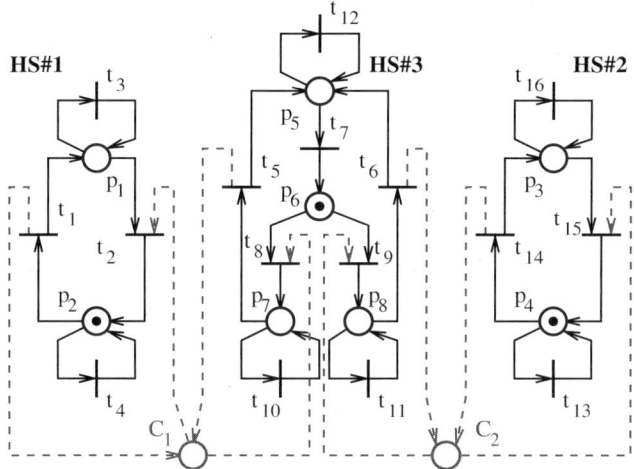

Fig. 9.11. Illustration of a difficulty arising in the supervision of abstractions.

when it has reached a state from which no further execution is possible. This is also called a global deadlock. A local deadlock corresponds to a deadlock that appears in only a part of the system. A system is *live* when no state of local deadlock can be reached. A transition t of a system is live if no state of local deadlock involving t can be reached. A system is $\mathcal{T}$-*live* if all transitions in the set $\mathcal{T}$ are live.

Assume that we have a number of linear-marking constraints on the total DES model of the hybrid systems. They could represent mode constraints, such as safety requirements. Let $\mathcal{N}$ be the total Petri net model of the hybrid systems and $\mathcal{N}_S$ the Petri net obtained by enforcing the linear-marking constraints on $\mathcal{N}$ using SBPI.

An important question is whether it is possible to have an execution such that at every time, for every component hybrid system, we may select at least one control input that insures the DES-level constraints are satisfied. Consider a state in which there is a component for which no such control inputs are desirable; in other words, for all applied inputs the next marking will not satisfy the marking constraints. Such a state is a deadlock for the supervisor. In fact, we can derive from $\mathcal{N}_S$ a Petri net which is in deadlock for such a state. A supervisor without such deadlocks is said to be **nonblocking**. Another question is whether it is possible to insure not only that the marking constraints are satisfied, but also that a number of actions can be repeated infinitely often. The actions correspond to transitions in the total Petri net. Therefore, this is a $\mathcal{T}$-liveness-enforcing problem. Thus, depending on the type of specifications we have, we may be interested in either a deadlock prevention supervisor or a $\mathcal{T}$-liveness-enforcing supervisor. (Note that $\mathcal{T}$-liveness implies deadlock-freedom.)

Two kinds of failures may occur in this approach. First, only admissible constraints can be enforced (see section 3.4). Inadmissible constraints can be transformed to admissible constraints, for instance, by using methods from [141, 142]. However, a failure occurs if we cannot obtain such admissible constraints. Second, deadlock prevention or the T-liveness enforcement approaches may fail. In both cases, the failures tend to occur because of excessive uncontrollability or unobservability of the Petri net. So, when such failures occur, we need to refine the hybrid system models, in order to reduce the nondeterminism of the model (manifested through excessive uncontrollability).

As discussed before, a supervisor should be *nonblocking*. This requirement can be formally described as follows.

Requirement 1. Let T_F be the set of transitions that can be forced to fire if enabled. The supervisor must ensure that for all markings μ reached under its supervision, there is a closed-loop-enabled firing vector q such that for all places $p \in \bullet T_F$ with $\mu(p) \geq 1$: (a) $\sum_{t \in p \bullet \cap T_F} q(t) \geq \mu(p)$, and (b) for all firing vectors $q_1, q_2, \ldots, q_k$ satisfying $q_1 + q_2 + \cdots q_k \leq q$, the sequence $q_1, q_2, \ldots, q_k$ is closed-loop enabled.

Here, a **supervisor** Ξ of a Petri net model is a map $\Xi : \mathcal{M} \to 2^{\mathbb{N}^{|T|}}$ that associates to each marking of $\mathcal{M}$ a set of firing vectors allowed to fire. Note that Requirement 1 is consistent with the SBPI, which ensures that any firing sequence $q_1, q_2, \ldots, q_k$ is enabled in the closed loop if $q_1 + q_2 + \cdots q_k \leq q$ and q is enabled in the closed loop. Ensuring this for other supervision techniques is necessary for the following reason. A concurrent firing command at the logic level may actually occur as a sequential firing at the physical level. Therefore, the specification should be satisfied by any firing sequence of Parikh vector less than or equal to q. Second, note that Requirement 1 is also adequate for Petri net models in which a place models the mode of several identical hybrid systems, and hence can have a marking greater than 1. Recall that if a place p corresponds to a certain mode of multiple identical hybrid systems, a marking $\mu(p)$ indicates that a number of $\mu(p)$ systems are in the mode of p. In this situation a supervisor should ensure that for all markings μ reached under its supervision, there is a valid control input that can be applied to each of the $\mu(p)$ systems.

Let $L_c \mu \geq b_c$ be marking constraints. We assume them to be admissible; if not, we can transform them to admissible constraints, using one of the approaches from [141, 142]. The two safety and liveness problems mentioned earlier in this section can be formally stated as follows.

Problem 1. Find a supervisor satisfying Requirement 1 such that the constraints $L_c \mu \geq b_c$ are enforced in $\mathcal{N}$.

Problem 2. Find a supervisor satisfying Requirement 1 such that the constraints $L_c\mu \geq b_c$ and $\mathcal{T}$-liveness are enforced in $\mathcal{N}$.

Problem 3. Find decentralized supervisors satisfying Requirement 1 for Problems 1 and 2, respectively.

Problems 1 and 2 are addressed first in sections 9.4.1 and 9.4.2. The approaches in these two sections can be generalized to Problem 3, discussed in section 9.4.3. As noted in the introduction, in the example of Fig. 9.3(b), deadlock prevention/$\mathcal{T}$-liveness enforcement can be problematic when the NFNI transitions are not guaranteed to eventually fire (when enabled). Section 9.4.4 addresses this problem.

9.4.1 Case 1: All Places Have Controllable or FNI Self-Loops

The Petri net models of this case have a controllable or FNI self-loop transition at every place. This case allows an immediate reduction of the problem to the design of supervisors in the traditional DES framework. Recall that the self-loops we consider involve transitions t such that $\bullet t = t\bullet = \{p\}$ and $W(t, p) = W(p, t) = 1$. In the traditional DES framework, the SBPI produces supervisors that enable the self-loop transitions at all times. This is also the case for the $\mathcal{T}$-liveness-enforcing procedure of the previous chapters, as it relies on the SBPI. Further, this is also true for most approaches in the literature for the enforcement of forbidden state or liveness specifications. Therefore, the design of the supervisor for Problem 1 or Problem 2 can be reduced to the design in the traditional DES framework of a supervisor enforcing $L_c\mu \leq b_c$ or $L_c\mu \leq b_c$ and $\mathcal{T}$-liveness. Requirement 1 is automatically satisfied, due to the fact that the self-loop transitions are always enabled by the supervisor, and so they can be forced to fire each time they are enabled by the plant. Further, note that a self-loop plays no role in the synthesis of a supervisor in the traditional DES setting, so it can be deleted during the design process. In summary, this is the procedure:

1. Let $\mathcal{N}' = (P, T', F', W')$ be $\mathcal{N}$ without self-loops. Let T_{uc} and T_{uo} be the sets of uncontrollable and unobservable transitions of $\mathcal{N}$, respectively.
2. Design in the traditional DES framework a supervisor $\mathcal{S}'$ of $\mathcal{N}'$ such that $L_c\mu \leq b_c$ (or $L_c\mu \leq b_c$ and $\mathcal{T}$-liveness) is enforced. The sets of uncontrollable and unobservable transitions of $\mathcal{N}'$ are $T_{uc} \cap T'$ and $T_{uo} \cap T'$.
3. The supervisor $\mathcal{S}$ of $\mathcal{N}$ is obtained from $\mathcal{S}'$ by enabling the self-loop transitions at all times.

Because Requirement 1 is satisfied, the supervisor $\mathcal{S}$ has at all times a valid control command for each of the component hybrid systems.

9.4.2 Case 2: Not All Places Have Controllable or FNI Self-Loops

In this case some places may have a NFNI self-loop transition or no self-loop at all. Two different approaches are proposed.

Solution 1: Increase the number of uncontrollable transitions

Consider the places p without self-loop transitions of types FI or FNI. The solution is to change the attributes of all transitions $t \in p\bullet$ to NI. Then, the supervisor $\mathcal{S}$ designed as in the Case 1 will never attempt to inhibit any transition in $p\bullet$. Therefore, Requirement 1 is satisfied. Finally, note that each time p contains a token, the supervisor $\mathcal{S}$ will have to force one of the forcible transitions in $p\bullet$ to fire. However, since the transitions in $p\bullet$ are taken of type NI, the supervisor design assumes the supervisor will not be able to guarantee that the transition it selects will actually fire. As this may not be the case (such as when all $t \in p\bullet$ are FI), this solution is suboptimal.

Solution 2: Transform the problem into a deadlock prevention problem

This solution assumes that the supervisor is significantly faster than the dynamics of the hybrid systems it controls. It also assumes ordinary Petri nets. Let $\mathcal{P}$ be the set of places p without self-loop transitions of types FI or FNI. To satisfy Requirement 1, a supervisor is to find a control decision for each token entering the places $\mathcal{P}$. The idea is that the supervisor can find the control decisions by considering the tokens separately, one at a time.

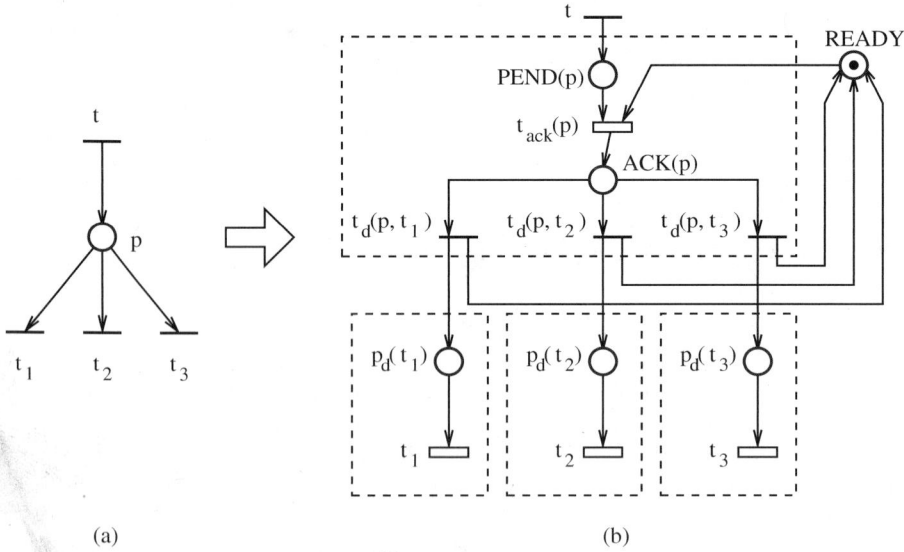

(a)   (b)

Fig. 9.12. Illustration of the transformation.

The solution is to transform the Petri net $\mathcal{N}$ into a new Petri net $\mathcal{N}'$ such that the nonblockingness requirement on the supervisor of $\mathcal{N}$ becomes a deadlock prevention specification in $\mathcal{N}'$. The transformation is illustrated first in Fig. 9.12. The transformation affects the places $p \in \mathcal{P}$ and the transitions $t \in \mathcal{P}\bullet$ that are controllable. (Throughout this description, the operator $\bullet$ is taken with respect to $\mathcal{N}$, not $\mathcal{N}'$.) In Fig. 9.12(a) $p \in \mathcal{P}$, as p does not have a self-loop. For this illustration, the transitions t_i, $i = 1, 2, 3$ are assumed controllable. In the transformation to $\mathcal{N}'$, a new place $READY$ is added. It is initially marked with one token. Further, p is replaced by $PEND(p)$, $ACK(p)$, $t_{ack}(p)$, $t_d(p, t_1)$, $t_d(p, t_2)$, and $t_d(p, t_3)$. $t_{ack}(p)$ is uncontrollable, while $t_d(p, t_1)$, $t_d(p, t_2)$, and $t_d(p, t_3)$ are controllable. Moreover, each transition t_i, $i = 1, 2, 3$, is enhanced with a place $p_d(t_i)$ and its controllability type is changed to NI. Note that a token entering $p_d(t_i)$ from $t_d(p, t_i)$ models the choice of the supervisor to fire t_i. For instance, in Fig. 9.12(b) the choice is between firing t_1, t_2, or t_3. In order to have deadlock when the supervisor has no choice $t_d(p, t_i)$ for which the specification will stay satisfied, self-loops are added as follows. The remaining transitions t of $\mathcal{N}$ that were not enhanced with a place p_d are connected to $READY$ through self-loop transitions (i.e., by both arcs $(READY, t)$ and $(t, READY)$).

The places added in the transformation have intuitive meanings. Thus, firing $t_{ack}(p)$ signifies that the supervisor acknowledges a new token has entered p. When the place $ACK(p)$ is marked, the supervisor works exclusively to select one of the possible choices $t_d(p, t_i)$. Once a choice has been made, a place $p_d(t_i)$ is marked and the supervisor is released to consider other requests ($READY$ is marked again).

The general description of the transformation is as follows. The places $p \in \mathcal{P}$ and the transitions $t \in \mathcal{P}\bullet$ that are controllable undergo the transformation of Fig. 9.12. As Fig. 9.12 makes the assumption that the transitions t_i are controllable and satisfy $|\bullet t_i| = 1$, we have the following enhancements. When some of the transitions $t \in p\bullet$ are uncontrollable, we do the construction above only for the controllable transitions, and we connect to each $p_d(t)$ a copy of each of the uncontrollable transitions in $p\bullet$. If all transitions of $p\bullet$ are uncontrollable, the transformation of Fig. 9.12(b) is not done for p. Now, in the case $|\bullet t_i| > 1$ for some controllable $t_i \in p\bullet$, the arcs $(p', t_d(p, t_i))$ or $(PEND(p'), t_d(p, t_i))$, as appropriate, are added for all $p' \in \bullet t_i \setminus \{p\}$. On the other hand, the case $|\bullet t_i| > 1$ is treated the same way as $|\bullet t_i| = 1$ if t_i is uncontrollable.

Finally, the specification on $\mathcal{N}$ is transformed as follows. For each p that is transformed (as in Fig. 9.12), $\mu(p)$ is substituted by the sum of the markings of $PEND(p)$, $ACK(p)$, and of the places $p_d(t)$. By construction, Problems 1 and 2 correspond in the transformed Petri net to the solution to the design of a supervisor for the enforcement of $L_c\mu \leq b_c$ with deadlock prevention, and the enforcement of $L_c\mu \leq b_c$ with $\mathcal{T}$-liveness enforcement, respectively.

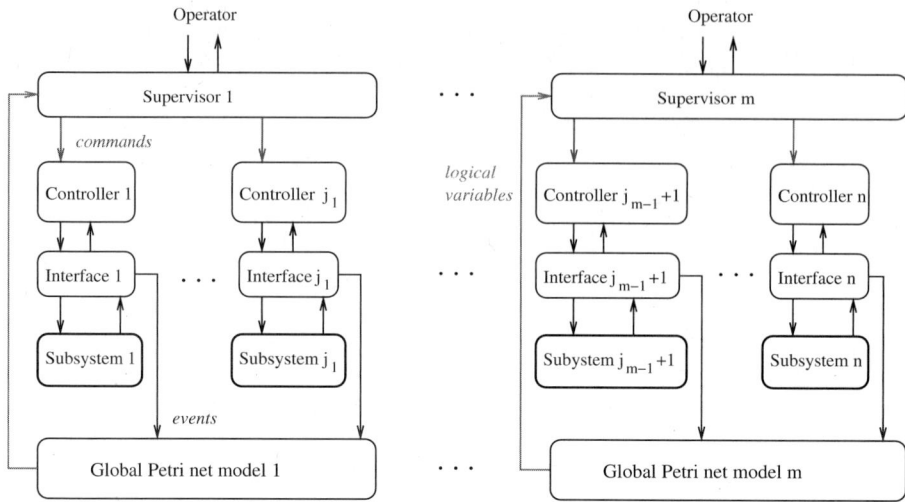

Fig. 9.13. Decentralized control architecture.

9.4.3 Decentralized Control

Decentralized control corresponds to the architecture shown in Fig. 9.13. In this case we have several global Petri net models which differ only by the fact that each has different sets of controllable and observable transitions. Thus we have groups of hybrid systems, and each group has one of the global Petri net models associated to it.

The decentralized control problem can be solved by using the approach of chapter 5 on the Petri net that resulted after either of the transformations of section 9.4.2. In the setting of chapter 5, each group of hybrid systems corresponds to a subsystem. Note that in the decentralized setting the deadlock prevention and $\mathcal{T}$-liveness enforcement problems are of increased difficulty. The deadlock prevention and $\mathcal{T}$-liveness enforcement procedures proposed in this book can be extended to the decentralized setting as follows. Each constraint for the control of a siphon is transformed to be d-admissible with respect to the Petri net model. This involves changing the transformation of section 8.4.3 to produce d-admissible instead of c-admissible constraints. Recall that d-admissibility is defined in chapter 5.

9.4.4 Dealing with NFNI Transitions

As shown in the example of Fig. 9.3(b) in the introduction, transitions of the type NFNI affect deadlock prevention and $\mathcal{T}$-liveness enforcement, as there is no guarantee they would eventually fire. Here, in the case of $\mathcal{T}$-liveness, it is assumed that no NFNI transition is in $\mathcal{T}$. For the procedures of this book, the dependence of the supervisor on the fireability of the NFNI transitions can be eliminated as follows. Let $\mathcal{N}$ be the Petri net model and $\mathcal{N}'$ be $\mathcal{N}$ without the NFNI transitions. Then, the deadlock prevention/$\mathcal{T}$-liveness enforcement supervisor of $\mathcal{N}$ can be computed by applying the procedures to $\mathcal{N}'$ and by ensuring that: (a) the generated constraints are admissible with respect to $\mathcal{N}$ (not $\mathcal{N}'$) and (b) in section 7.5.2, page 167, the check $C\bullet \subseteq \bullet S$ is done with respect to $\mathcal{N}$ (not $\mathcal{N}'$). Note that the admissibility requirement can easily be achieved by including the NFNI transitions in D_{uc} and D_{uo}. The requirement (b) means that the arcs between the NFNI transitions and the places of $\mathcal{N}$ are introduced in $\mathcal{N}_i$ when $C\bullet \subseteq \bullet S$ is tested (and C is computed).

To see that the supervisors produced this way prevent deadlock/enforce $\mathcal{T}$-liveness, note the following. Let T_u be the set of the NFNI transitions and $L\mu \geq b$ and $L_0\mu \geq b_0$ be the sets of constraints produced by the procedure. Given a firing sequence of $\mathcal{N}$, let $\mathcal{P}_u(\sigma)$ denote the projection of σ that removes the transitions in T_u. The requirement (a) ensures that no firing of a transition in T_u can decrease $L\mu$. (This results from the fact that the transformation to admissible constraints ensures that no arc exists from a control place to an uncontrollable transition.) The requirement (b) ensures that if $t \in T_u$, $L_0\mu \geq b_0$, and $\mu \xrightarrow{t} \mu_x$, then $L_0\mu_x \geq b_0$. Given the initial marking μ_0 such that $L\mu_0 \geq b$ and $L_0\mu_0 \geq b_0$, these two facts imply that for all reachable markings of the closed loop, the plant $\mathcal{N}$ satisfies $L\mu \geq b$ and $L_0\mu \geq b_0$. Since the closed loop of $\mathcal{N}'$ with $L\mu \geq b$ is deadlock-free/$\mathcal{T}$-live, it follows that regardless of the transitions in T_u that may fire in $\mathcal{N}$, the following is true of the closed loop of $\mathcal{N}$: in the case of deadlock prevention no reachable state is a deadlock state, and in the case of $\mathcal{T}$-liveness enforcement there is no reachable state with dead transitions in $\mathcal{T}$.

10

Hybrid System Level Control

10.1 Introduction

This chapter proposes a two-level approach for the supervision of concurrent hybrid systems. This approach has been illustrated in Fig. 9.2. Thus, at the upper level, a supervisor controls the DES behavior of the system, where the supervisor is designed based on the DES abstractions of the lower level. At the lower level, each hybrid subsystem of the concurrent system has a controller. The controllers implement the control decisions taken at the upper level. Note that the DES abstraction of the lower level is formed by composing the abstractions of the controlled hybrid subsystems. (A controlled hybrid subsystem is the closed-loop of a hybrid subsystem with its controller.)

The DES modeling of the lower level and the design of the supervisor of the upper level has been approached in chapter 9. This chapter approaches the design of the controllers of the lower level, and the DES abstraction of the controlled hybrid subsystems. The goal of the DES abstraction is to obtain DES models within the extended DES framework described in section 9.3 of the previous chapter. The role of the controllers is to ensure that the lower level behaves according to its abstraction, as long as the supervisor outputs commands consistent with the DES abstraction. The hybrid subsystems are assumed to be given in the form of hybrid automata [133]. Then, the DES abstraction and controllers can be obtained based on methods for the computation of predecessor sets, controlled invariant sets, and controllable invariant sets. Note that the concept of controllable invariant sets is introduced here, in this chapter.

While the computation of predecessor sets and controlled invariant sets has been approached in the literature in several settings, this chapter will focus on the discrete-time setting. Thus, the dynamics of each hybrid system mode will be assumed to be linear in the state variable, input, and disturbance. In this setting, the computation of the maximal controlled invariant sets has been approached and solved in [204]. The computation of the predecessor sets appears also in [104, 204]. Unfortunately, the predecessor computation may not terminate. Nonetheless, the computation of the predecessor set over a finite time horizon terminates, though it may not scale well with the length of the horizon. Some preliminary results on suboptimal but scalable approaches will be considered in this chapter. Further, this chapter also defines the controllable invariant sets and presents a method for their computation.

The controllable invariant sets are interesting for several reasons. First, they are a class of controlled invariant sets. As this chapter will show that controllable invariants can be computed in a nonrecursive fashion, a natural implication is that we can compute controlled invariant sets nonrecursively. However, this is generally not true of the maximal controlled invariant sets, because they may not be controllable invariants. Another benefit of controllable invariance is that it has the property that from any "large enough" connected region of the set it is possible to reach any such other region of the set, regardless of disturbances. Note that in general controlled invariants do not have this property.

While this chapter suggests the discrete-time setting for the computation of predecessor sets, controlled invariants, and controllable invariants, note that the proposed abstraction approach is general. The abstraction approach can be applied in any setting in which the predecessors and the controlled/controllable invariants are computable. This assumes also that it is possible to perform basic set operations, such as union and complement. The abstraction approach receives as input a hybrid automaton with control inputs, and outputs a state machine. This approach has the property that any switching sequence that can be induced in the abstraction can be induced also in the hybrid automaton, regardless of disturbances. While the abstraction procedure does not have a guaranteed termination, it can be terminated before its convergence, in which case the abstraction will still be valid, though incomplete (not all regions of interest of the state space would be mapped into the abstraction).

The chapter is organized as follows. Related work is presented in section 10.2. A description of the hybrid automata from which the DES abstraction is extracted is given in section 10.3. Definitions necessary in the subsequent developments are included there as well. An approach for the DES abstraction is proposed in section 10.4. Section 10.5 presents a method for the computation of the controllable invariant sets. Then, in section 10.6, the computation of the predecessor sets is considered.

10.2 Related Work

A significant amount of work has been done in the area of hybrid systems [62, 7, 3, 8, 6, 10, 11, 5]. Some of the methods that use DES abstractions for the control of hybrid systems appear in the following references. In [179, 180, 112], the controller is designed in the Ramadge–Wonham framework [159] based on a DES abstraction of the plant. The control architecture consists of the plant, the interface, and the controller. The interface consists of an actuator that maps the discrete commands into (possibly time-varying) control inputs, and a generator that generates events depending on the continuous state of the system. Other similar approaches appear in [38, 132, 157]. A different approach of discrete-event abstraction is to use bisimulations [68, 118, 4]. The goal of the bisimulation approach is to map the hybrid dynamics into DES dynamics described by a finite transition system (finite automaton) such that certain properties of interest are preserved by the transformation. This is achieved by a semidecidable procedure, proposed in [102, 27], that builds a bisimilar system. While most work so far has used bisimulations for verification purposes, there is also more recent work on the use of bisimulations for abstractions fit for supervisory control. Thus, in [188] supervision can be applied to abstractions of controllable discrete-time linear systems.

Compared to the DES abstraction approaches existing in the literature, the approach presented here is designed for supervisory control. Thus, the main concern here has been to be able to generate physically all switching sequences that can be induced in the abstraction, regardless of disturbances or uncertainties in the physical model. In this setting uncertainties could arise from the reset maps; at this stage, discrete disturbances have not been considered. Since the construction of the abstraction has a supervisory control purpose, this work is related to [179, 180, 112]. On the other hand, since reachability analysis is used to derive the abstraction, the works on bisimulations are also related [68, 118, 4]. Moreover, the fact that invariants are also computed in the process of abstraction relates this work to the controlled invariance literature. Note also that a distinguishing feature of this approach is that the DES abstractions model the closed loop of a controller with a hybrid system, and not just the hybrid system. Here, the goal of the abstractions is to facilitate the enforcement of higher level specifications. Thus, in our approach the controller is not designed based on the DES abstraction, but rather at the same time as the abstraction, and the abstraction is based on the operation of the controller.

In the area of hybrid systems, the study of controlled invariant sets has been done mostly in the context of the enforcement of safety specifications. Safety specifications describe a set of forbidden states that a hybrid system should not reach. Thus, the maximal controlled invariant set is of interest in order to compute the least restrictive controller. In [134] a dynamic game theory approach is used for this; the game is between the controller (to be designed) and the disturbance. Such an approach involves challenging problems, including the computational problem of obtaining steady state solutions of Hamilton–Jacobi equations. A new approach appeared in [170], in which it is shown that for some classes of hybrid systems with linear dynamics, the synthesis of the least restrictive controller is semidecidable. In [170], solving Hamilton–Jacobi equations is avoided. There are approaches for the computation of controlled invariants which take advantage of particular types of systems. For instance, for linear discrete-time systems, a method for invariant computation appears in [64]. For the same class of systems, the computation of the maximal controlled invariant set is shown to be decidable under certain conditions [204]. A decidable procedure for some classes of linear *continuous-time* systems for the computation of the maximal controlled invariant set appears also in [169].

Note that the approach of this chapter will propose the use of controllable invariant sets. The controllable invariant sets, which are defined in this chapter, are subsets of the maximal controlled invariant sets. They have the property that every "region" of the controllable set can be reached from any other "region" of the set. This invariant concept resembles the idea that a system could gradually progress from the neighborhood of an equilibrium point to another (e.g., section 6.2 in [137]). The computation approach we propose applies to linear discrete-time dynamics with disturbances. This approach is related to the approaches used in [104, 203] for predecessor operator computations and in [47, 204] for the computation of the maximal controlled invariant set. Related approaches in a nonhybrid context include [103, 24]. Note that some model uncertainties could also be incorporated in this framework [130].

There are also other methods for the synthesis of controllers for hybrid systems, such as the following. In the context of the viability theory, [44] shows a controller synthesis method for hybrid systems in which the mode dynamics is described by differential inclusions. A procedure for the synthesis of controllers has been shown also in [209] for linear hybrid automata. Note that in a linear hybrid automaton the continuous dynamics requires the derivative of the state to satisfy a finite and mode-dependent set of linear inequalities. A very different approach to the control of hybrid systems appears in [22], which uses discrete-time modeling and integer programming for controller design. The controllability of hybrid systems has also been studied [21, 201].

10.3 The Hybrid Automaton Model

In this chapter, the hybrid systems are assumed to be given as hybrid automata. The definition of hybrid automata below corresponds to that of [133].

A **hybrid automaton** is $H = (Q, X, V, Init, f, Inv, Edg, G, Res, \phi)$, where

1. Q is the set of modes (discrete states).
2. $X \subseteq \mathbb{R}^n$ is the domain of the continuous state variable, denoted by x.
3. $V = U \times D \times \Sigma_C$ is the domain of the input, where U is the domain of the control input, D is the domain of the disturbances, and Σ_C is the set of controllable events (or discrete inputs). A control input is denoted by u, a disturbance by d, and a controllable event by α. The null event is denoted by ϵ, and we consider that $\epsilon \notin \Sigma_C$.

4. *Init* is the set of initial states (modes).
5. $f : Q \times X \times U \times D \to \mathbb{R}^n$ is the right-hand side of the continuous state equation

$$\dot{x} = f(q, x, u, d). \tag{10.1}$$

6. $Inv : Q \to 2^X$ maps to each q a set in which the x must be when the system is in the state q. (For instance, $Inv(q)$ may be the set of x for which the dynamics of the continuous state is described by (10.1).)
7. $Edg \subseteq Q \times Q$ is the set of transitions (edges) between modes; (Q, Edg) is a state machine.
8. $G : Edg \to 2^{X \times U \times \Sigma_C}$ maps to each transition a guard, meaning that a transition $e \in Edg$ may occur only if $(x, u, \alpha) \in G(e)$. In particular, when $G(e)$ does not depend on u and α, the transition e is *uncontrollable*. It will be assumed that a transition e occurs if $(x, u, \alpha) \in G(e)$ (where (u, α) is the input applied to the system). Note that *nondeterminism* arises when $G(q, q_1) \cap G(q, q_2) \neq \emptyset$ and $q_1 \neq q_2$, as the system may switch to either of q_1 or q_2 when $(x, u, \alpha) \in G(q, q_1) \cap G(q, q_2)$.
9. $Res : Edg \times X \times U \to 2^X$ is the reset map, mapping (e, x, u), $(x, u) \in G(e)$, to the set in which x may be after the transition e occurs.
10. $\phi : Q \times X \to 2^{U \times D}$ identifies the admissible inputs at every state.

The following notation is used:

- *Pre* represents the *controlled predecessor*. That is, $Pre(M)$ is the set of continuous states from which M can be robustly reached. In other words, $\forall x_0 \in Pre(M)$ there is a feedback control policy which, regardless of disturbances, leads the continuous state x from x_0 to some $x_f \in M$.
- The preset/postset symbol $\bullet$ is used as in the Petri net notation: $\bullet q = \{e : e = (q', q) \text{ for } (q', q) \in Edg, \}$, $q\bullet = \{e : e = (q, q') \text{ for } (q, q') \in Edg\}$, $\bullet e = \{q : e = (q, q')\}$, and $e\bullet = \{q : e = (q', q)\}$.
- For some $q \in Q$, we say that $I \subseteq Inv(q)$ is a **controlled invariant set** if for all $x \in I$ there is an admissible feedback control law taking values in $U \times \Sigma_C$ such that for all subsequent times t: $x(t) \in I$, regardless of the disturbance input.
- Let $Reach : X \to \mathcal{P}(\mathcal{P}(X))$, where for $M \subset X$ we have $M \in Reach(x)$ if it is possible to robustly reach M starting from x (i.e., regardless of disturbances, it is possible to reach M from x).[1] In other words $Reach(x)$ is the collection of sets M with the property that it is possible to robustly reach M from x.
- Given a set $\Omega \subset \mathbb{R}^n$, let $\Omega_x = \{z \in \mathbb{R}^n : \exists y \in \Omega : z = y + x\}$. Further, let Ω^o denote the interior of Ω. For some $q \in Q$ we say that $I \subseteq Inv(q)$ is a **controllable invariant set** if a connected compact set $\Omega \subset \mathbb{R}^n$ exists such that $0 \in \Omega^o$ and
  1. $\forall x \in I$: Ω_x is a controlled invariant set;
  2. $\bigcup_{x \in I} \Omega_x \subseteq Inv(q)$;

---

[1] $\mathcal{P}(Y) = \{E : E \subseteq Y\}$ denotes the collection of all subsets of Y.

3. $\forall x_1, x_2 \in I, \forall x \in \Omega_{x_1}: \Omega_{x_2} \in Reach(x)$.

Note that we have defined *Pre*, *Reach* and the controlled and controllable invariant sets with respect to the dynamics of a single mode. The definitions could be extended by using $Q \times X$ instead of X.

A class of hybrid systems for which the interesting problems are more tractable computationally are the systems in which equation (10.1) is discrete-time and affine in x, u, and d. Then, the dynamics in any mode q can be described by

$$x(t+1) = A(q)x(t) + B(q)u(t) + E(q)d(t). \tag{10.2}$$

10.4 Extracting the DES Abstraction

This section approaches the extraction of DES dynamics from controlled hybrid dynamics. The extraction process produces a model in the DES framework of section 9.3. This model also represents the (feasible) specification for the design of a hybrid system controller. Indeed, referring to the illustration of Fig. 9.2, the DES model obtained here abstracts the behavior of the closed loop of a controller and a hybrid subsystem.

A hybrid subsystem is assumed to be given in the form of a hybrid automaton. If the abstraction process does not attempt to refine this model, the DES abstraction will result in a subnet of the state machine (Q, Edg) of the hybrid automaton. Note that the notation of section 10.3 is used here.

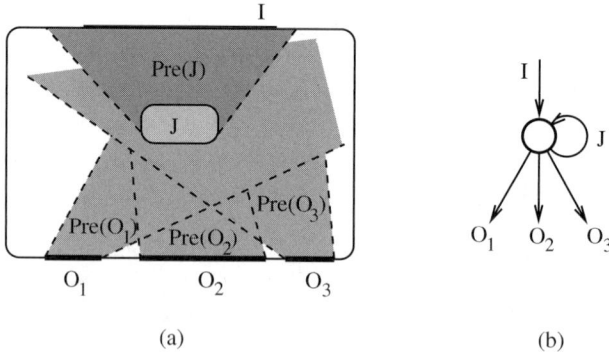

(a)   (b)

Fig. 10.1. Illustration of a desirable situation in the controlled behavior of a hybrid system. (a) A hybrid system mode with input set I and output sets O_1, O_2, and O_3 corresponding to the thick lines, controlled invariant set J, and $Pre(O_1)$, $Pre(O_2)$, $Pre(O_3)$, and $Pre(J)$ represented through the shaded areas. (b) Equivalent DES abstraction of the mode, where the self-loop corresponds to J and the other transitions to the transitions exiting O_1, O_2, and O_3.

The process of DES abstraction has two favorable situations which we consider below. First we define for every mode $q \in Q$ the following sets:

(i) $J_q \subseteq Inv(q) \cap Safe(q)$, where $Safe(q)$ is the set specifying the safety specification for the mode q (i.e., $Inv(q) \setminus Safe(q)$ is the forbidden state set of the mode q).

(ii) For every $(q, q') \in Edg$, let $O_{q \to q'} \subseteq Inv(q) \cap Safe(q)$ denote the continuous states for which there is an input leading the system from q to q', regardless of disturbances.

(iii) Let I_q be the set of continuous states in which the mode q may be entered from the modes q_c such that $(q_c, q) \in Edg$.

Note that the set I_q could be reduced by an appropriate control law. An ideal situation for the DES abstraction is when for all $q \in Q$ there is a J_q such that:

(a) J_q is a controlled invariant set.
(b) $I_q \subseteq Pre(J_q)$.
(c) $J_q \subseteq \bigcap_{q' \in q \bullet \bullet} Pre(O_{q \to q'})$.

This situation is illustrated in Fig. 10.1, together with the DES abstraction of the mode. Thus, once we have the sets I_q and $O_{q \to q'}$, we are interested in computing the maximal controlled invariant set J_q satisfying (i) and (c). Indeed, if the maximal controlled invariant set does not satisfy (b), there is no controlled invariant set J_q satisfying (a)–(c). However, even when (b) is not satisfied, we may still be able to reduce the set I_q (through a control law) such that (b) is satisfied. Note that $Pre(J_q)$ at (b) does not need to be calculated when J_q is the *maximal* controlled invariant set, since $J_q = Pre(J_q)$. An interesting variant of the requirements (a)–(c) is given below:

(a') J_q is a controllable invariant set with a set Ω such that $\bigcup_{x \in J_q} \Omega_x \subseteq Inv(q) \cap Safe(q)$.
(b') $I_q \subseteq Pre(J_q)$.
(c') $\forall q' \in q \bullet \bullet \; \exists x \in J_q \colon \Omega_x \subseteq Pre(O_{q \to q'})$.

This situation is illustrated in Fig. 10.2, together with the DES abstraction of the mode. Again, once we have the sets I_q and $O_{q \to q'}$, we are interested in computing a controllable invariant set J_q satisfying (a') and (c'). This can be achieved by computing a maximal controllable invariant set satisfying (a'). Then, if (c') is not satisfied and J_q is maximal no solution to (a'-c') exists, but if (b') is not satisfied, we may still be able to reduce the set I_q.

Note that the conditions (a')–(c') are a variant of (a)–(c). Indeed, by the definition of the controllable invariant set, (c) implies $J_q \subseteq Pre(O_{q \to q'})$. Thus $J_q \subseteq \bigcap_{q' \in q \bullet \bullet} Pre(O_{q \to q'})$. Further, every controllable invariant set is a controlled invariant set. (However, the converse is not true.) The (a')–(c') variant may be computationally advantageous when it is not easy to compute $Pre(O_{q \to q'})$; then we do not need to compute the whole sets $Pre(O_{q \to q'})$, but only to show that they intersect J_q as shown at (c'). This quality may be of interest especially in the discrete-time case, in which the computation of

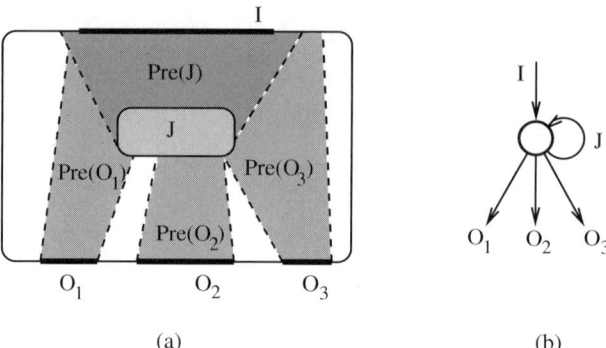

(a)   (b)

Fig. 10.2. Illustration of another desirable situation in the controlled behavior of a hybrid system. (a) A hybrid system mode with input set I and output sets O_1, O_2, and O_3 corresponding to the thick lines, controllable invariant set J, and $Pre(O_1)$, $Pre(O_2)$, $Pre(O_3)$, and $Pre(J)$ represented through the shaded areas. (b) Equivalent DES abstraction of the mode, where the self-loop corresponds to J and the other transitions to the transitions exiting O_1, O_2, and O_3.

the predecessor is iterative and may not terminate. Note also that here the controllable invariant set is computed first, and then the predecessor sets. On the other hand, in the previous situation the maximal controlled invariant set was computed only after the computation of the predecessor sets.

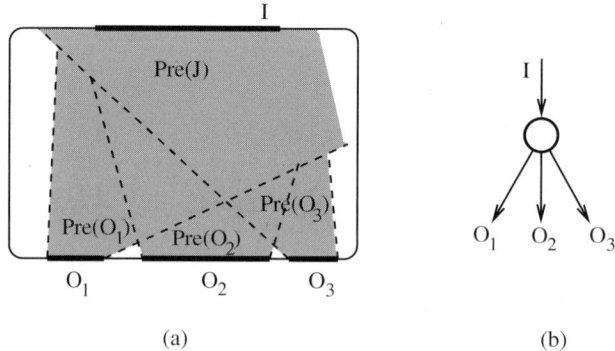

(a)   (b)

Fig. 10.3. Illustration of the situation in which no adequate controllable or controlled invariant set exists. (a) A hybrid system mode with input set I and output sets O_1, O_2, and O_3 corresponding to the thick lines, and $Pre(O_1)$, $Pre(O_2)$, and $Pre(O_3)$ represented through the shaded areas. (b) Equivalent DES abstraction of the mode. Note that the abstraction has no self-loop.

When no invariant set J_q satisfying (a)–(c) or (a′)–(c′) can be found, we could still abstract the mode by considering the inclusion relations between I_q on one side and $Pre(O_{q \to q'})$ on the other, or among I_q, $Pre(J_q)$, J_q and $Pre(O_{q \to q'})$. A situation in which a mode has no invariant sets is illustrated in Fig. 10.3.

In principle, the following abstraction procedure could be used. The procedure assumes there are no discrete disturbances (refer to point 8 of the definition of the hybrid automaton; note that continuous disturbances are allowed). This assumption could be removed in future work. First, the notation is defined.

- A hybrid automaton $H = (Q, X, V, Init, f, Inv, Edg, G, Res, \phi)$ is assumed to be given. This is the input of the procedure.
- The output of the procedure is a state machine $(S, \to)$, where S is the set of states and $\to \subseteq S \times S$ is the transition relation.
- Similar to $Inv(q)$ in the hybrid automaton, let's also define $Inv : S \to X$ for the states q' of the abstraction. Further, the map $C : S \to Q$ is defined, to associate a mode $q \in Q$ to each $q' \in S$. Thus, each $q' \in S$ corresponds to a region $(C(q'), Inv(q'))$ of the hybrid system, where $Inv(q') \subseteq Inv(C(q'))$.
- Let $Res_{q' \to q}$ denote the area in which the state is reset when switching from q' to q for $q', q \in Q$. Technically, let ϕ_u be the restriction of ϕ to 2^U (that is, $\phi_u(p, x)$ is the set of inputs that can be applied when the mode is p and the state x.) Then $Res_{q' \to q} = \bigcup_{(x,u) \in V} Res((q', q), x, u)$ for $V = \{(x, u) \mid (\exists \alpha \in \Sigma_C : (x, u, \alpha) \in G(q', q)) \wedge x \in Inv(q') \wedge u \in \phi_u(q', x)\}$.
- Given $I \subseteq Inv(q)$, the set $G^{-1}_{q' \to q}(I)$, $q', q \in Q$, denotes the set of states x in mode q' from which there is an input leading to the mode q with the state x reset within I. Formally, $G^{-1}_{q' \to q}(I) = \{x \in Inv(q') \mid \exists u \in \phi_u(q', x), \exists \alpha \in \Sigma_C : Res((q', q), x, u) \subseteq I \wedge (x, u, \alpha) \in G(q', q)\}$.
- The predecessor Pre is defined with respect to each mode $q \in Q$. Let $\psi(t)$ denote the solution to $\dot{x} = f(q, x, u, d)$ or $x(t+1) = f(q, x(t), u, d)$, assuming it exists. Let ψ_t denote $\psi(t)$. Formally, $Pre(M) = \{z \in X : \exists t > 0, \exists u : [0, t] \times X \to \mathbb{R}^m, \forall d : [0, t] \to \mathbb{R}^p : [\forall \tau \in [0, t], u(\tau, \psi_\tau) \in \phi_u(q, \psi_\tau)] \wedge [(\forall \tau \in [0, t], d(\tau) \in \phi_d(q, \psi_\tau)) \Rightarrow (\exists \tau \in [0, t], \psi_\tau \in M)] \wedge \psi(0) = z\}$. Similarly to ϕ_u, ϕ_d denotes the restriction of ϕ to 2^D. In other words, $Pre(M)$ denotes the set of states x from which it is possible to reach M regardless of disturbances, assuming the dynamics of the mode q.

The procedure starts with a number of sets of interest (q, J) specified by the user, where $q \in Q$, $J \subseteq Inv(q)$, and J represents a set of interest in the mode q. The abstraction procedure is defined next.

1. Initialize $S = \emptyset$ and $\to = \emptyset$.
2. For all sets of interest (q, J), create a state $q' \in S$ with $C(q') = q$ and let $Inv(q') := Pre(J) \cap Inv(q)$. If J is a controlled invariant, add (q', q') to $\to$.
3. Initialize $ModeList := S$.

4. While $ModeList \neq \emptyset$ do
   (a) For all $q \in ModeList$
      (i) Compute $I_{p \to q} = Inv(q) \cap Res_{p \to C(q)}$ for all $p \in Q$ such that $(p, C(q)) \in Edg$.
      (ii) For each $I_{p \to q}$ computed above find $O_{p \to q} = G^{-1}_{p \to C(q)}(I_{p \to q})$.
      (iii) For all $O_{p \to q}$ computed above, if $O_{p \to q} \neq \emptyset$, add $O_{p \to q}$ to the list $L(p)$.
   (b) Set $ModeList = \emptyset$.
   (c) For all $p \in Q$ with $L(p) \neq \emptyset$ do:
      (i) For all $p \in Q$, $O_{p \to q} \in L(p)$, and $q' \in S$, if $C(q') = p$ and $Inv(q') \subseteq Pre(O_{p \to q}) \cap Inv(p)$, then add (q', q) to $\to$.
      (ii) Distribute the states q with $O_{p \to q} \in L(p)$ into k disjoint groups $\Gamma_1 \dots \Gamma_k$, such that $Inv_i \neq \emptyset$ for $i = 1 \dots k$ and $Inv_i \neq Inv_j$ for $i \neq j$, where $Inv_i := Inv(p) \cap (\bigcap_{q \in \Gamma_i} Pre(O_{p \to q}))$.
      (iii) Let $q'_1 \dots q'_{2k}$ be new discrete states.
      (iv) Let $cinv(Inv_i)$ denote the maximal controlled invariant set included in Inv_i.
      (v) For all $i = 1 \dots k$, let $Inv(q'_i) = cinv(Inv_i)$ and $Inv(q'_{i+k}) = Inv_i$.[2]
      (vi) For all $i = 1 \dots 2k$, if $Inv(q'_i) = Inv(q)$ for some $q \in S$, then set $q'_i \equiv q$.
      (vii) Let $\Gamma = \{i : Inv(q'_i) \neq \emptyset\}$. For all $i \in \Gamma$, if $q'_i \notin S$ then add q'_i to S and $ModeList$, and set $C(q'_i) = p$.
      (viii) For all $i \in \Gamma$ with $i \leq k$, add (q'_i, q'_i) to $\to$ and (q'_i, q) to $\to$ for all $q \in \Gamma_i$.
      (ix) For all $i \in \Gamma$ with $i > k$, for all $q \in \Gamma_{i-k}$, add (q'_i, q) to $\to$.
   (d) Set $L(p) = \emptyset$ for all $p \in Q$.

The procedure is graphically illustrated in Fig. 10.4. Note that the abstraction procedure does not assume a set of initial states. Rather, the abstraction could be used to determine the states in which the system could be initialized.

The abstraction is done in such a way that its (controllable) transitions can always be enforced by a controller. Formally, a controller is defined as follows.

Let $\gamma : Q \times X \to \mathbb{R}^m$ be the observation of the continuous state x available to the controller. Let Y be the range of γ. The controller is viewed as a set-valued map

$$\kappa : Q_C \times Q \times Q \times Y \to Q_C \times 2^{U \times \Sigma_C},$$

where Q_C is a finite set of discrete states of the controller. For a current state observation y, mode q, current controller state q_c, and next desired mode q', we have

$$\kappa(q_c, q, q', y) = (q'_c, Z),$$

where q'_c is the next state of the controller (which may be the same as q_c), and $Z \in 2^{U \times \Sigma_C}$ can be decomposed in $Z = Z_U \times Z_\Sigma$; the controller requests

---

[2] However, if $cinv(Inv_i) = Inv_i$, we set q'_{i+k} to be identical to q'_i: $q'_{i+k} \equiv q'_i$.

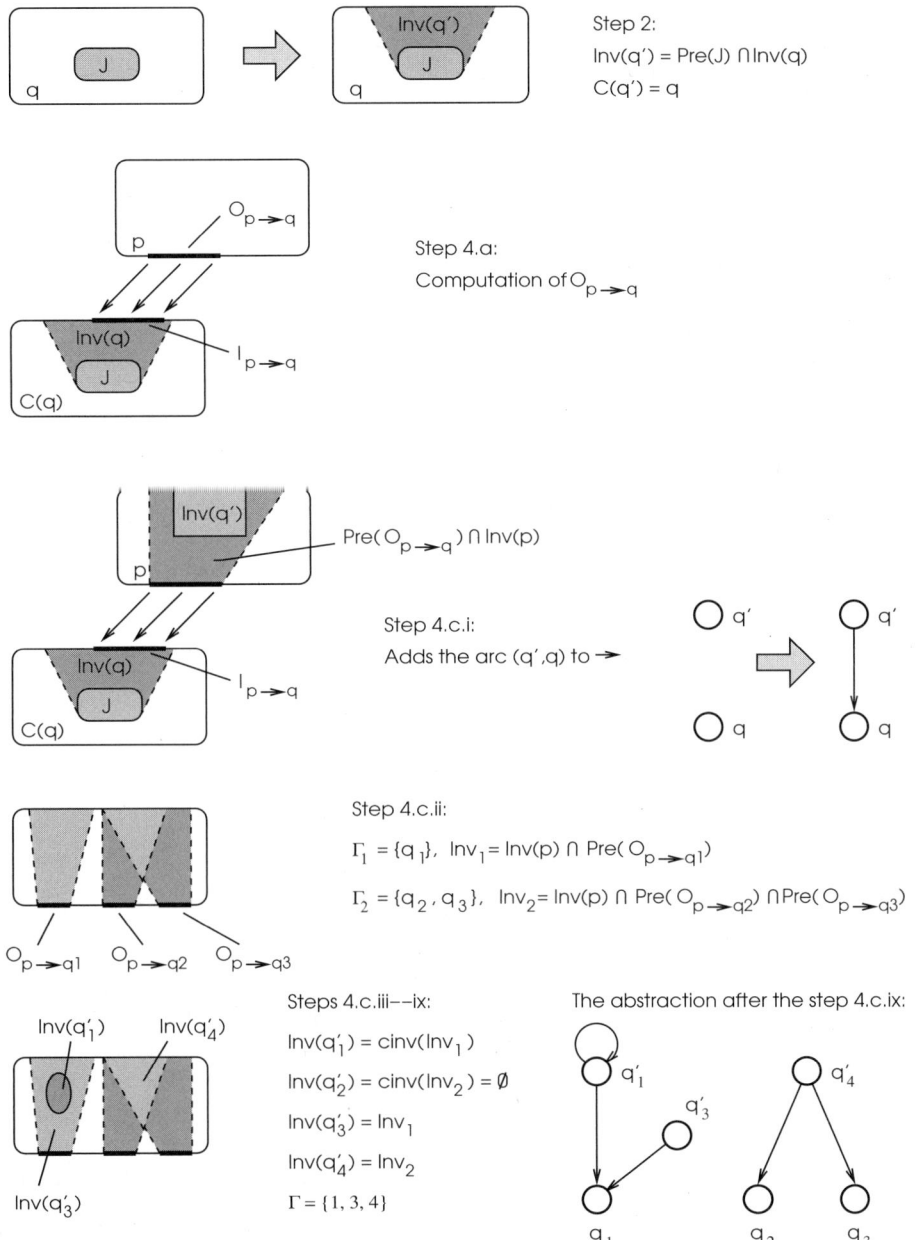

Fig. 10.4. Illustration of the abstraction procedure.

a control input $u \in Z_U$ and discrete input $\alpha \in Z_\Sigma$. When we want to stay in a mode q, we let $q' = q$. Note that q' is an external input, as it is not generated by the hybrid system or the controller. Through this input the user can steer the operation of the controlled hybrid system.

10.5 Computation of the Controllable Invariant Sets

10.5.1 The Controlled Invariance Context

Controllable invariant sets are a class of controlled invariant sets. For this reason, a brief discussion of controlled invariance is included here. In the literature, recursive approaches have been proposed for the computation of the maximal controlled invariant sets. This recursive computation is more tractable when done for single modes (rather than for the whole state space of the hybrid system) and for particular types of dynamics. Of particular interest in this section are mode dynamics of the form

$$x(t + 1) = Ax(t) + Bu(t) + Ed(t), \tag{10.3}$$

where u denotes the control input and d the disturbance input, and mode invariants of the form

$$Rx \leq r, \tag{10.4}$$

Thus, we are interested in the maximal controlled invariant set that is a subset of the set given by (10.4). For mode dynamics (10.3), the recursive computation of the maximal controlled invariant set is guaranteed to terminate under certain assumptions [204]. However, each iteration of the recursive computation suffers from the double-exponential complexity of the Fourier–Motzkin elimination of variables.

In this section we will present a nonrecursive method for the computation of controllable invariant sets. The mode dynamics (10.3) will be assumed, where the input u takes values in a convex domain $\mathcal{U}$, and the disturbance d takes values in a bounded set $\mathcal{D}$. Since disturbances are allowed, the model (10.3) is not very particular, as nonlinearities could be incorporated in the disturbance term. In fact, future work may incorporate piecewise linear dynamics and some of the polytopic uncertainties of [130].

Note that compared to controlled invariants, the maximal controlled invariant may not be a controllable invariant. Nonetheless, it may be that in practice we are only interested in the part of the maximal controlled invariant that is "controllable." This may be especially the case when we are interested in "reversible" behavior, where reversibility here refers to the ability to return to a neighborhood around some initial state.

10.5.2 The Idea of the Approach

Considering dynamics again of the form (10.3), if the system is stabilizable, there is a state feedback controller $u = Kx$ such that the system is stable. Furthermore, for each $u = Kx + r$, where r is a constant, there is an equilibrium point x^* to which (in the absence of disturbances) the state converges. Intuitively it is clear that around that equilibrium point there is a region of attraction, such that no matter of the bounded disturbances, that region is invariant. Furthermore, if each equilibrium point has such a region of attraction, we can go from one region to another. Indeed, if we are in the region of (x_1^*, r_1), by applying the control $u = Kx + r_2$ we move to the region of (x_2^*, r_2). Also, in order to keep the control $u = Kx + r$ within its admissible domain $\mathcal{U}$, we can "slowly" change r from r_1 to r_2. Therefore the controllable invariant set would correspond to the set of equilibrium points x^*. While linear state feedback was used in this illustration, we are not going to refer to it in what follows. We consider a more general state feedback solution. Also, we will continue to use the notation x^*, but this may no longer denote an equilibrium point.

10.5.3 The Computation

We consider the dynamics of equation (10.3) and sets Ω (see the definition of controllable invariant sets) of the form $\Omega = \{x : |x| \leq b\}$ where $b \in \mathbb{R}^n$ and $b > 0$. Recall that given x^*, $\Omega_{x^*} = \{x : |x - x^*| \leq b\}$. Let $\mathcal{U}$ denote the domain of the control input and $\mathcal{D}$ the (bounded) domain of the disturbance.

Given x_1^*, the set of points x_2^* satisfying that $\exists u(t) \in \mathcal{U} \ \forall x(t) \in \Omega_{x_1^*}$: $x(t+1) \in \Omega_{x_2^*}$ can be expressed as

$$\exists u \in \mathcal{U}, \forall x \in \Omega_{x_1^*} : \begin{cases} Ax + Bu + d^+ \leq x_2^* + b, \\ Ax + Bu - d^- \geq x_2^* - b, \end{cases} \tag{10.5}$$

where $x = x(t)$, $u = u(t)$, $d^+ = \max_{d \in \mathcal{D}} Ed$, and $d^- = -\min_{d \in \mathcal{D}} Ed$ and the maximum/minimum is taken separately on each row of Ed. Note that the requirement that $\Omega_{x_1^*}$ be invariant corresponds to (10.5) when $x_2^* = x_1^*$.

Assuming a convex domain $\mathcal{U} = \{u : L_u u \leq b_u\}$, the input u can be eliminated from (10.5) using the Fourier–Motzkin elimination (FME) [144]. The result is of the form

$$\forall x \in \Omega_{x_1^*} : Gx + Hb + Mx_2^* \leq g \tag{10.6}$$

or

$$\forall \alpha \in [-b, b] : (G + M)x_1^* + G\alpha + Hb + M\beta \leq g, \tag{10.7}$$

where $\alpha = x - x_1^*$ and $\beta = x_2^* - x_1^*$. Note that α can also be eliminated, as $\max_{\alpha \in [-b, b]} G\alpha = |G|b$, where the maximum is taken separately on each row of $G\alpha$, and $|G| = [|G_{ij}|]$ denotes the absolute value of G. We obtain:

$$(G + M)x_1^* + (|G| + H)b + M\beta \leq g. \tag{10.8}$$

To satisfy (10.8) for all $\beta \in [-\delta, \delta]$, where $\delta \in \mathbb{R}^n$, $\delta \geq 0$, is given, the following constraint is obtained:

$$(G + M)x_1^* + (|G| + H)b + |M|\delta \leq g. \tag{10.9}$$

Note that (10.9) describes the set of points x_1^* such that $\Omega_{x_1^*}$ is a controlled invariant and from all points $x \in \Omega_{x_1^*}$ it is possible to reach any $\Omega_{x_2^*}$ with $|x_2^* - x_1^*| \leq \delta$ in one time step. Obviously, we would like this set of points x_1^* to be as large as possible. At the same time, we are also interested in having the sets Ω_{x^*} as small as possible (i.e., b as small as possible). In view of (10.5) the minimum value of b is

$$b \geq \frac{d^+ + d^-}{2}. \tag{10.10}$$

On the other hand, the minimum value of δ is 0. From (10.9) with $\delta = 0$ we derive the controllable invariant set

$$(G + M)x + (|G| + H)b < g. \tag{10.11}$$

Note that $<$ denotes strict inequality on all elements, that is, $y < z \Rightarrow y_i < z_i$ for all indices i.

EXAMPLE 10.1

Assume a system described by the dynamics

$$x(t + 1) = ax(t) + u(t) + d(t), \tag{10.12}$$

where $a \in \mathbb{R}$. Assume $d^+ = d^- = d_0$ and the control input domain $-u_0 \leq u \leq u_0$. The relation (10.5) can be written as

$$\exists u \in \mathcal{U}, \forall x \in \Omega_{x_1^*} : \begin{cases} ax + u + d_0 \leq b + x_2^*, \\ -ax - u + d_0 \leq b - x_2^*. \end{cases} \tag{10.13}$$

Then (10.6) becomes

$$\forall x \in \Omega_{x_1^*} : \begin{cases} d_0 \leq b, \\ ax + d_0 \leq b + x_2^* + u_0, \\ -ax + d_0 \leq b - x_2^* + u_0, \end{cases} \tag{10.14}$$

while (10.9) is

$$\begin{cases} d_0 \leq b, \\ (a - 1)x_1^* + (|a| - 1)b + \delta + d_0 \leq u_0, \\ (-a + 1)x_1^* + (|a| - 1)b + \delta + d_0 \leq u_0. \end{cases} \tag{10.15}$$

We see that there is no solution unless $|a|d_0 < u_0$ or $|a| < 1$. Once these conditions are satisfied, the controllable invariant is given by

$$\begin{cases} (a-1)x + (|a|-1)b + d_0 < u_0, \\ (-a+1)x + (|a|-1)b + d_0 < u_0 \end{cases} \qquad (10.16)$$

for a b such that $b \geq d_0$ and $(|a|-1)b + d_0 < u_0$.

The following results establish properties of the controllable invariant sets computed this way. Notably, we prove that (10.11) describes a controllable invariant set and that, with the possible exception of (some of) its boundary, it coincides with the maximal controllable invariant set with $\Omega = [-b, b]$. For the moment, $Inv(q) = \mathbb{R}^n$ is assumed in the definition of the controllable invariant sets.

Let $J_\delta = \{x : (G + M)x + (|G| + H)b + |M|\delta \leq g\}$, where the notation of (10.9) is used.

Proposition 10.1. *The set J_δ is a controllable invariant of set $\Omega = [-b, b]$.*

Proof. The proof is divided in three parts. Part (a) shows that $\forall x_1^* \in J_\delta$, $\forall x_2^* \in [x_1^* - \delta, x_1^* + \delta]$, $\forall x(t) \in \Omega_{x_1^*}$, $\exists u \in \mathcal{U}$ $\forall d \in \mathcal{D}$: $x(t+1) \in \Omega_{x_2^*}$. Part (b) shows that Ω_{x^*} is a controlled invariant for all $x^* \in J_\delta$. Part (c) shows that $\forall x_1^*, x_2^* \in J_\delta$, $\forall x \in \Omega_{x_1^*}$: $\Omega_{x_2^*} \in Reach(x)$.

(a) Let $\alpha = x - x_1^*$ and $\beta = x_2^* - x_1^*$. From $\beta \leq \delta$ we get that $M\beta \leq |M|\delta$. Since x_1^* satisfies (10.9), it follows that (10.8) is also satisfied. Similarly, we derive $(G+M)x_1^* + G\alpha + Hb + M\beta \leq g$, and so $Gx + Hb + Mx_2^* \leq g$. However, this is the projection of

$$\begin{cases} Ax + Bu + d^+ \leq x_2^* + b, \\ Ax + Bu - d^- \geq x_2^* - b, \end{cases} \qquad (10.17)$$

that removes the variable $u \in \mathcal{U}$. Therefore, there is a $u \in \mathcal{U}$ such that (10.17) is satisfied for the given x and x_2^* $\forall d \in \mathcal{D}$. However, (10.17) is precisely the condition that some $x(t+1) \in \Omega_{x_2^*}$ is reached from $x(t) = x$ by applying the input u.

(b) This results from (a) for $x_1^* = x_2^* = x^*$.

(c) Let $x_1^*, x_2^* \in J_\delta$ be chosen arbitrarily. Let $n > 0$ be an integer such that $|x_2^* - x_1^*| \leq n\delta$. Let $z_0^*, z_1^*, \ldots, z_n^*$ be such that $z_k^* = \frac{k}{n}x_1^* + \frac{n-k}{n}x_2^*$ for $k = 0\ldots n$. Since J_δ is convex, $z_k^* \in J_\delta$ for all $k = 0\ldots n$. Further, $|z_{k+1}^* - z_k^*| \leq \delta$ for $k = 0\ldots n-1$. Then, in view of (a), we reach $\Omega_{x_2^*}$ in n steps by going from $x(t) \in \Omega_{x_1^*}$ to $\Omega_{z_1^*}$, then to $\Omega_{z_2^*}$, and so on to $\Omega_{z_n^*}$. $\qquad \square$

Proposition 10.2. *x^* satisfies $(G + M)x^* + (|G| + H)b \leq g$ if and only if Ω_{x^*} is a controlled invariant.*

Proof. "$\Rightarrow$" Let $x \in \Omega_{x^*}$. From $(G+M)x^* + (|G|+H)b \leq g$ and $G(x - x^*) \leq |G|b$ we get $Gx + Hb + Mx^* \leq g$. Since $Gx + Hb + Mx^* \leq g$ is the projection of

$$\begin{cases} Ax + Bu + d^+ \leq x^* + b, \\ Ax + Bu - d^- \geq x^* - b \end{cases} \qquad (10.18)$$

that removes the variable $u \in \mathcal{U}$, it follows that there is a $u \in \mathcal{U}$ such that when $x(t) = x \in \Omega_{x^*}$, $\forall d \in \mathcal{D}$: $x(t+1) \in \Omega_{x^*}$.

"$\Leftarrow$" If Ω_{x^*} is a controlled invariant, then (10.5) is satisfied for $x_1^* = x_2^* = x^*$. This is also true of (10.6) and (10.8) with $\beta = 0$. So the conclusion follows.

□

Proposition 10.3. *The set* $J = \{x : (G + M)x + (|G| + H)b < g\}$ *is a controllable invariant.*

Proof. By Proposition 10.2, Ω_x is a controlled invariant for all $x \in J$. It remains to show that any $x_2^* \in J$ can be reached from any $x_1^* \in J$. Let $x_1^*, x_2^* \in J$. Note that $\exists \delta_1, \delta_2 > 0$: $(G + M)x_1^* + (|G| + H)b + |M|\delta_1 \leq g$ and $(G + M)x_2^* + (|G| + H)b + |M|\delta_2 \leq g$. Let $\delta = \min(\delta_1, \delta_2)$. It follows that $x_1^*, x_2^* \in J_\delta$, and so the conclusion follows by Proposition 10.1.

□

Proposition 10.4. *All controllable invariant sets of set* $\Omega = [-b, b]$ *are subsets of the set* $\overline{J} = \{x : (G + M)x + (|G| + H)b \leq g\}$.

Proof. For any controllable invariant set I, the set Ω_x for $x \in I$ should be a controlled invariant. Therefore, the conclusion follows immediately from Proposition 10.2.

□

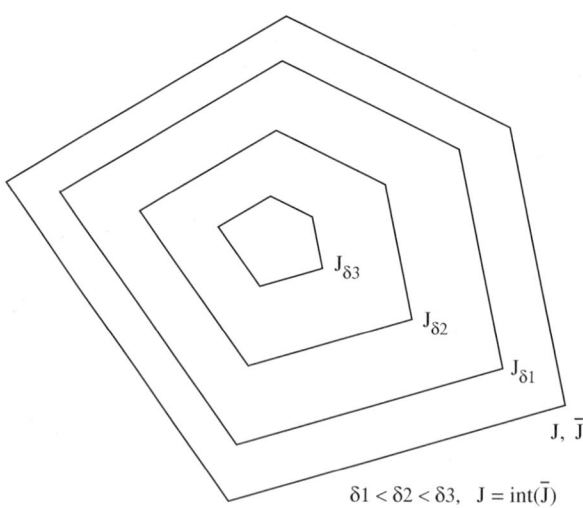

$$\delta 1 < \delta 2 < \delta 3, \quad J = \text{int}(\overline{J})$$

Fig. 10.5. Illustration of the (inclusion) relation among the sets J_δ, J, and $\overline{J}$.

Propositions 10.3 and 10.4 indicate that the construction of the controllable set J in (10.11) is nearly optimal, as all controllable invariant sets I of set Ω satisfy $I \subseteq \overline{J}$. Further, if the maximal controllable set J_m exists, it satisfies $J \subseteq J_m \subseteq \overline{J}$. Note that J is the interior of $\overline{J}$. So J is a very tight approximation of the optimum.

The computation of the set J has been done assuming $Inv(q) = \mathbb{R}^n$. In the general case, a controllable invariant set can be computed as follows. Let $M = \{x \in Inv(q) : \Omega_x \subseteq Inv(q)\}$. Assuming M to be connected, note that a controllable invariant set is $J_0 = J \cap M$. This construction ensures that regardless of the current state x, as long as $x \in \Omega_{x'}$ for some $x' \in J_0$, the state can be kept inside $Inv(q)$.

Finally, note that the computation steps have polynomial complexity, except for the step from (10.5) to (10.6), involving the Fourier–Motzkin elimination. That step, in the worst case, has double-exponential complexity in the number of controls (i.e., the size of u). However, in practice intensive computation can be avoided by removing all redundant constraints (which has polynomial complexity in the number of constraints) after each component u_i of u is eliminated.

10.6 Computation of the Predecessor

The computation of the predecessor sets for dynamics of the form (10.3) or piecewise linear dynamics has been described at length in [104]. We propose to use the approach of [104]. There, the one-step predecessor Pre^1 can be computed by quantifier elimination. The one-step predecessor is defined by $x \in Pre^1(I)$ if $\exists u(t) \in \mathcal{U}$, $\forall d(t) \in \mathcal{D}$, $x(t+1) \in I$. Of course, we are more interested in an N-step predecessor, recursively defined as $Pre^N(I) = Pre^1(Pre^{N-1}(I))$, if not on $Pre(I) = \lim_{N \to \infty} Pre^N(I)$. The main difficulty here is that the number of constraints for computing Pre^N may increase exponentially with N. Therefore, it is of interest to have a suboptimal computation of Pre^N that has a polynomial complexity in N. Preliminary results are presented in this section. In future work, the algorithms presented here could be further refined for improved performance. The idea is to fix a number m, and then whenever we have a convex set described by $n > n_1$ linear inequalities, where $n_1 \geq m$ is a fixed number, we approximate it with m linear inequalities. A possible underapproximation algorithm is given below. Although the algorithm starts with a convex set, it can be easily extended to nonconvex sets.

ALGORITHM 10.1

Input: *A convex set I, an integer $N_0 > 0$, and an integer $n_1 > 0$*

Let $V = I$ and $N = 1$

Compute $J = Pre^1(I)$ (the predecessor of I in one step)

Remove[3] the redundant constraints of J.

While $N < N_0$ and $J \not\subseteq V$ **do**

  1. **Let** $V = J$, $N \rightarrow N + 1$, and $J \rightarrow Pre^1(J)$

  2. *Remove the redundant constraints of J.*

  3. **Let** n *be the number of constraints of J.*

  4. **If** $n > n_1$ **then** *find a convex set $K \subseteq J$ described by $m < n_1$ inequalities and set $J = K$.*

Note that by limiting the number of inequalities to n_1, the computation is linear in N. There are several ways in which we can find a subset of J described by a given number of linear inequalities. In what follows we propose two simple methods involving little computational effort. The methods assume J to be bounded, but they may be extendable to the case when J is unbounded. Let $L_J x \leq b_J$ be the set of inequalities describing the convex set J. The first method computes a set K of the form of an interval $[a_1, c_1] \times [a_2, c_2] \times \ldots [a_v, c_v]$ and is given by the linear program below, in which we let $a = [a_1 \ldots a_v]^T$ and $c = [c_1 \ldots c_v]^T$:

$$\max \sum_{i=1}^{v} (c_i - a_i) \text{ subject to}$$

$$\begin{cases} (L_J + |L_J|)c + (L_J - |L_J|)a \leq 2b_J, \\ a \leq c. \end{cases}$$

Given the number of variables v, the second approach finds $v + 1$ vertices of $L_J x \leq b_J$. Then the set K is the polytope having the $v + 1$ vertices, and each of the $v + 1$ facets can easily be computed. The algorithm is given below.

ALGORITHM 10.2

Find $v + 1$ vertices. This is done as follows:

  1. *Select[4] $v + 1$ distinct groups of v linear independent rows of L_J, $\{i_1, \ldots, i_v\}$.*

  2. **For** *each group restrict L_J and b_J to the rows $\{i_1, \ldots, i_v\}$.* **Let** $L_{J,k}$ *and $b_{J,k}$ be the restrictions for each group $k = 1 \ldots v + 1$.*

  3. **For** $k = 1 \ldots v + 1$ *solve $L_{J,k} x = b_{J,k}$. The solutions are $v + 1$ vertices $x_1 \ldots x_{v+1}$.*

For $u = 1 \ldots v + 1$ *let H_u be $[x_1 \ldots x_{v+1}]$ without the column x_u.*

For $u = 1 \ldots v + 1$ *let $[f_u^T, d_u]$ be a nonzero solution of $f_u^T H_u + d_u \mathbf{1}^T = 0$, where $f_u \in \mathbb{R}^v$, $d_u \in \mathbb{R}$ and $\mathbf{1}^T$ is a row vector of elements 1 and appropriate dimension.*

---

[3] For n constraints this operation involves at most n feasibility problems (which are solvable via linear programming)

[4] The $v+1$ groups exist since J is assumed to be nonempty and bounded. Finding the $v + 1$ groups can be carried out with polynomial complexity.

The set K is given by the inequalities in x

$$(f_u^T x + d_u)(f_u^T x_u + d_u) \geq 0 \qquad (10.19)$$

for $u = 1 \ldots v + 1$.

In the algorithm above note that by construction every half-space in (10.19) contains all vertices $x_1 \ldots x_{v+1}$, therefore K, which is the intersection of the half-spaces, is nonempty.

The algorithms presented in this section represent preliminary results toward a more efficient computation of the predecessor sets. Their application or extension to a precise approximation of predecessor sets is a matter of further research.

Finally, note that these algorithms only attempt to speed up the (recursive) computation of the predecessor sets. An approach detecting when it is no longer useful to continue the recursive computation is given in [104]. There, the termination condition uses a quantization grid. The termination test first underapproximates the current predecessor set as a union of quantization cells. Then, it compares this underapproximation to underapproximations of past iterations. When no more differences are observed, the computation is terminated. Note that underapproximations are not used for the computation of the actual predecessor sets. They are only used to evaluate when the recursive computation could be terminated.

References

1. Z. Achour, N. Rezg, and X. Xie. Supervisory control of partially observable marked graphs. *IEEE Transactions on Automatic Control*, 49(11):2007–2011, 2004.

2. R. Alur, C. Courcoubetis, N. Halbwachs, T. A. Henzinger, P.-H. Ho, X. Nicollin, A. Oliveiro, J. Sifakis, and S. Yovine. The algorithmic analysis of hybrid systems. *Theoretical and Computer Science*, 138:3–34, 1995.

3. R. Alur, T. A. Henzinger, and E. Sontag, editors. *Hybrid Systems III*, volume 1066 of *Lecture Notes in Computer Science*. Springer-Verlag, Berlin, 1996.

4. R. Alur, T. A. Henzinger, G. Lafferriere, and G.J. Pappas. Discrete abstractions of hybrid systems. *Proceedings of the IEEE*, 88(7):971–984, 2000.

5. P. J. Antsaklis, editor. *Proceedings of the IEEE, Special Issue on Hybrid Systems: Theory and Applications*, volume 88(7), July 2000.

6. P. J. Antsaklis, W. Kohn, M. D. Lemmon, A. Nerode, and S. Sastry, editors. *Hybrid Systems V*, volume 1567 of *Lecture Notes in Computer Science*. Springer-Verlag, Berlin, 1999.

7. P. J. Antsaklis, W. Kohn, A. Nerode, and S. Sastry, editors. *Hybrid Systems II*, volume 999 of *Lecture Notes in Computer Science*. Springer-Verlag, Berlin, 1995.

8. P. J. Antsaklis, W. Kohn, A. Nerode, and S. Sastry, editors. *Hybrid Systems IV*, volume 1273 of *Lecture Notes in Computer Science*. Springer-Verlag, Berlin, 1997.

9. P. J. Antsaklis and X. D. Koutsoukos. Hybrid system control. In *Encyclopedia of Physical Science and Technology*, volume 7, pages 445–458. Academic Press, New York, third edition, 2002. Appears also as ISIS Technical Report ISIS-2001-003, February 2001.

10. P. J. Antsaklis and X. D. Koutsoukos. Hybrid dynamical systems: Review and recent progress. In T. Samad and G. Balas, editors, *Software-Enabled Control: Information Technologies for Dynamical Systems*. John Wiley & Sons, New York, 2003.

11. P. J. Antsaklis, X. D. Koutsoukos, and J. Zaytoon. On hybrid control of complex systems: A survey. *European Journal of Automation*, 32(9–10):1023–1045, 1998.

12. A. Balluchi, L. Benvenuti, M. D. Di Benedetto, G. M. Miconi, U. Pozzi, T. Villa, H. Wong-Toi, and A. L. Sangiovanni-Vincentelli. Maximal safe set

computation for idle speed control of an automotive engine. In N. Lynch and B. H. Krogh, editors, *Hybrid Systems: Computation and Control*, volume 1790 of *Lecture Notes in Computer Science*, pages 32–44. Springer-Verlag, Berlin, 2000.

13. Z. Banaszak and B. H. Krogh. Deadlock avoidance in flexible manufacturing systems with concurrently competing process flows. *IEEE Transactions on Robotics and Automation*, 6(6):724–734, 1990.

14. K. Barkaoui and I. Abdallah. Deadlock avoidance in fmss based on structural theory of Petri nets. In *IEEE Symposium on Emerging Technologies and Factory Automation*, 1995.

15. K. Barkaoui and J. F. Pradat-Peyre. On liveness and controlled siphons in Petri nets. In *Lecture Notes in Computer Science: 17th International Conference in Application and Theory of Petri Nets (ICATPN'96), Osaka, Japan*, volume 1091, pages 57–72, Springer-Verlag, Berlin, June 1996.

16. G. Barrett and S. Lafortune. Decentralized supervisory control with communicating controllers. *IEEE Transactions on Automatic Control*, 45(9):1620–1638, 2000.

17. G. Barrosso, A. Lima, and A. Perkusich. Supervision of discrete event systems using Petri nets and supervisory control theory. In *Proceedings of 1st International Workshop on Manufacturing and Petri Nets, 17th International Conference on Application and Theory of Petri Nets*, pages 77–96, 1998.

18. F. Basile, P. Chiacchio, and A. Giua. On the choice of suboptimal monitor places for supervisory control of Petri nets. In *Proceedings of the IEEE International Conference on Systems, Man, and Cybernetics*, pages 752–757, 1998.

19. F. Basile, P. Chiacchio, and A. Giua. Supervisory control of Petri nets based on suboptimal monitors places. In *Proceedings of the 4th International Workshop on Discrete Event Systems*, pages 85–87, 1998.

20. F. Basile, P. Chiacchio, and A. Giua. Optimal control of Petri net monitors with control and observation costs. In *Proceedings of the 39th IEEE International Conference on Decision and Control*, pages 424–429, 2000.

21. A. Bemporad, G. Ferrari-Trecate, and M. Morari. Observability and controllability of piecewise affine and hybrid systems. *IEEE Transactions on Automatic Control*, 45(10):1864–1876, 2000.

22. A. Bemporad and M. Morari. Control of systems integrating logic, dynamics, and constraints. *Automatica*, 35(3):407–427, 1999.

23. J. Billington. Protocol specification using P-graphs, a technique based on coloured Petri nets. In Reisig, W. and Rozenberg, G., editors, *Lectures on Petri Nets II: Applications*, volume 1492 of *Lecture Notes in Computer Science*, pages 293–330, Springer-Verlag, Berlin, 1998.

24. F. Blanchini and W. Ukovich. A linear programming approach to the control of discrete-time periodic system with state and control bounds in the presence of disturbance. *Journal of Optimization Theory and Applications*, 73(3):523–539, 1993.

25. J. Blazewicz, D. Bovet, and G. Gambosi. Deadlock-resistant flow control procedures for store-and-forward networks. *IEEE Transactions on Communications*, 32(8):884–887, 1984.

26. R. K. Boel, L. Ben-Naoum, and V. Van Breusegem. On forbidden state problems for a class of controlled Petri nets. *IEEE Transactions on Automatic Control*, 40(1):1717–1731, 1995.

27. A. Bouajjani, J.-C. Fernandez, and N. Halbwachs. Minimal model generation. In *Computer-Aided Verification*, volume 531 of *LNCS*, pages 197–203, Springer-Verlag, Berlin, 1990.

28. B. A. Brandin and W. M. Wonham. Supervisory control of timed discrete-event systems. *IEEE Transactions on Automatic Control*, 39(2):329–342, 1994.

29. U. A. Buy and R. H. Sloan. Automatic real-time analysis of reactive systems with the PARTS toolset. *Automated Software Engineering*, 23(4):227–273, 2001.

30. H. Chen. Net structure and control logic synthesis of controlled Petri nets. *IEEE Transactions on Automatic Control*, 43(10):1446–1450, 1998.

31. H. Chen. Control synthesis of Petri nets based on s-decreases. *Discrete Event Dynamic Systems: Theory and Applications*, 10(3):233–250, 2000.

32. H. Chen and B. Hu. Distributed control of discrete event systems described by a class of controlled Petri nets. In *Preprints of IFAC International Symposium on Distributed Intelligence Systems*, 1991.

33. H. Chen and B. Hu. Control of discrete event systems with their dynamics and legal behavior specified by Petri nets. In *Proceedings of the 32nd IEEE Conference on Decision and Control*, pages 239–240, 1993.

34. H. Chen and B. Hu. Monitor-based control of a class of controlled Petri nets. In *Proceedings of the 3rd International Conference on Automation, Robotics and Computer Vision*, 1994.

35. R. Cieslak, C. Desclaux, A. Fawaz, and P. Varayia. Supervisory control of discrete-event processes with partial observations. *IEEE Transactions on Automatic Control*, 33(3):249–260, 1988.

36. E. Coffman, M. Elphick, and A. Shaoshani. System deadlocks. *Computing Surveys*, 3:67–78, 1971.

37. J.-P. Courtiat, J. M. Ayache, and B. Algayres. Petri nets are good for protocols. In *ACM, SIGCOMM'84 Tutorials and Symposium, Communications Architectures and Protocols*, pages 66–74, 1984.

38. J. Cury, B. H. Krogh, and T. Niinomi. Synthesis of supervisory controllers for hybrid systems based on approximating automata. *IEEE Transactions on Automatic Control*, 43(4):564–568, 1998.

39. P. Darondeau and X. Xie. Linear control of live marked graphs. *Automatica*, 39(3):429–440, 2003.

40. R. David and H. Alla. Continuous Petri nets. In *8th European Workshop on Application and Theory of Petri Nets*, 1987.

41. R. David and A. Hassane. Petri nets for modeling of dynamic systems—a survey. *Automatica*, 32(2):175–202, 1994.

42. J. Desel and J. Esparza. *Free Choice Petri nets*. Number 40 in Cambridge Tracts in Theoretical Computer Science. Cambridge University Press, 1995.

43. J. Desel and W. Reisig. Place/transition Petri nets. *Lecture Notes in Computer Science: Lectures on Petri Nets I: Basic Models*, 1491:122–173, 1998.

44. A. Deshpande and P. Varaiya. Viable control of hybrid systems. In P. J. Antsaklis et al., editor, *Hybrid Systems II*, volume 999 of *Lecture Notes in Computer Science*, pages 128–147, Springer-Verlag, Berlin, 1995.

45. A.A. Desrochers and R.Y. Al'Jaar. *Applications of Petri nets in Manufacturing Systems: Modelling, Control and Performance Analysis*. IEEE Press, Piscatway, NJ, 1995.

46. F. DiCesare, G. Harhalakis, J.M. Proth, M. Silva, and F.B. Vernadat. *Practice of Petri Nets in Manufacturing*. Chapman and Hall, London, 1993.

47. C. Dórea and J. Hennet. (A,B)-Invariant polyhedral sets of linear discrete time systems. *Journal of Optimization Theory and Applications*, 103(3):521–542, 1999.

48. J. Ezpeleta, J. M. Colom, and J. Martínez. A Petri net based deadlock prevention policy for flexible manufacturing systems. *IEEE Transactions on Robotics and Automation*, 11(2):173–184, 1995.

49. M. Fanti, B. Maione, S. Mascolo, and B. Turchiano. Event-based feedback control for deadlock avoidance in flexible production systems. *IEEE Transactions on Robotics and Automation*, 13(3), 1997.

50. H. J. Genrich, K. Lautenbach, and P. S. Thiagarajan. Elements of general net theory. In W. Brauer, editor, *Net Theory and Applications*, volume 84 of *Lecture Notes in Computer Science*, pages 21–163, Springer-Verlag, Berlin, 1980.

51. A. Ghaffari, N. Rezg, and X. Xie. Design of a live and maximally permissive Petri net controller using the theory of regions. *IEEE Transactions on Robotics and Automation*, 19(1):137–142, 2003.

52. A. Ghaffari, N. Rezg, and X. Xie. Feedback control logic for forbidden-state problems of marked graphs: application to a real manufacturing system. *IEEE Transactions on Automatic Control*, 48(1):2–17, 2003.

53. C. Girault and R. Valk. *Petri Nets for Systems Engineering*. Springer-Verlag, Berlin, 2003.

54. A. Giua and F. DiCesare. Supervisory design using Petri nets. In *Proceedings of the 30th IEEE International Conference on Decision and Control*, pages 92–97, 1991.

55. A. Giua and F. DiCesare. Blocking and controllability of Petri nets in supervisory control. *IEEE Transactions on Automatic Control*, 39(4):818–823, 1994.

56. A. Giua and F. DiCesare. Decidability and closure properties of weak Petri net languages in supervisory control. *IEEE Transactions on Automatic Control*, 40(5):906–910, 1995.

57. A. Giua, F. DiCesare, and M. Silva. Generalized mutual exclusion constraints on nets with uncontrollable transitions. In *Proceedings of the IEEE International Conference on Systems, Man and Cybernetics*, pages 974–979, 1992.

58. A. Giua and C. Seatzu. Supervisory control of railway networks with Petri nets. In *Proceedings of the 40th IEEE Conference on Decision and Control*, pages 5004–5009, 2001.

59. A. Giua and C. Seatzu. Observability of place/transition nets. *IEEE Transactions on Automatic Control*, 47(9):1424–1437, 2002.

60. C. H. Golaszewski and P. J. Ramadge. Control of discrete event processes with forced events. In *Proceedings of the 26th IEEE Conference on Decision and Control*, pages 247–251, 1987.

61. U. Goltz. Synchronic distance. In W. Brauer, W. Reisig, and G. Rozenberg, editors, *Petri Nets: Central Models and Their Properties, Advances in Petri Nets 1986, Part I*, volume 254 of *Lecture Notes in Computer Science*, pages 338–358, Springer-Verlag, Berlin, 1987.

62. R. Grossman, A. Nerode, A. Ravn, and H. Rischel, editors. *Hybrid Systems*, volume 736 of *Lecture Notes in Computer Science*. Springer-Verlag, Berlin, 1993.

63. X. Guan and L. E. Holloway. Control of distributed discrete event systems modeled as Petri nets. In *Proceedings of the 1997 American Control Conference*, pages 2342–2347, 1997.

64. P. Gutman and M. Cwikel. An algorithm to find maximal state constraint sets for discrete-time linear dynamical systems with bounded controls and states. *IEEE Transactions on Automatic Control*, 32(3):251–254, 1987.

65. K. X. He and M. D. Lemmon. Modelling hybrid control systems using programmable timed Petri nets. *European Journal of Automation*, 32(9–10):1187–1208, 1998.

66. K. X. He and M. D. Lemmon. Liveness verification of discrete event systems modeled by *n*-safe ordinary Petri nets. In M. Nielsen and D. Simpson, editors, *21st International Conference on Application and Theory of Petri Nets (ICATPN 2000), Aarhus, Denmark, June 2000*, volume 1825 of *Lecture Notes in Computer Science*, pages 227–243, Springer-Verlag, Berlin, 2000.

67. K. X. He and M. D. Lemmon. Liveness-enforcing supervision of bounded ordinary Petri nets using partial order methods. *IEEE Transactions on Automatic Control*, 47(7):1042–1055, 2002.

68. T. A. Henzinger. Hybrid automata with finite bisimulations. In Z. Füllöp and G. Gécgeg, editors, *ICALP'95: Automata, Languages, and Programming*, Springer-Verlag, Berlin, 1995.

69. Y.C. Ho, editor. *Discrete Event Dynamic Systems: Theory and Application*, Special Issue on Hybrid Petri Nets, volume 11(1/2). Kluwer Academic Press, Norwell, MA, 2001.

70. L. E. Holloway, B. H. Krogh, and A. Giua. A survey of Petri net methods for controlled discrete event systems. *Discrete Event Dynamic Systems*, 7(2):151–190, 1997.

71. L. E. Holloway, X. Guan, and L. Zhang. A generalization of state avoidance policies for controlled Petri nets. *IEEE Transactions on Automatic Control*, 41(6):804–816, 1996.

72. L. E. Holloway and B. H. Krogh. Synthesis of feedback control logic for a class of controlled Petri nets. *IEEE Transactions on Automatic Control*, 35(5):514–523, 1990.

73. L. E. Holloway and B. H. Krogh. On closed-loop liveness of discrete-event systems under maximally permissive control. *IEEE Transactions on Automatic Control*, 37(5):692–697, 1992.

74. F.-S. Hsieh and S.-C. Chang. Dispatching-driven deadlock avoidance controller synthesis for flexible manufacturing systems. *IEEE Transactions on Robotics and Automation*, 10(2):196–209, 1994.

75. A. Ichikawa and K. Hiraishi. Analysis and control of discrete event systems represented by Petri nets. In P. Varaiya and A. B. Kurzhanski, editors, *Discrete Event Systems: Models and Applications*, volume 103 of *Lecture Notes in Control and Information Sciences*, pages 115–134, Springer-Verlag, Berlin, 1988.

76. A. Ichikawa, K. Yokoyama, and S. Kurogi. Reachability and control of discrete event systems represented by conflict-free Petri nets. In *Internationl Symposium on Circuits and Systems, Proceedings*, pages 487–490, 1985.

77. M. V. Iordache and P. J. Antsaklis. Decentralized control of Petri nets. In *Proceedings of the Workshop on Discrete Event Systems Control, of the International Conference on the Application and Theory of Petri Nets (ATPN 2003)*, pages 143–158, 2003.

78. M. V. Iordache. Deadlock Prevention in Discrete Event Systems Using Petri Nets. Master's thesis, University of Notre Dame, Notre Dame, IN, 1999.

79. M. V. Iordache and P. J. Antsaklis. $\mathcal{T}$-liveness enforcement in Petri nets based on structural net properties. In *Proceedings of the 40th IEEE International Conference on Decision and Control.*, pages 4984–4989, December 2001.

80. M. V. Iordache and P. J. Antsaklis. Generalized conditions for liveness enforcement and deadlock prevention in Petri nets. In J.M. Colom and M. Koutny, editors, *22nd International Conference on Application and Theory of Petri Nets (ICATPN 2001), Newcastle upon Tyne, UK, June 2001,* volume 2075 of *Lecture Notes in Computer Science,* pages 184–203, Springer-Verlag, Berlin, 2001.

81. M. V. Iordache and P. J. Antsaklis. Decentralized control of DES using Petri nets. Technical report isis-2002-005, University of Notre Dame, Notre Dame, IN, September 2002.

82. M. V. Iordache and P. J. Antsaklis. Software tools for the supervisory control of Petri nets based on place invariants. Technical report isis-2002-003, University of Notre Dame, Notre Dame, IN, April 2002.

83. M. V. Iordache and P. J. Antsaklis. Synthesis of supervisors enforcing general linear vector constraints in Petri nets. In *Proceedings of the 2002 American Control Conference,* pages 154–159, 2002.

84. M. V. Iordache and P. J. Antsaklis. Synthesis of supervisors enforcing general linear vector constraints in Petri nets. Technical report isis-2002-002, University of Notre Dame, Notre Dame, IN, February 2002.

85. M. V. Iordache and P. J. Antsaklis. Admissible decentralized control of Petri nets. In *Proceedings of the 2003 American Control Conference,* pages 332–337, 2003.

86. M. V. Iordache and P. J. Antsaklis. Decentralized control of Petri nets with constraint transformations. In *Proceedings of the 2003 American Control Conference,* pages 314–319, 2003.

87. M. V. Iordache and P. J. Antsaklis. On a class of controlled invariant sets. In *Proceedings of the 2004 American Control Conference,* pages 2522–2527, 2004.

88. M. V. Iordache and P. J. Antsaklis. Supervision based on place invariants: A survey. Technical report isis-2004-003, University of Notre Dame, Notre Dame, IN, July 2004. Accepted for publication in *Discrete Event Dynamic Systems.*

89. M. V. Iordache and P. J. Antsaklis. A structural approach to the enforcement of language and disjunctive constraints. In *Proceedings of the 2005 American Control Conference,* pages 3920–3925, 2005.

90. M. V. Iordache, J. O. Moody, and P. J. Antsaklis. A method for deadlock prevention in discrete event systems using Petri nets. Technical report of the isis group, isis-99-006, University of Notre Dame, Notre Dame, IN, July 1999.

91. M. V. Iordache, J. O. Moody, and P. J. Antsaklis. Automated synthesis of deadlock prevention supervisors using Petri nets. Technical report of the isis group, isis-2000-003, University of Notre Dame, Notre Dame, IN, May 2000.

92. M. V. Iordache, J. O. Moody, and P. J. Antsaklis. Automated synthesis of liveness enforcement supervisors using Petri nets. Technical report of the isis group, isis-2000-004, University of Notre Dame, Notre Dame, IN, September 2000.

93. M. V. Iordache, J. O. Moody, and P. J. Antsaklis. A method for the synthesis of deadlock prevention controllers in systems modeled by Petri nets. In *Proceedings of the 2000 American Control Conference,* pages 3167–3171, 2000.

94. M. V. Iordache, J. O. Moody, and P. J. Antsaklis. A method for the synthesis of liveness enforcing supervisors in Petri nets. In *Proceedings of the 2001 American Control Conference*, pages 4943–4948, 2001.

95. M. V. Iordache, J. O. Moody, and P. J. Antsaklis. Synthesis of deadlock prevention supervisors using Petri nets. *IEEE Transactions on Robotics and Automation*, 18(1):59–68, February 2002.

96. M. V. Iordache and P. J. Antsaklis. Design of T-liveness enforcing supervisors in Petri nets. *IEEE Transactions on Automatic Control*, 48(11):1962–1974, 2003.

97. M. V. Iordache and P. J. Antsaklis. Synthesis of supervisors enforcing general linear vector constraints in Petri nets. *IEEE Transactions on Automatic Control*, 48(11):2036–2039, 2003.

98. S. S. Isloor and T. A. Marsland. The deadlock problem: An overview. *Computer*, 13(9):58–78, 1980.

99. M. Jantzen and R. Valk. Formal properties of place/transition nets. In W. Brauer, editor, *Net Theory and Applications, Proceedings of the Advanced Course on General Net Theory of Processes and Systems, Hamburg, 1979*, volume 84 of *Lecture Notes in Computer Science*, pages 165–212, Springer-Verlag, Berlin, 1980.

100. S. Jiang, V. Chandra, and R. Kumar. Decentralized control of discrete event systems with multiple local specilizations. In *Proceedings of 2001 American Control Conference*, pages 959–964, 2001.

101. S. Jiang and R. Kumar. Decentralized control of discrete event systems with specializations to local control and concurrent systems. *IEEE Transactions on Systems, Man and Cybernetics, Part B*, 30(5):653–660, 2000.

102. P.C. Kanellakis and S.A. Smolka. CCS expressions, finite-state processes, and three problems of equivalence. *Information and Computation*, 86:43–68, 1990.

103. S. Keerthi and E. Gilbert. Computation of minimum-time feedback control laws for discrete-time systems with state-control constraints. *IEEE Transactions on Automatic Control*, 32(5):432–435, 1987.

104. X. D. Koutsoukos. *Analysis and Design of Piecewise Linear Hybrid Dynamical Systems*. Ph.D. thesis, University of Notre Dame, Notre Dame, IN, 2000.

105. X. D. Koutsoukos and P. J. Antsaklis. Supervisory control design of hybrid systems modeled by timed Petri nets based on invariant properties. In *Proceedings of the 8th IFAC/IFORMS/IMACS/IFIP/ Symposium on Large Scale Systems: Theory and Applications LSS'98*, Rio, Greece, July 1998.

106. X. D. Koutsoukos and P. J. Antsaklis. Hybrid control of a robotic manufacturing system. In *Proceedings of the 7th IEEE Mediterranean Conference on Control and Automation*, pages 144–159, 1999.

107. X. D. Koutsoukos and P. J. Antsaklis. Hybrid control systems using timed Petri nets: Supervisory control design based on invariant properties. In P. J. Antsaklis, W. Kohn, M. D. Lemmon, A. Nerode, and S. Sastry, editors, *Hybrid Systems V*, volume 1567 of *Lecture Notes in Computer Science*, pages 142–162, Springer-Verlag, Berlin, 1999.

108. X. D. Koutsoukos, P. J. Antsaklis, K. X. He, and M. D. Lemmon. Programmable timed Petri nets in the analysis and design of hybrid control systems. In *Proceedings of the 37th IEEE Conference on Decision and Control*, pages 1617–1622, 1998.

109. X. D. Koutsoukos, K. X. He, M. D. Lemmon, and P. J. Antsaklis. Timed Petri nets in hybrid systems: Stability and supervisory control. *Journal of Discrete Event Dynamic Systems: Theory and Applications*, 8(2):137–173, 1998.

110. X. D. Koutsoukos, K. X. He, M. D. Lemmon, and P. J. Antsaklis. Timed Petri nets in hybrid systems: Stability and supervisory control. *Journal of Discrete Event Dynamic Systems: Theory and Applications*, 8(2):137–173, 1998.

111. X. D. Koutsoukos, F. Zhao, H. Haussecker, J. Reich, and P. Cheung. Fault modeling for monitoring and diagnosis of sensor-rich hybrid systems. In *Proceedings of the 40th IEEE Conference on Decision and Control*, pages 793–801, 2001.

112. X. D. Koutsoukos, P. J. Antsaklis, J. A. Stiver, and M. D. Lemmon. Supervisory control of hybrid systems. *Proceedings of the IEEE*, pages 1026–1049, July 2000.

113. P. Kozak and W. M. Wonham. Fully decentralized solutions of supervisory control problems. *IEEE Transactions on Automatic Control*, 40(12):2094–2097, 1995.

114. B. H. Krogh. Controlled Petri nets and maximally permissive feedback logic. In *Proceedings of the 25th Annual Allerton Conference, University of Illinois, Urbana*, pages 317–326, 1987.

115. B. H. Krogh and L. E. Holloway. Synthesis of feedback control logic for manufacturing systems. *Automatica*, 27(4):641–651, 1991.

116. R. Kumar and L. E. Holloway. Supervisory control of Petri nets with regular specification languages. *IEEE Transactions on Automatic Control*, 41(2):245–249, 1996.

117. R. Kumar and M. Shayman. Formulae relating controllability, observability, and co-observability. *Automatica*, 34(2):211–215, 1998.

118. G. Lafferriere. Hybrid systems with finite bisimulations. In P. J. Antsaklis, W. Kohn, M. D. Lemmon, A. Nerode, and S. Sastry, editors, *Hybrid Systems V*, volume 1567 of *Lecture Notes in Computer Science*, pages 186–203, Springer-Verlag, Berlin, 1999.

119. H. Lamouchi and J. Thistle. Effective control synthesis for des under partial observations. In *Proceedings of the 39th IEEE Conference on Decision and Control*, pages 22–28, 2000.

120. K. Lautenbach and H. Ridder. The linear algebra of deadlock avoidance—a Petri net approach. Technical report, University of Koblenz, Institute for Computer Science, 1996.

121. K. Lautenbach and P. S. Thiagarajan. Analysis of a resource allocation problem using Petri nets. In *Proceedings of the 1st European Conference on Parallel and Distributed Processing*, pages 260–266. Cepadues Editions, 1979.

122. M. D. Lemmon, K. X He, and I. Markovsky. Supervisory hybrid systems. *IEEE Control Systems Magazine*, 19(4):42–55, 1999.

123. M. D. Lemmon, K. X. He, and C. J. Bett. Modeling Hybrid Control Systems Using Programmable Timed Petri Nets. In *3rd International Conference ADMP'98, Automation of Mixed Processes: Dynamic Hybrid Systems, (ADPM'98)*, pages 177–184, Rheims, France, March 1998.

124. Y. Li and W. M. Wonham. Control of vector discrete-event systems I—The base model. *IEEE Transactions on Automatic Control*, 38(8):1214–1227, 1993.

125. Y. Li and W. M. Wonham. Control of vector discrete-event systems II—Controller synthesis. *IEEE Transactions on Automatic Control*, 39(3):512–530, 1994.

126. Y. Li and W. M. Wonham. Concurrent vector discrete-event systems. *IEEE Transactions on Automatic Control*, 40(4):628–638, 1995.

127. Y. Li and W. M. Wonham. Controllability and observability in the state feedback control of discrete-event systems. In *Proceedings of the 27th IEEE Conferenece on Decision and Control*, pages 203–208, 1988.

128. F. Lin and W. M. Wonham. On observability of discrete-event systems. *Information Sciences*, 44(3):173–198, 1988.

129. F. Lin and W. M. Wonham. Decentralized control and coordination of discrete-event systems with partial observation. *IEEE Transactions on Automatic Control*, 35(12):1330–1337, 1990.

130. H. Lin and P. J. Antsaklis. Controller synthesis for a class of uncertain piecewise linear hybrid dynamical systems. In *Proceedings of the 41st IEEE Conference on Decision and Control*, pages 3188–3193, 2002.

131. J. Lunze. A Petri-net approach to qualitative modelling of continuous dynamical systems. *Systems Analysis, Modelling, Simulation*, 9:89–111, 1992.

132. J. Lunze, B. Nixdorf, and J. Schroder. Deterministic discrete-vent representations of linear continuous-variable systems. *Automatica*, 35(3):396–406, 1999.

133. J. Lygeros, D. Godbole, and S. Sastry. Verified hybrid controllers for automated vehicles. *IEEE Transactions on Automatic Control*, 43(4):522–539, 1998.

134. J. Lygeros, C. Tomlin, and S. Sastry. Controllers for reachability specifications for hybrid systems. *Automatica*, 35(3):349–370, 1999.

135. D. Mandrioli, R. Zicari, C. Ghezzi, and F. Tisato. Modeling the Ada task system by Petri nets. *Computer Languages*, 10(1):43–61, 1985.

136. B. J. McCarragher and H. Asada. The discrete event modeling and trajectory planning of robotic assembly tasks. *Transactions of the ASME–Journal of Dynamic Systems, Measurement and Control*, 117(3):394–400, 1995.

137. N. H. McClamroch, I. Kolmanovsky, and M. Reyhanoglu. Hybrid closed loop systems: A nonlinear control perspective. In *Proceedings of the 36th Conference on Decision and Control*, pages 114–119, 1997.

138. A. Merlin and S. Schweitzer. Deadlock avoidance in store-and-forward networks–I: Store-and-forward deadlock. *IEEE Transactions on Communications*, 28(3):345–354, 1980.

139. A. Merlin and S. Schweitzer. Deadlock avoidance in store-and-forward networks–II: Other deadlock types. *IEEE Transactions on Communications*, 28(3):355–360, 1980.

140. J. O. Moody and P. J. Antsaklis. Deadlock avoidance in Petri nets with uncontrollable transitions. In *Proceedings of the 1998 American Control Conference*, pages 1257–1258, 1998.

141. J. O. Moody and P. J. Antsaklis. *Supervisory Control of Discrete Event Systems Using Petri Nets*. Kluwer Academic Publishers, Norwell, MA, 1998.

142. J. O. Moody and P. J. Antsaklis. Petri net supervisors for DES with uncontrollable and unobservable transitions. *IEEE Transactions on Automatic Control*, 45(3):462–476, 2000.

143. J. O. Moody, M. V. Iordache, and P. J. Antsaklis. Enforcement of event-based supervisory constraints using state-based methods. In *Proceedings of the 38th IEEE Conference on Decision and Control*, pages 1743–1748, 1999.

144. T. Motzkin. *The theory of linear inequalities*. Rand Corp., Santa Monica, CA, 1952.

145. T. Murata. Petri nets: Properties, analysis and applications. In *Proceedings of the IEEE*, pages 541–580, April 1989.

146. A. Overkamp and J. H. van Schuppen. Maximal solutions in decentralized supervisory control. *SIAM Journal of Control and Optimization*, 39(2):492–511, 2000.

147. J. Park and S. Reveliotis. Deadlock avoidance in sequential resource allocation systems with multiple resource acquisitions and flexible routings. *IEEE Transactions on Automatic Control*, 46(10):1572–1583, 2001.

148. J. Park and S. Reveliotis. Liveness-enforcing supervision for resource allocation systems with uncontrollable behavior and forbidden states. *IEEE Transactions on Robotics and Automation*, 18(2):234–240, 2002.

149. J. L. Peterson. *Petri Net Theory and the Modeling of Systems*. Prentice Hall, Englewood Cliffs, NJ, 1981.

150. C. A. Petri. *Kommunikation mit Automaten*. Bonn: Institut für Instrumentelle Mathematik, Schriften des IIM Nr. 2, 1962.

151. C. A. Petri. Fundamentals of a theory of asynchronous information flow. In *Proceedings of IFIP Congress 62*, pages 386–390, North Holland Publ. Comp., Amsterdam, 1963.

152. C. A. Petri. Kommunikation mit automaten. *New York: Griffiss Air Force Base, Technical Report RADC-TR-65-377*, 1:Suppl. 1, 1966. English translation.

153. C. A. Petri. *Interpretations of Net Theory*. St. Augustin: Gesellschaft für Mathematik und Datenverarbeitung Bonn, Interner Bericht ISF-75-07, Second Edition, December 1976.

154. J. H. Prosser, M. Kam, and H.G. Kwanty. Decision fusion and supervisor synthesis in decentralized discrete event systems. In *Proceedings of the 1997 American Control Conference*, pages 2251–2255, 1997.

155. J.-M. Proth and X. Xie. *Petri Nets: A Tool for Design and Management of Manufacturing Systems*. John Wiley & Sons, New York, 1997.

156. A. Puri, S. Tripakis, and P. Varaiya. Problems and examples of decentralized observation and control for discrete event systems. In *Symposium on the Supervisory Control of Discrete Event Systems*, 2001.

157. J. Raisch and S. O'Young. Discrete approximation and supervisory control of continuous systems. *IEEE Transactions on Automatic Control*, 4(43):568–573, 1998.

158. P. J. Ramadge and W. M. Wonham. Supervisory control of a class of discrete event processes. *SIAM Journal on Control and Optimization*, 25(1):206–230, 1987.

159. P. J. Ramadge and W. M. Wonham. The control of discrete event systems. *Proceedings of the IEEE*, 77(1):81–98, 1989.

160. W. Reisig. *Petri Nets*, volume 4 of *EATCS Monographs on Theoretical Computer Science*. Springer-Verlag, Berlin, 1985.

161. W. Reisig. *Elements of Distributed Algorithms: Modeling and Analysis with Petri nets*. Springer-Verlag, Berlin, 1998.

162. W. Reisig, E. Kindler, T. Vesper, and H. Völzer. Distributed algorithms for networks of agents. In Reisig, W. and Rozenberg, G., editors, *Lectures on Petri Nets II: Applications*, volume 1492 of *Lecture Notes in Computer Science*, pages 331–385, Springer-Verlag, Berlin, 1998.

163. E. Roszkowska and R. Wojcik. Problems of process flow feasibility in fas. In *IFAC CIM in Process and Manufacturing Industries*, pages 115–121, 1992.

164. K. Rudie. The current state of decentralized discrete-event control systems. In *Proceedings of the 10th Mediterranean Conference on Control and Automation*, 2002.

165. K. Rudie, S. Lafortune, and F. Lin. Minimal communication in a distributed discrete-event system. In *Proceedings of the 1999 American Control Conference*, pages 1965–1970, 1999.

166. K. Rudie and J. C. Willems. The computational complexity of decentralized discrete-event control problems. *IEEE Transactions on Automatic Control*, 40(7):1313–1319, 1995.

167. K. Rudie and W. M. Wonham. Think globally, act locally: Decentralized supervisory control. *IEEE Transactions on Automatic Control*, 37(11):1692–1708, 1992.

168. A. van der Schaft and H. Schumacher. *An Introduction to Hybrid Dynamical Systems*, volume 251 of *Lecture Notes in Control and Information Sciences*. Springer Verlag, London, 2000.

169. O. Shakernia, G. Pappas, and S. Sastry. Decidable controller synthesis for classes of linear systems. In N. Lynch et al., editor, *Hybrid Systems: Computation and Control*, number 1790 in *Lecture Notes in Computer Science*, pages 407–420, Springer-Verlag, Berlin, 2000.

170. O. Shakernia, G. Pappas, and S. Sastry. Semidecidable controller synthesis for classes of linear hybrid systems. In *Proceedings of the 39th IEEE Conference on Decision and Control*, pages 1834–1839, 2000.

171. M. Silva, E. Teruel, and J. M. Colom. Linear algebraic and linear programming techniques for the analysis of place/transition net systems. *Lecture Notes in Computer Science: Lectures on Petri Nets I: Basic Models*, 1491:309–373, 1998.

172. R. S. Sreenivas. On a free-choice equivalent of a Petri net. In *Proceedings of the 36th IEEE Conference on Decision and Control*, pages 4092–4097, 1997.

173. R. S. Sreenivas. On commoner's liveness theorem and supervisory policies that enforce liveness in free-choice Petri nets. *Systems & Control Letters*, pages 41–48, 1997.

174. R. S. Sreenivas. On the existence of supervisory policies that enforce liveness in discrete event dynamic systems modeled by controlled Petri nets. *IEEE Transactions on Automatic Control*, 42(7):928–945, July 1997.

175. R. S. Sreenivas. An application of independent, increasing, free-choice Petri nets to the synthesis of policies that enforce liveness in arbitrary Petri nets. *Automatica*, 44(12):1613–1615, December 1998.

176. R. S. Sreenivas. On supervisory policies that enforce liveness in a class of completely controlled Petri nets obtained via refinement. *IEEE Transactions on Automatic Control*, 44(1):173–177, 1999.

177. R. S. Sreenivas. On supervisory policies that enforce liveness in completely controlled Petri nets with directed cut-places and cut-transitions. *IEEE Transactions on Automatic Control*, 44(6):1221–1225, 1999.

178. R. S. Sreenivas. On a minimally restrictive supervisory policy that enforces liveness in partially controlled free choice Petri nets. In *Proceedings of the 39th IEEE Conference on Decision and Control*, pages 2651–2656, 2000.

179. J. A. Stiver, P. J. Antsaklis, and M. D. Lemmon. Interface and controller design for hybrid control systems. In P. J. Antsaklis, W. Kohn, A. Nerode, and S. Sastry, editors, *Hybrid Systems II*, volume 999 of *Lecture Notes in Computer Science*, pages 462–492, Springer-Verlag, Berlin, 1995.

180. J. A. Stiver, P. J. Antsaklis, and M. D. Lemmon. A logical des approach to the design of hybrid control systems. *Mathematical and Computer Modelling*, pages 55–76, 1996.

181. G. Stremersch. *Supervision of Petri Nets*. Kluwer Academic Publishers, Norwell, MA, 2001.

182. G. Stremersch and R. K. Boel. Enforcing k-safeness in controlled state machines. In *Proceedings of the 38th IEEE Conference on Decision and Control*, pages 1737–1742, 1999.

183. G. Stremersch and R. K. Boel. Reduction of the supervisory control problem for Petri nets. *IEEE Transactions on Automatic Control*, 45(12):2358–2363, 2000.

184. G. Stremersch and R. K. Boel. Decomposition of the supervisory control problem for Petri nets under preservation of maximal permissiveness. *IEEE Transactions on Automatic Control*, 46(9):1490–1496, 2001.

185. G. Stremersch and R. K. Boel. Structuring acyclic Petri nets for reachability analysis and control. *Discrete Event Dynamic Systems*, 12(1):7–41, 2002.

186. Z. Suraj. Resource allocation problem. In *Proceedings of the 3rd Symposium on the Mathematical Foundations of Computer Science, Zaborow 1980, ICS PAS Reports*, pages 83–86, 1980.

187. T. Suzuki, S. M. Shatz, and T. Murata. A protocol modeling and verification approach based on a specification language and Petri nets. *IEEE Transactions on Software Engineering*, 16(5):523–536, 1990.

188. P. Tabuada and G. Pappas. Model checking LTL over controllable linear systems is decidable. In *Hybrid Systems: Computation and Control*, volume 2623 of *LNCS*, pages 498–513, Springer-Verlag, Berlin, 2003.

189. S. Takai and S. Kozama. Decentralized state feedback control of discrete event systems. *Systems & Control Letters*, 22(5):369–375, 1994.

190. S. Takai and T. Ushio. Reliable decentralized supervisory control of discrete event systems. *IEEE Transactions on Systems, Man, and Cybernetics—Part B: Cybernetics*, 30(5):661–667, 2000.

191. J. G. Thistle and W. M. Wonham. Control of infinite behavior of finite automata. *SIAM Journal on Control and Optimization*, 32(4):1075–1097, 1994.

192. J. G. Thistle and W. M. Wonham. Supervision of infinite behavior of discrete-event systems. *SIAM Journal on Control and Optimization*, 32(4):1098–1113, 1994.

193. M. Tittus and B. Egardt. Hierarchical supervisory control for batch processes. *IEEE Transactions on Control Systems Technology*, 7(5):542–554, 1999.

194. F. Tricas, F. Garcia-Valles, J. M. Colom, and J. Ezpeleta. New methods for deadlock prevention and avoidance in concurrent systems. *Actas de las Jornadas de Concurrencia 2000*, pages 97–110, June 2000.

195. S. Tripakis. Undecidable problems of decentralized observation and control. In *Proceedings of the 40th IEEE Conference on Decision and Control*, pages 4104–4109, 2001.

196. S. Tripakis. Decentralized control of discrete event systems with bounded or unbounded delay communication. In *6th International Workshop on Discrete Event Systems*, 2002.

197. J. N. Tsitsiklis. On the control of discrete event dynamical systems. *Mathematics of Control, Signals and Systems*, 2(2):95–107, 1989.

198. T. Ushio. Maximally permissive feedback and modular control synthesis in Petri nets with external input. *IEEE Transactions on Automatic Control*, 35(7):844–848, 1990.

199. R. Valk and M. Jantzen. The residue of vector sets with applications to decidability problems in Petri nets. *Acta Informatica*, 21:643–674, 1985.

200. J. H. van Schuppen. Decentralized supervisory control with information structures. In *Proceedings of the International Workshop on Discrete Event Systems (WODES98)*, pages 36–41, 1998.

201. J. H. van Schuppen. A sufficient condition for controllability of a class of hybrid systems. In *Hybrid Systems: Computation and Control*, volume 1386 of *LNCS*, pages 374–383, Springer-Verlag, Berlin, 1998.

202. L. Vandenberge and S. Boyd. Semidefinite programming. *SIAM Review*, 38(1):49–95, 1996.

203. R. Vidal, S. Schaffert, J. Lygeros, and S. Sastry. Controlled invariance of discrete time systems. In N. Lynch and B. H. Krogh, editors, *Hybrid Systems: Computation and Control*, volume 1790 of *Lecture Notes in Computer Science*, pages 437–450, Springer-Verlag, Berlin, 2000.

204. R. Vidal, S. Schaffert, O. Shakernia, J. Lygeros, and S. Sastry. Decidable and semi-decidable controller synthesis for classes of discrete time hybrid systems. In *Proceedings of the 40th IEEE Conference on Decision and Control*, pages 1243–1248, 2001.

205. H. P. Williams. Linear and integer programming applied to the propositional calculus. *International Journal of Systems Research and Information Science*, 2:81–100, 1987.

206. H. P. Williams. *Model building in mathematical programming*. John Wiley & Sons, New York, 1993. (3rd ed.).

207. Y. Willner and M. Heymann. Supervisory control of concurrent discrete-event systems. *International Journal of Control*, 54(5):1143–1169, 1991.

208. K. C. Wong and J. H. van Schuppen. Decentralized supervisory control of discrete-event systems with communication. In *Proceedings International Workshop on Discrete Event Systems (WODES96)*, pages 284–289, 1996.

209. H. Wong-Toi. The synthesis of controllers for linear hybrid automata. In *Proceedings of the 36th IEEE Conference on Decision and Control*, pages 4607–4612, 1997.

210. K. Xing, B. Hu, and H. Chen. Deadlock avoidance policy for Petri net modeling of flexible manufacturing systems with shared resources. *IEEE Transactions on Automatic Control*, 41(2):289–295, February 1996.

211. A. Yakovlev, L. Gomes, and L. Lavagno, editors. *Hardware Design and Petri Nets*. Kluwer Academic Publishers, Norwell, MA, 2000.

212. A. Yakovlev and A. Koelmans. Petri nets and digital hardware design. In W. Reisig and G. Rozenberg, editors, *Lectures on Petri Nets II: Applications*, volume 1492 of *Lecture Notes in Computer Science*, pages 154–236, Springer-Verlag, Berlin, 1998.

213. E. Yamalidou and J. Kantor. Modeling and optimal control of discrete-event chemical processes using Petri nets. *Computers and Chemical Engineering*, 15(7):503–519, 1991.

214. E. Yamalidou, J. O. Moody, P. J. Antsaklis, and M. D. Lemmon. Feedback control of Petri nets based on place invariants. *Automatica*, 32(1):15–28, 1996.

215. T. Yoo and S. Lafortune. New results on decentralized supervisory control of discrete-event systems. In *Proceedings of the 39th IEEE Conference on Decision and Control*, pages 1–6, 2000.

216. T. Yoo and S. Lafortune. A general architecture for decentralized supervisory control of discrete-event systems. *Discrete Event Dynamic Systems: Theory and Applications*, 12(3):335–377, 2002.

217. L. Zhang and L. E. Holloway. Forbidden state avoidance in controlled Petri nets under partial observation. In *Proceedings of the 33rd Annual Allerton Conference on Communications, Control, and Computing*, pages 146–155, 1995.

218. M. Zhou and K. Venkatesh. *Modeling, Simulation, and Control of Flexible Manufacturing Systems: A Petri Net Approach*, volume 6 of *Series in Intelligent Control and Intelligent Automation*. World Scientific Publishing Company, River Edge, NJ, 1999.

Index

Systems & Control: Foundations & Applications

Series Editor
Tamer Başar
Coordinated Science Laboratory
University of Illinois at Urbana-Champaign
1308 W. Main St.
Urbana, IL 61801-2307
U.S.A.

Systems & Control: Foundations & Applications

Aims and Scope

The aim of this series is to publish top quality state-of-the art books and research monographs at the graduate and post-graduate levels in systems, control, and related fields. Both foundations and applications will be covered, with the latter spanning the gamut of areas from information technology (particularly communication networks) to biotechnology (particularly mathematical biology) and economics.

Readership

The books in this series are intended primarily for mathematically oriented engineers, scientists, and economists at the graduate and post-graduate levels.

Types of Books

Advanced books, graduate-level textbooks, and research monographs on current and emerging topics in systems, control and related fields.

Preparation of manuscripts is preferable in LATEX. The publisher will supply a macro package and examples of implementation for all types of manuscripts.

Proposals should be sent directly to the editor or to: Birkhäuser Boston, 675 Massachusetts Avenue, Cambridge, MA 02139, U.S.A. or to Birkhäuser Publishers, 40-44 Viadukstrasse, CH-4051 Basel, Switzerland

A Partial Listing of Books Published in the Series

Identification and Stochastic Adaptive Control
Han-Fu Chen and Lei Guo

Viability Theory
Jean-Pierre Aubin

Representation and Control of Infinite Dimensional Systems, Vol. I
A. Bensoussan, G. Da Prato, M. C. Delfour, and S. K. Mitter

Representation and Control of Infinite Dimensional Systems, Vol. II
A. Bensoussan, G. Da Prato, M. C. Delfour, and S. K. Mitter

Mathematical Control Theory: An Introduction
Jerzy Zabczyk

H_∞-Control for Distributed Parameter Systems: A State-Space Approach
Bert van Keulen

Disease Dynamics
Alexander Asachenkov, Guri Marchuk, Ronald Mohler, and Serge Zuev

Theory of Chattering Control with Applications to Astronautics,
Robotics, Economics, and Engineering
Michail I. Zelikin and Vladimir F. Borisov

Modeling, Analysis and Control of Dynamic Elastic
Multi-Link Structures
J. E. Lagnese, Günter Leugering, and E. J. P. G. Schmidt

First-Order Representations of Linear Systems
Margreet Kuijper

Hierarchical Decision Making in Stochastic Manufacturing Systems
Suresh P. Sethi and Qing Zhang

Optimal Control Theory for Infinite Dimensional Systems
Xunjing Li and Jiongmin Yong

Generalized Solutions of First-Order PDEs: The Dynamical
Optimization Perspective
Andreĭ I. Subbotin

Finite Horizon H_∞ and Related Control Problems
M. B. Subrahmanyam

Control Under Lack of Information
A. N. Krasovskii and N. N. Krasovskii

H^∞-Optimal Control and Related Minimax Design Problems:
A Dynamic Game Approach
Tamer Başar and Pierre Bernhard

Control of Uncertain Sampled-Data Systems
Geir E. Dullerud

Robust Nonlinear Control Design: State-Space and
Lyapunov Techniques
Randy A. Freeman and Petar V. Kokotović

Adaptive Systems: An Introduction
Iven Mareels and Jan Willem Polderman

Sampling in Digital Signal Processing and Control
Arie Feuer and Graham C. Goodwin

Ellipsoidal Calculus for Estimation and Control
Alexander Kurzhanski and István Vályi

Minimum Entropy Control for Time-Varying Systems
Marc A. Peters and Pablo A. Iglesias

Chain-Scattering Approach to H^∞-Control
Hidenori Kimura

Output Regulation of Uncertain Nonlinear Systems
*Christopher I. Byrnes, Francesco Delli Priscoli,
and Alberto Isidori*

High Performance Control
Teng-Tiow Tay, Iven Mareels, and John B. Moore

Optimal Control and Viscosity Solutions of
Hamilton–Jacobi–Bellman Equations
Martino Bardi and Italo Capuzzo-Dolcetta

Stochastic Analysis, Control, Optimization and Applications:
A Volume in Honor of W.H. Fleming
*William M. McEneaney, G. George Yin,
and Qing Zhang, Editors*

Mutational and Morphological Analysis: Tools for Shape
Evolution and Morphogenesis
Jean-Pierre Aubin

Stabilization of Linear Systems
Vasile Dragan and Aristide Halanay

The Dynamics of Control
Fritz Colonius and Wolfgang Kliemann

Optimal Control
Richard Vinter

Advances in Mathematical Systems Theory:
A Volume in Honor of Diederich Hinrichsen
*Fritz Colonius, Uwe Helmke, Dieter Prätzel-Wolters,
and Fabian Wirth, Editors*

Nonlinear and Robust Control of PDE Systems:
Methods and Applications to Transport-Reaction Processes
Panagiotis D. Christofides

Foundations of Deterministic and Stochastic Control
Jon H. Davis

Partially Observable Linear Systems Under Dependent Noises
Agamirza E. Bashirov

Switching in Systems and Control
Daniel Liberzon

Matrix Riccati Equations in Control and Systems Theory
Hisham Abou-Kandil, Gerhard Freiling, Vlad Ionescu, and Gerhard Jank

The Mathematics of Internet Congestion Control
Rayadurgam Srikant

H^{∞} Engineering and Amplifier Optimization
Jeffery C. Allen

Advances in Control, Communication Networks, and Transportation Systems:
In Honor of Pravin Varaiya
Eyad H. Abed

Convex Functional Analysis
Andrew J. Kurdila and Michael Zabarankin

Max-Plus Methods for Nonlinear Control and Estimation
William M. McEneaney

Uniform Output Regulation of Nonlinear Systems:
A Convergent Dynamics Approach
Alexey Pavlov, Nathan van de Wouw, and Henk Nijmeijer

Current Trends in Nonlinear Systems and Control:
In Honor of Petar Kokotović and Turi Nicosia
Laura Menini, Luca Zaccarian, Chaouki T. Abdallah, Editors

Supervisory Control of Concurrent Systems: A Petri Net Structural Approach
Marian V. Iordache and Panos J. Antsaklis